MORPHOGENESIS

MORPHOGENESIS

By
Dr. D.R. Khanna
Reader in Zoology
Gurukul Kangri University
Haridwar (Uttaranchal)

DISCOVERY PUBLISHING HOUSE
NEW DELHI-110002

First Published-2004

ISBN 81-7141-771-X

Published by

DISCOVERY PUBLISHING HOUSE
4831/24, Ansari Road, Prahlad Street,
Darya Ganj, New Delhi-110002 (India)
Phone: 23279245 • Fax: 91-11-23253475
E-mail:dphtemp@indiatimes.com

Printed at:

Tarun Offset Printers, Delhi-53

Preface

The fundamentals of Morphogenesis are presented within the framework of scientific discovery. Researches in Embryology have made almost increadible strides in the past few decades. Conse-quently, existing concepts of the Morpho-genesis have expanded. There has been a revolution indeed in this direction.

The text integrates the descriptive, experimental and biochemical approaches into a conceptual framework for the analysis of development. All important points are illustrated diagramatically. The title is not intended to be comprehensive nor could it be at length, but it concentrates as putting across the basic principles of the subject as briefly and lucidly as possible. It does this with the aid of careful selected examples—some recent and other classic of the field and with numerous illustrations. The aim is to enthuse the reader with this active and exciting area of research and to lay a solid foundation on which further study of its various facets may be based.

The present title is designed primarily for undergraduate students though it may also serve an introductory text to those preparing for their master's degree. The title is not intended to be comprehensive, nor could it be at this length, but it concentrates on putting across the basic principles of the subject as briefly and lucidly as possible. It does this with the aid of carefully selected examples, some recent and other classic of the field, and with numerous illustrations.

In the preparation of this book large number of books and research papers have been consulted. So no authenticity is claimed.

Text book can not be written without the support and professional contributions of many people. This book is no exception. The author is grateful to those teachers and colleagues whose stimulating discussions have clarified certain vague points for him, but all errors, omissions, and corrections needed are solely his responsibility.

The author tried hard to be accurate and upto date in statement and realises the impossibility of completely avoiding errors therefore, the author will greatly appreciate having his attention called to any questionable statement.

The author expresses his gratitute to Mr. Wasan and staff of M/s Discovery Publishing House for their whole hearted co-operation in the publication of this book.

Author

Contents

1

Primordial Germ Cells

The primordial germ cells PGCs are large compared with the surrounding cells, are roundish, oval or pear-shaped and have a distinct cell boundary and a diameter varying from 10 to 20 µm depending on shape, stage and species. The voluminous nucleus is roundish, oval or lobular in shape and has a distinct unclear membrane. It is often situated eccentrically; its diameter may vary between 6 and 10 µm. The chromatin is fairly evenly distributed throughout the nucleus contains one or two prominent nucleoli. Depending on the stage of development and the taxonomic group the cytoplasm contains varying amounts of yolk granules, fat droplets and pigment granules. In general the cytoplasm has a dense, granular appearance and contains a roughly hemispherical cytoplasmic body called variously the centre sphere cytoplasmic crescent, mitochondrial crescent or attraction sphere. This consists of one or two centrioles, numerous mitochondria, and small yolk platelets.

The unequal distribution of yolk and lipid inclusions and the position of the cytoplasmic body result in the eccentric location of the nucleus and the oval or pear-shaped form of the cell. The PGCs are often also characterized by the presence of lobopodia or filopodia, particularly during their migratory period. The general features of the PGCs have been described in teleost fishes, anuran amphibians, urodele amphibians, coecilians, chelonian reptiles, birds and in mammals. It is clear that none of the features of PGCs described above is specific for these cells, so that an unequivocal identification can only be made on the basis of a combination of all or most of them. Moreover, the earlier the stage of development, the more yolk the cells usually contain and the less pronounced are the various cell characteristics.

A large amount of yolk also often leads to deformation' of both cell and nucleus, and a consequent blurring of several of the above features. The early workers were therefore unable to trace the PGCs back to the earliest stages of development. The discovery by Bounoure in anurans of a special cytoplasmic structure, the so-called 'germinal plasm', which is originally located in the subcortical layer of the vegetal region of the fertilized egg and later in the PGCs of the tadpole, marked an important step forward in the analysis of the origin of the PGCs in the anuran amphibians. Cytological and cytochemical studies indicated that the germinal plasm contained, besides small yolk platelets and pigment granules, mitochondria and RNA. However, it was the electron microscope which revealed the true nature of the germinal plasm.

In recent years numerous investigations have been devoted to the study of this structure that is specific to germ cells. It must however be emphasized that not all chordate or vertebrate PGCs are characterized by the presence of a typical germinal plasm and that moreover it is not found during all developmental stages. Since nothing is known about the ultrastructure of the PGCs in the Cephalochordata and Tunicata the following description will only deal with the ultrastructural and cytochemical characteristics of the PGCs in the various groups of the vertebrates.

The Ultrastructure and Biochemistry of the PGCs in Anamnia

Amphibia

In the amphibians extensive ultrastructural and cytochemical analyses have been carried on the PGCs of the anurans, few on those of the urodeles, and none on those of the coedlians.

Anura

In the anurans electron-microscopic analysis has revealed, apart from the usual cell organelles found in almost every cell of the embryo, a number of special cytoplasmic and nuclear structures which may be considered as being characteristic of PGCs and their descendants. The first and most striking of these is the 'germinal plasm', which has now been described in different anuran species by many investigators. The germinal plasm is a well-circumscribe structure occurring in the form of patches 5-12 μm in diameter. These consist of numerous mitochondria, many containing protein crystals, interspersed with so-called 'germinal granules' (electron-dense bodies). These electron-dense bodies are composed of small electron-dense foci which seem to be embedded in a matrix of extremely fine fibrils. It is not clear whether the structure of the germinal granules is essentially granular, fibrillar or fibrillo-granular.

The electron-dense bodies are not bounded by a membrane and are frequently found in contact with mitochondrial membranes ('intermitochondrial cement') or with the nuclear membrane ('nuage material'). The germinal plasm also contains pigment granules and some glycogen, as well as a few lipid droplets and a varying amount of small yolk platelets. Blackler (1958) thought that the islands of germinal plasm found in cleaving eggs arise only after fertilization, but Czolowska (1969) demonstrated the presence of small patches of germinal plasm in ovarian oocytes. Williams and Smith (1971) found small electron-dense areas devoid of ribosomes within clusters of mitochondria just beneath the vegetal pole cortex of ovarian oocytes.

Around the time of germinal vesicle breakdown larger electron-dense areas are seen which are now associated with ribosomes since typical germinal granules also appear in progesterone-treated oocytes which are subsequently enucleated, this 'condensation' process does not depend upon germinal vesicle contributions. The characteristic germinal granules therefore seem to coalesce from precursors during oogenesis with a possible synthesis of components *de novo* during maturation. According to Mahowald and Hennen (1971) the precursors of the germinal plasm in *Rana pipiens*, described as cytoplasmic islands in the unfertilized and fertilized egg and in early cleavage stages, consist essentially of the same components as the germinal plasm in the PGCs of the tadpole. However, a certain structural evolution seems to occur in the germinal plasm during development.

In the fertilized egg the germinal plasm is present in the form of small cytoplasmic patches with germinal granules 0.2-0.3 μm in diameter. This is in agreement with Czolowska's (1969) observations on *Xenopus*, where a large number of very small patches was found in the cortical and subcortical cytoplasm of the vegetal, region of ovarian oocytes. In the unfertilized egg the electron-dense foci of the germinal granules measure about 20 nm. At the late blastula stage the germinal granules have acquired a diameter of 0.4-.05 μm, while the electron-dense foci now have a diameter of up to 50 nm. A condensation process has apparently occurred. In the unfertilized and fertilized egg and at early cleavage stages the germinal plasm seems to contain some RNA in the form of clusters of ribosomes increase upon fertilization.

At the early tadpole stage the germinal granules are in the form of amorphous bodies of a clearly fibrous nature. They are situated in close proximity to the nuclear membrane or are attached to mitochondria, but are no longer associated with clusters of ribosomes. Mahowald and Hennen (1971) ascribe a morphogenetic significance to the temporary association of clusters of ribosomes with the germinal granules, mainly in analogy to similar events observed in the development of the polar granules in the pole cells of certain insects. Protein synthesis has been demonstrated in the metabolically active germinal plasm of the *Xenopus* egg during cleavage. According to Ikenishi and Kotani (1975) the fibrillo-granular germinal granules found in *Xenopus* at early stages of development first change into irregular string-like bodies at the late neurula to early tail-bud stages and then into granular material at the feeding tadpole stage.

In *Xenopus* Kalt (1973) distinguishes three different structures which are found in both the male and the female germ line but in no other cells. The first of these is the germinal plasm, which takes the form of large proteinaceous masses of electron-dense material, which he calls 'nuage material'. The second structure, also cytoplasmic in nature, in the electron-dense so-called chromatoid or nucleolus-like body. This is found in germ cells of the indifferent gonad and throughout the female germ line, as well as in the early stages of the male germ line, whether it dissociates into coated vesicles during spermatogenesis. The third structure is a fibrillar nuclear region (fibrillar meshwork) which is free of DNA. A similar structure has been described in *Xenopus*. Opinions differ on the homology of the various electron-dense structures found in the cytoplasm of PGCs. Some authors, for example Kalt (1973) and Kalt, Pinney and Graves (1975), make a distinction between chromatoid or nucleolus-like bodies on the one hand and germinal granules or nuage material on the other, on the basis of differential sensitivity to various cytochemical treatments.

Kalt (1973), Eddy (1974), and Webb (1976) emphasized the great-similarity in ultrastructure between germinal granules and nuage material as well as their resemblance to nucleolar components in the nucleus. Other authors, for example Fawcett (1972), consider the various electron-dense components of the cytoplasm of germ cells as belonging to one and the same family. It must however be emphasized that neither is the homology of the various structures established nor is their functional significance sufficiently understood. The germinal plasm original present in the fertilized egg does not change in amount during cleavage.

It is distributed unevenly over the exponentially increasing number of blastomeres. Although the number of blastomeres containing germinal plasm initially increased, their total number later remains constant as a consequence of the transfer of the germinal plasm to one daughter cell of a division only (pPGCs). During gastrulation and neurulation a small number of clonal divisions occur, as a result of which the number of PGCs increases about fourfold; there are followed by one more clonal division before stage 35. It is evident that during these clonal divisions the amount of germinal plasm much increase, since otherwise dilution would occur.

The frequently observed passage of electron-dense material through nuclear pores may be associated with the formation of new germinal plasm. However, the static electron micrographs do not allow a distinction between a transfer of material from the nucleus to the cytoplasm or vice versa. During the clonal divisions the PGCs also synthesis RNA. It seems of importance also to mention some negative features of the PGCs in anurans. Gipoulous (1967) was unable to demonstrate either glycogen or alkaline phosphatase in anuran PGCs. The former is at variance with Kessel's (1971) election microscope observations, but the latter is in accordance with Chiquoine and Rothenberg's (1957) studies on PGCs in urodele amphibians.

The distinct cell boundary of PGCs already by the earlier workers in the various groups of vertebrates turns out to be due to the presence of larger intercellular spaces between the PGCs and the surrounding somatic cells. This emphasizes the non-participation of the PGCs in embryonic tissues. The PGCs show pseudopodia and lobopodia which protrude into the intercellular spaces. The demonstration of microtubules and microfilaments in the cytoplasm of spermatogonia constitutes additional evidence for their migratory activity.

Urodela

In the urodele amphibians only very scanty and partially conflicting data on the ultrastructure of PGCs are available. Recent electron-microscopic observations of Ikenishi and Nieuwkoop (1978) on the ultrastructure of the PGCs in *Ambystoma mexicanum* reveal that the PGCs of the urodele larva at stage 46 are characterized by the presence of germinal plasm consisting of clusters of mitochondria and electron-dense bodies of a fibrillo-granular nature. However, the germinal plasm only appears at early larval stages (from stage 40 onwards), at which time the passage of electron-dense material through nuclear pores is also observed. The amount of electron-dense nuclear pores is also observed.

The amount of electron-dense material increases rapidly during subsequent developmental stages, until the definitive complement is reached by stage 46. The late appearance of the germinal plasm in the urodeles seems to be at variance with Williams and Smith's (1971) brief and unconfirmed statement on the presence of germinal granules in the subcortical cytoplasm of the equatorial region of the fertilized, uncleaved egg of *Ambystoma mexicanum*. Intranuclear fibrillar regions as found by Kalt (1973) in anuran PGCs, have recently also been observed by Ikenishi and Nieuwkoop (1978) in *Ambystoma mexicanum* PGCs but nothing is known about their functional significance.

Fishes

Little is known about the ultrastructure of the PGCs in fishes and the relevant literature is virtually restricted to the Teleostomi. The PGCs in the Agnatha. Chondrichthyes and Osteichthyes have so far only been studied with the light microscope, as described in the first part of this chapter. Gamo (1961*b*) emphasized the fact that the PGCs in meroblastic embryos contain less yolk than those in holoblastic embryos. This may explain their tendency to associate themselves with nutritional sources, e.g., with the periblast in teleost embryos. Pala (1970) mentions the presence of small PAS-positive granules in the perinuclear cytoplasm of teleost PGCs, particularly in the attraction sphere Nedelea and Steopoe (1970), were the first to describe the presence in the cytoplasm of PGCs of electron-dense bodies, which they call nuage material.

Satoh (1974) gives a more detailed description of the electron-dense bodies. They are up to 1 μm in diameter, not bounded by a membrane, and consist of an interwoven meshwork of very fine fibrils. They are often found in close association with clusters of mitochondria, in direct contact with their outer membranes. This description fits that of germinal plasm very well. Nothing is known, however of the time of appearance of these structures in the development of the teleost PGCs.

The Ultrastructure and Biochemistry of PGCs in Amniota

Aves

So far electron-microscopic studies have only revealed the presence of common cell organelles. No mention has made of structures specific to germ cells, such as germinal plasm or nuage material. However, some cytochemical features can be used to characterize the PGCs in birds. Although alkaline phosphatase activity had been demonstrated in chick PGCs by Chiquonie and Rothenberg (1975), the activity is too low to characterize them properly, since a similar activity was found in other cells. The most prominent feature is their high content of PAS-positive material.

According to McCallion and Wong (1956) it is not an absolute criterion, since PAS-positive material is also found in other cells. We may say, however, that the combination of the various cytological and cytochemical features adequately characterizes the PGCs in birds. The PGCs in birds also go through some form of evolution during development. Whereas in early stages PGCs contain a large amount of yolk and lipids and only a small amount of glycogen, in later stages they contain less yolk and lipids, but abundant glycogen. The amount of glycogen drops again during migration. Clawson and Domn (1963*a*) and Fujimoto *et al.* (1976*a*) also observed changes in the ultrastructure and regional distribution of the glycogen: it is non-granular and evenly distributed in early stages, but granular and concentrated at one pole of the cell in later stages.

Mammalia

In the mammals McKay *et al.* (1953) were the first to demonstrate high alkaline phosphatase activity in the cytoplasmic rim of PGCs. Clark and Eddy (1975) and Fujimoto et. al. (1977) mention that the alkaline phosphatase activity along the cell periphery is only faint in early stages but increases markedly in later stages. Asayama (1963) emphasizes the fact that high alkaline phosphatase activity is not a feature specific to germ cells; since other cells may also show it. It is moreover stage dependent. During germ cell migration as well as during gametogenesis the germ cells seem to associate themselves with other tissues due to a lack of adequate endogenous energy sources. The alkaline phosphatase activity may thus be connected with an active intercellular exchange of material.

Asayama (1965) observed the presence of rather large amounts of PAS-positive material in mammalian PGCs. This feature is not specific either, since other cells also contain much glycogen. It disappears, moreover, with the onset of meiosis in the development of the female gametes, and at a later stage in spermatogenesis. Fujimoto *et al*. also observed a larger number of lipid droplets in human PGCs. Eddy (1974) described in detail the presence of electron-dense granulofibrous material, called 'nuage material', in the cytoplasm of mammalian primordial germ cell, which generally poor in inclusions. It was found either in the form of free, discrete bodies or as intermitochondrial cement, and was sometimes spatially associated with pores in the nuclear membrane. Eddy found these structures in young PGCs but also in the PGCs of the indifferent gonad and in the germ cells of neonatal and adult animals. These obervations were corroborated by Motta and Van Blerkom (1974), who found nucleolus-like bodies in the cytoplasm of large oocytes as well as in PGCs of early developmental stages.

During spermatogenesis Comings and Okada (1972) had found similar structures probably formed by the extrusion of nucleolar components into the cytoplasm, which they called *chromatoid bodies*. Fawcen, Eddy and Phillips (1970), who had seen similar structures, did not assume them to be of nuclear origin. It is not yet clear whether these structures really are a constant feature of mammalian germs cells. While Eddy points out the strong resemblance of the nuage material of the germinal plasm of the anuran egg. Sud (1961) and Motta and Van Blerkom (1974) moreover call attention to its similarity to the chromatoid body found during spermatogenesis. However in the latter RNA could not be demonstrated. It may finally be mentioned that during their migratory period mammalian PGCs show cytoplasmic lobopodia and filopodia containing numerous microfilaments.

Reptilia

Milaire (1957) mentioned that in the Chelonia both the extra-gonadal and the gonadal PGCs contain large amounts of PAS-positive material. The PGCs of Lacertilia besides some glycogen, also contain numerous yolk granules and lipid-droplets. The ultrastructure of the PGCs has been fairly extensively studied only in the Squamata Hubert (1970*b*) describes the ultrastructural organization of the PGCs in the Lacertilia. Structures specific to germ cells are not found in the cytoplasm but only in the nucleus. He mentions the presence of a paranucleolar mass as well as of a small, dense fibrillar corpuscle ('ring-shaped' in cross section): both are situated in the vicinity of the large 'ring shaped' nucleolus. When there are two nucleoli both structures are also present in duplicate. The paranuclear mass consists of a network of the fine fibrils and very fine particles.

It is probably of a proteinaceous nature, since it is Feulgen-negative but Fast Green-positive and stains green with Methyl Green Pyronine. A similar structure was described as fibrillar meshwork' in PGCs of anuran and urodele amphibians. No general plasm or nuage material has hitherto been demonstrated. In *Lacerta* the paranuclear mass is a constant feature of PGCs, of oogonia and oocytes, and of spermatogonia. Hubert (1968) found the same structure in the snake *Vipera* but not in the slow-worm *Anguis*, so that it may actually not be a constant feature in the squamata. The small fibrillar corpuscle is also found in spermatogonia of *Lacerta*. During the migratory period the PGCs show micropseudopodia containing bundles of microfibrils, which provide evidence of amoeboid motility.

Conclusions

Summarizing, it may be stated that the *general cytological features* of the PGCs are very much alike in the various groups of chordates. Although any of these features in itself is not specific for germ cells, the combination of various features quite satisfactorily serves for the unequivocal identification of PGCs. It must however by emphasized that during certain periods of development, particularly in the very early stages and during rapid multiplication, several of these features are rather indistinct or even unrecognizable. This had led to the suggestion of discontinuity in the germ line and in formation of germ cells *de novo* during certain period of development particularly during gametogenesis.

Ultrastructural analysis which unfortunately is still rudimentary or even lacking in certain groups, suggests quite strongly that the electron-dense fibrillo-granular bodies or the intermitochondrial cement found in the germinal plasm are structures specific to germ cells. They have not been found in all groups, however, so that they certainly do not represent a universal feature of germ cells. It is intriguing that the PGCs, which show almost identical general characteristics in all chordate groups, nevertheless show pronounced differences in ultrastructure or cytochemistry.

We have already mentioned the non-universality of the germinal plasm. Glycogen, which is so abundant in avian germ cells, is also found in mammalian and reptilian germ cells, but in smaller amounts, while it is virtually absent from the PGCs of amphibians and fishes. A high alkaline phosphatase activity of the cell periphery seems to characterize the mammalian PGCs. A special nuclear structure in the form of a fibrillar meshwork inside the nucleus, which is free of DNA but RNA-positive is found in the PGCs of anuran and urodele amphibians as well as in those of certain reptiles. During the last decades attention has been focussed particularly on the cytoplasmic features of the PGCs, so that specific nuclear structures may have been overlooked in a number of vertebrate groups. In our opinion it is not unlikely that nuclear structure may represent more constant and more specific features of the PGCs than the well-known germinal plasm.

2

Origin of Primordial Germ Cells

Before considering the site and mode of Origin of primordial germ cells, a clear cut distinction must be made between descriptive and experimental evidence. In the older literature the place where PGCs were first identified during embryonic development was often taken as synonymous with their actual site of origin. We know now that the PGCs are formed extra-gonadally and have amoeboid motility, migrating though various tissues towards their final location in the gonadal anlagen. Although migration usually occurs during later stages of development (e.g., early larval stages), the possibility cannot be excluded that active displacements also take place during early stages. This may for instance be the reason for the often-observed accumulation of PGCs in the endo-mesodermal interspace in the Anamnia. Descriptive evidence may therefore only be considered as circumstantial. It is clear that the earlier the stage of development at which the PGCs can be recognized, the more likely it is that the place where they are first identified corresponds to the actual site of origin.

However, conclusive evidence for the site of origin of the PGCs requires experimental analysis. We may only learn about the causal factors responsible for germ cell formation (in other words, their mode of origin) from adequately planned experiments. These are virtually lacking in the lower planned experiments. These are virtually lacking in the lower chordates, and are mainly restricted in the Anamnia to the amphibians, and in the Amniota to the birds. The anuran and urodele amphibians will therefore be discussed first, followed by the various groups of fishes and the cephalochordates and tunicates. Among the amniotes the birds will be treated first, followed by the mammals and reptiles.

THE ANAMNIA

Amphibia

Anura

The history of the germinal plasm. Since the discovery of the so-called 'germinal plasm' in the egg of *Rana temporaria* by Bounoure in 1931 and the elucidation of its subsequent history through embryogenesis, the anuran amphibians have served as the classical example of the presence of an uninterrupted germ line in the vertebrates. These observation agree very well with the results of similar studies in several insects and other invertebrates (for further information). The correspondence in germ cell history, as well as the striking similarity in ultrastructure of the germinal plasm in groups of the animal kingdom which otherwise differ so much, has led to the idea that the presence of an uninterrupted germ line from the fertilized egg to the adult organism is a basic mechanism for maintaining organismal identity through successive generations. We must however test whether such a postulate is actually justified, and decide to what extent other mechanisms may also be represented in the chordates. The fate of the germinal plasm during cleavage, gastrulation, neurulation and further embryogenesis was carefully studied for the first time by Bounoure in the years 1931 to 1934.

In the vicinity of the vegetal pole of *Rana temporaria* eggs he found small subcortical patches of a special cytoplasm which he called '*germinal plasm*'. During the first two cleavages these are more or less evenly distributed among the four blastomeres and are partially pulled inwards along the cleavage furrows. In subsequent cleavage cycles they are usually transferred to only one of the two daughter cells (differential division), and consequently are distributed among a restricted number of vegetal blastomeres. As a consequence of the pre-gastrulation movements the blastomeres containing germinal plasm are displaced together with other endodermal cells towards the centre of the vegetal yolk mass and sometimes even as far as the floor of the blastocoel.

At late blastula or early gastrula stages the germinal plasm, which was hitherto situated at the periphery of the cell, is displaced intracellularly to a juxta-nuclear position, which is its final location in the PGC. Subsequently the PGCs are displaced along with the vegetal yolk mass by the gastrulation and neurulation movements, and are ultimately found among the vegetal blastomeres in the caudal portion of the embryo. In the anurans a distinction must be made between true PGCs and their forerunners, the presumptive PGCs (pPGCs).

During the first two cleavage divisions and the succeeding differential blastomere divisions one can only speak of pPGCs. One may define true PGCs as cells which are characterized by the presence of germinal plasm and which divide by clonal divisions. When the PGCs are first formed they do not divide very frequently, so that a considerable time may elapse between the last 'unequal' division and the first clonal division. It is interesting that this transition from presumptive to true PGCs coincides with the centripetal displacement of the germinal plasm. 'Unequal' divisions occur as long as the germinal plasm is situated peripherally, while 'equal' divisions take place when the germinal plasm has reached its juxta-nuclear position. It is evident from the literature that the anurans great significance is attributed in the germinal plasm for the determination and differentiation of the PGCs. However, what is the present evidence for this assumption? To test this question the following approaches have been made (a) removal or destruction of the germinal plasm or its transplantation to an egg lacing functional germinal plasm, and (b) analysis of the chemical composition and possible function of the germinal plasm.

The significance of the germinal plasm for germ cell formation. As already described above. PGCs develop from, blastomeres which contain germinal plasm. A number of workers have tried to remove the germinal plasm surgically. Nieuwkoop and Suminski (1959) were among the first to try to remove the germinal plasm at the 2 to 4-cell stage of *Xenopus* by superficial pricking of the vegetal pole region without causing a considerable outflow of cytoplasm. In general this did not lead to a marked reduction in the number of PGCs in the developing larvae. Fischiarolo (1960), using *Discoglossue* eggs found normal gonads with germ cells in metamorphosed larvae after exovation of about 40% of the egg contents as a consequence of pricking the vegetal region. Librera (1964) obtained varying results after withdrawal of about one-quarter of the egg contents with a micropipette inserted into the vegetal region of uncleaved eggs, 4-cell, blastula and early gastrula stages of *Discoglossus*.

Of the 49 animals which reached metamorphosis, 32 had normal gonads, 5 had gonads with only a few germ cells, and 12 were completely sterile. Buehr and Blackler (1970) found either complete sterility or a marked reduction in the number of PGCs at the tadpole stage after pricking the vegetal pole at the 2 to 4-cell stage in *Xenopus*. They checked that some or nearly all of the germinal plasm was present in the exovate and found a quantitative relationship between the amount of germinal plasm remaining and the number of PGCs in the tadpole. From these experiments they concluded that the germinal plasm is indispensable for germ cell formation and that at least upto the tadpole stage germ cells are not formed from secondary sources.

Gipoulous (1971, 1975) confirmed these results in *Discoglossus*, applying numerous superficial punctures to the vegetal pole region at the 2- to 4-cell stage. He ascribed the negative results of Nieuwkoop and Suminski to the experimental technique used, which did not lead to any substantial outflow of cytoplasm. In 1937 Bounoure described for the first time the effect of UV-irradiation on the germinal plasm in the vegetal hemisphere of the fertilized anuran egg. He observed a dramatic reduction (up to c. 90%) in the number of PGCs in the treated animals at metamorphosis. Although some animals had very few germ cells, he never actually obtained complete sterility. The gonads, though reduced in size, showed a normal somatic structure. Aubry (1953a, b) obtained still better results with UV-irradiation combined with animal-vegetal compression of the eggs.

Padoa (1963b) studied UV sensitivity at different stages of development and found it to be highest in the period from shortly before the initiation of first cleavage till the beginning of second cleavage. He ascribed this to the accumulation of the germinal plasm in the vicinity of the vegetal pole shortly before first cleavage. The decrease in sensitivity after second cleavage seems to be due to the inward displacement of part of the germinal plasm along the cleavage furrows. This interpretation is supported by the observation that irradiation of the vegetal pole of the blastula no longer has any effect, since nearly all the germinal plasm is by now situated in the interior of the embryo.

Smith (1966) obtained complete sterility in nearly 100% of the developing larvae after UV-irradiation with 8000 erg/mm^2 of the vegetal hemisphere of *Rana pipiens* eggs at the beginning of first cleavage, whereas similar irradiation of the animal pole had no effect. By testing the effect of different wavelengths he obtained a UV spectrum for PGC inactivation. The maximally effective wavelength of 254 nm corresponds to the UV sensitivity of nucleic acids. Tanabe and Kotani (1974) also found complete sterility after irradiation of the vegetal hemisphere of the 2- cell stage of *Xenopus* with UV of wavelength 253.7 nm and dose c. 6000 erg mm^2. When the stage of irradiation was gradually shifted from the 2- to the 4-cell stage an increasing number of PGCs was found in the larvae. When the dose of UV-irradiation at the 2-cell stage was decreased from 6000 to 750 erg/mm2 the number of germ cells found in tadpoles increased proportionally.

The number of germ cells was also inversely proportional to the irradiated area. In *Xenopus* Ikenishi, Kotani and Tanabe (1974) observed a swelling and vacuolation of the mitochondria and a fragmentation of the germinal granules as a direct effect of UV-irradiation of the germinal plasm. They concluded that maintenance of the normal structure of the germinal plasm is indispensable for its role as germ cell determinant, however, warn against the conclusion that the effect of UV-irradiation represents a specific inactivation of the germinal plasm. They observed an inhibition of cleavage in the vegetal hemisphere in *Xenopus*, leading to a markedly abnormal segregation of the germinal plasm into particulars blastomeres, and suggest that the sterilizing effect of the UV-irradiated is more likely to be due to a derangement of the normal cell lineage than to a direct deteriorating effect on the germinal plasm.

Grant, Wacaster and Turner (1976) observed a disturbance of the cleavage process in the UV-irradiated vegetal half of the anuran egg and a subsequent delay in the gastrulation process. The most elegant experiment indicating the great significance of the germinal plasm in PGC determination and differentiation was done by Smith (1966). He found that he could restore the germ line in *Rana pipiens* eggs previously sterilized by UV-irradiation by injecting cytoplasm from the vegetal pole of normal eggs, but that cytoplasm from the animal hemisphere was ineffective. However, whole vegetal cytoplasm was injected, not pure germinal plasm, so that this is not absolute proof of the indispensability of the germinal plasm.

Nevertheless the restoration of the PGC-forming capacity by injection of certain fractions of a crude homogenate of vegetal pole plasm into UV-irradiated Rana eggs suggests that germinal granules rather than mitochondria are indeed involved. Very recently Dixon (Personal communication) found that in contrast to all previous observations in *Xenopus* early PGC formation is probably not affected at all by UV-irradiation, but that the migration of the PGCs toward the genital ridges is markedly retarded, so that the gonadal anlagen become colonized only at a late stage of development. This finding, which requires further confirmation, seems to call for a reinterpretation of several of the results mentioned above. A positive regulation after UV-irradiation towards the normal germ cell number at metamorphosis was noted by Bounoure et al. (1954). This may explain the negative results of Fischiarolo (1960) as well as the varying results obtained by Librera (1964).

In Gipouloux's (1972) experiments negative regulation occurred, i.e., a restoration of the normal number of germ cells after an initial excess in a parabiont with one set of gonadal anlagen and two endodermal masses as sources of PGCs. Positive or negative regulation must be due to an increases or decrease in the number of mitoses in the PGCs between the tadpole stage and metamorphosis. Bounoure et. al. (1954) actually observed a much higher mitotic rate in the PGCs of animals with a markedly reduced number of germ cells. The fact that Bounoure and Librera obtained metamorphosing tadpoles with only a few PGCs suggests that regulation may be in impaired when the number of germ cells falls below a certain minimum level. This lack of regulation may, however, also be due to the remaining germ cells being damaged too much by the treatment used.

Composition and possible function of the germinal plasm. Gipouloux (1975) distinguishes three successive phases in germ cell formation. The first phase is characterized by the peripheral position of the germinal plasm during cleavage (pPGCs). In the second phase the germinal plasm occupies a juxta-nuclear position in contact with the nuclear membrane (PGCs). According to Gipouloux germ cells determination occurs in this phase, which is moreover characterized by an arrest of mitosis. The third phase is that of cellular-differentiation of the PGCs in the gonadal anlagen.

Several authors have studied the composition of the germinal plasm in order to analyze its possible function in germ cell determination. The UV sensitivity of the vegetal hemisphere of the anuran egg seems to correspond to that of nucleic acids (Smith, 1966). However, these nucleic acids are not necessarily in the germinal granules themselves, but may be those in the ribosomes and polysomes that are associated with the germinal granules at the unfertilized egg to early cleavage stages or even the mitochondrial DNA, the germinal plasm being very rich in mitochondria.

Smith (1965) investigated the capacities of germ cell and somatic nuclei for supporting normal development when injected into enucleated eggs of *Rana pipiens*. Eggs with germ cell nuclei showed a much higher incidence of cleavage than eggs with somatic nuclei (43% compared with 18%), while 40%, compared with 10% of completely cleaved eggs formed normal embryos. He concludes that the nuclei of PGCs are apparently still totipotent, whereas the somatic nuclei have lost part of their potency. He therefore makes the suggestion that the germinal plasm in effect protects the nucleus from becoming channelled into a somatic pathway of differentiation. Smith and Ecker (1970), reviewing the relevant literature, came to the conclusion that nucleic acids in the form of stable templates are present in the germinal plasm and act as germ cell determinants.

Using autoradiography, Eddy and Ito (1971) failed to demonstrate any RNA synthesis in the electron-dense bodies of ovarian oocytes but found a rapid incorporation and turnover of amino acids. They also failed to detect any RNA cytochemically. They pointed out the great similarity between the electron-dense bodies found in oogonia and oocytes and the germinal plasm found in PGCs. Wallace, Morray and Langridge (1971), studying gene amplification during gametogenesis, inferred the presence of self-replicating ribosomal DNA particles in the nucleolar core of oocyte nuclei. They suggest that during maturation these particles pass into the cytoplasm and end up in the germinal plasm. It must however be stressed that DNA has never been demonstrated in germinal plasm. Deuchar (1972), who also considers a large store of ribosomal DNA as a characteristic of the germ line, assumes that during embryonic development and early larval life this DNA is transcribed into RNA, which is sequestered in the germinal plasm.

Webb (1976) demonstrated ribosomal RNA synthesis associated with nuage material in oogonia. He suggests that the negative results of Eddy and Ito (1971) may be due to a protective effect resulting from the predominantly proteinaceous nature of the germinal plasm. Dziadek and Dixon (1977) found three phases of DNA synthesis in PGCs of *Xenopus laevis*, viz. In the stages 10 to 20, 22 to 24, and 37 to 39 no DNA synthesis occurring in the intervening periods. The PGCs distinguish themselves from surrounding endoderm cells as early as the mid-gastrula, since in the endoderm cells DNA synthesis is not temporarily arrested. In the PGCs RNA synthesis begins in the mid-gastrula (stage 11) and ceases at the early tail-bud (stage 24). Dziadek and Dixon did not find RNA synthesis in the germinal plasm and conclude that it is probably confined to the nucleus.

Zust and Dixon (1977) state that the germ cell lineage consists of a number of successive cell generations, so that the process of germ cell differentiation is a stepwise process. This may also hold for the process of germ cell determination. Both the descriptive and the experimental evidence point to a crucial role for the germinal plasm in the determination and differentiation of the PGCs in the anuran amphibians. Only those blastomeres which contain a sufficient (?) amount of intact germinal plasm, or which receive it after previous UV treatment, develop into PGCs.

Smith (1966) therefore calls the germinal plasm of the anuran egg *a germ cell determinant*. This means that the presence or absence of this plasm–more particularly of us nucleic acid component-determines whether a cell will become a PGC or a somatic cell. The present authors feel, however, that one should be cautious in drawing such a far-reaching conclusion. So long as the germinal plasm or its active component has not been isolated and tested separately, Smith's conclusion must be considered premature. On the basis of the available evidence it can be said only that the vegetal pole region of the anuran egg contains factors which apparently act in germ cell determination.

Urodela

Descriptive evidence. The descriptive evidence concerning the site of origin of the PGCs in the urodeles is somewhat scanty. On the basis of their yolk content Abramowicz (1913) concluded that in *Triturus* the PGCs arise from the endoderm and migrate into the mesoderm at a relatively advanced stage of development. In

Hemidactylium and other urodeles, however, Humphrey (1925) identified the PGCs at an early tail-bud stage (from the 7-somite stage onwards) in the medial and later in the dorsal region of the lateral plate mesoderm. He stated that there is no evidence for Abramowicz's conclusion that their initial position is in the endoderm. Humphrey's observations were confirmed by Nieuwkoop in Triturus embryos at older tail-bud stages and recently in *Ambystoma* by Ikenish and Nieuwkoop (1978), who could not, however, trace the PGCs back to stages earlier than 23 (H).

Experimental evidence. Humphrey (1929) confirmed the localization of the PGCs in the medial to dorsal lateral plate mesoderm by unilateral extirpation any by transplantation of this portion of the mesoderm to the ventro-lateral side of host embryos at stages 21 to 25. The former experiments led to the complete absence of PGCs on the operated side and the latter to additional germ cell formation. Nieuwkoop (1947) carried out a more systematic analysis of the origin of the PGCs in the urodeles. Neither removal of vegetal cytoplasm at the 1-cell stage, nor removal of the central portion of the vegetal blastomeres at the early gastrula stage in *Ambystoma* and *Triturus*, respectively, affected the number of PGCs in the developing larvae. Removal of the entire endoderm from early neurulae led to endoderm-free larvae containing a restricted number of PGCs situated in that portion of the mesoderm which had already invaginated and had apparently come into contact with the caudal endoderm prior to operation.

Transplantation of the entire endoderm between neurulae of different *Triturus* species and between *Triturus* and *Ambystoma* showed that in the chimaeric embryos the PGCs always belonged to the species which had furnished the ecto-mesodermal envelope. Removal of the presumptive lateral plate mesoderm at the early neurula stage led either to complete sterility or to a very marked reduction in the number of PGCs, the remaining PGCs being situated in the unaffected, most cranial section of the fertile region. Heteroplastic transplantation of ventro-lateral marginal Zone (presumptive lateral plate mesoderm), at an early yolk-plug stage (11-11 H.) gave rise to chimaeric larvae in which the regional distribution of the PGCs of the two species corresponded to the regional composition of the chimaeric lateral plate mesoderm. These experiments demonstrated unequivocally that in the urodeles the PGCs are of mesodermal origin and are evenly distributed throughout the presumptive lateral plate mesoderm. Contrary to Nieuwkoop, Takamoto (1953) observed complete absence of PGCs after removal of entire endoderm from late neurulae of *Triturus pyrrhogaster*, but in Amanuma's (1958) experiments extirpation of the dorso-caudal endoderm of early neurulae of *T. pyrrhogaster* did not affect the number of PGCs.

Amanuma (1957) moreover found that extirpation of presumptive lateral plate mesoderm at mid-gastrula or early neurula stages of *T. pyrrhogaster* and *Hynobius nebulosus* yielded larvae with a varying number of PGCs. He assumed that regulation occurred and concluded that the determination of the PGCs is not yet completed at the neurula stage. Orthotopic and heterotopic transplants of lateral plate mesoderm at tail-bud stages of *T. pyrrhogaster* by Asayama (1950, 1961) led to similar conclusions. Lacroix and Capuron (1966) concluded from grafts of intermediate mesoderm of a tail-bud stage (stage 26) between Iberian and Moroccan races of *Pleurodeles* (which are characterized by a different lampbrush chromosome structure) that at this stage the PGCs are actually localized in the intermediate mesoderm.

Maufroid and Capuron (1972) used defect experiments to determine the exact location of the PGCs in the lateral plate mesoderm at neurula and tail-bud stages of *Pleurodeles*, and found that in the early neurula the PGCs are located in the neighbourhood of the blastopore, from where they are displaced anteriorly (along with the lateral plate mesoderm) during mesodermal mantle formation. At tail-bud stages (stages 21 to 23, G and D) the PGCs move inside the lateral plate mesoderm from a ventral and lateral position towards the vicinity of the Wolffian ducts. Ablation of lateral plate mesoderm containing PGCs did not lead to any formation *de novo* from adjacent mesoderm. The seemingly contradictory results of Asayama and Amanuma can be adequately explained on the basis of the exact location of the PGCs at successive stages of development.

Maufroid and Capuron (1973) transplanted presumptive ventral and lateral marginal zone of early gastrulae of *Pleurodeles* into the blastocoel of host embryos of the same age, and observed the formation of extra ventral mesoderm with PGCs. They concluded that the PGCs are of mesodermal origin and are distributed throughout the entire ventral and ventro-lateral marginal

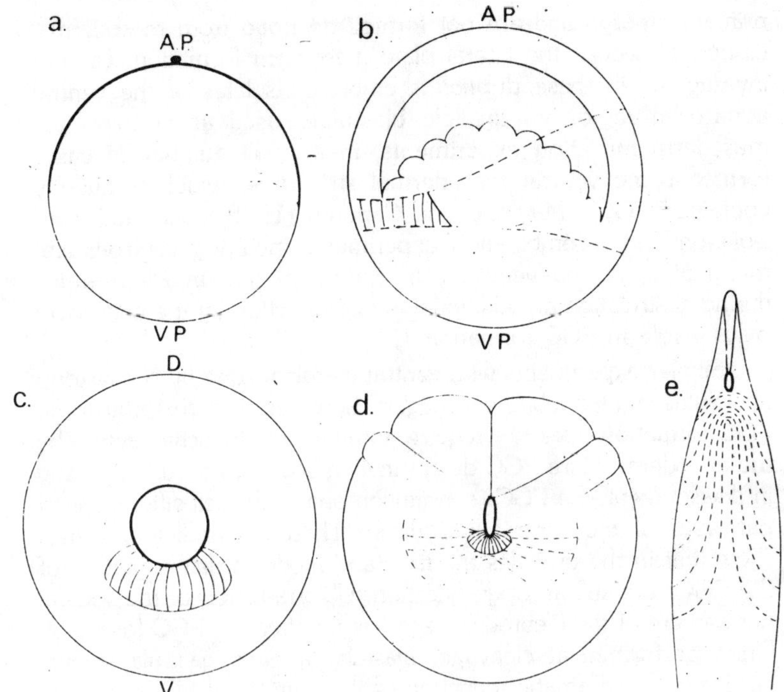

Fig. 2.1. Diagrammatic representation of the development of lateral plate mesoderm in the urodele embryo. (a) Distinction of animal and vegetal moieties in the uncleaved egg. (b) Location of presumptive lateral plate mesoderm in the ventral and lateroventral marginal zone of the blastula. (c) External location of presumptive lateral plate mesoderm at yolk-plug stage. (d) Formation of left and right wings of lateral plate mesoderm. (e) Cranial and ventral extension of lateral plate mesoderm at tail-bud stage; ventral view. A.P., animal pole; V.P., vegetal pole.

zone at the early gastrula stage. Asayama and Amanuma (1957) grafted an extra dorsal blastoporal lip into the ventral marginal zone of a host embryo, and concluded from the distribution of PGCs in the induced secondary anlage that PGCs are induced *de novo* in the secondary embryo.

The conclusion seems to be at variance with the observation by Capuron (1968) that in *duplicatas* embryos the total number of PGCs never exceeded that of a normal embryo and was often considerably lower. This would indicate that the PGCs of the secondary embryo come from the same stock as those of the primary embryo and are not formed *de novo* from mesodermal tissues. However, the lateral plate mesoderm formation was not investigated in these *duplicitas* embryos. Isolates of the ventral equatorial region of urodele blastulae, used as controls for mesoderm induction experiments, in a small number of cases formed some ventral mesodermal structures, which frequently contained PGCs. Maufroid and Capuron (1978) concluded from isolation and recombination experiments involving ventro-lateral marginal zone and ventral yolk endoderm that in *Pleurodeles* during gastrulation an additional inductive action by the endoderm plays a role in PGC formation.

Similar experiments with ventral marginal zone by Nieuwkoop and Sutasurya (unpublished) suggest, however, that in *Ambystoma* PGC formation does not require an additional inductive action by the endoderm. Here PGC determination seems to occur very early in development, the PGCs constitution one of the self-differentiation pathways of the ventral mesoderm. These contradictory results show that in the urodeles the time and mode of determination of the PGCs are still an open question. UV-irradiation of the vegetal hemisphere of the Pleurodeles egg has no effect on PGC formation when performed at cleavage, blastula or early gastrula stages, but leads to a dramatic reduction of the number of PGCs or even to complete sterility when applied at a late yolk-plug stage. At that stage the presumptive ventro-caudal mesoderm—the source of the PGCs—is localized in the ventral blastoporal lip and is therefore accessible to the irradiation.

In our opinion this result indicates that the PGCs pass through a sensitive period at the late gastrula stage, which possibly represents a (first) decisive step in their determination. Attention must now be drawn to a different line of approach. Kotani (1957) replaced presumptive lateral plate mesoderm of early gastrulae of *T. pyrrhogaster* by presumptive epidermis and found PGC formation in the grafts. He concluded that the presumptive epidermis of early gastrulae possesses germ-cell-forming potencies Smith (1966) replaced ventro-later marginal zone of stage 11 gastrulae by presumptive epidermis of early gastrulae, using white and black axolotls. When testing the progeny of the operated embryos, he did not find PGC formation of graft origin.

Kotani's results were later confirmed by Kocher-Becker and Tiedemann (1971), who found PGC formation in partially mesodermised and endodermised 'ectodermal' explants from

Triturus early gastrulae treated with a vegetalising factor isolated from chick embryos. In the course of the analysis of mesoderm induction in the urodeles many workers found that PGCs are a normal constituent of induced ventral mesoderm. When studying the mesoderm inducing capacities of the dorsal, lateral and ventral portions of the yolk mass found that both the lateral and ventral portions induce predominantly ventral mesoderm, while the dorsal portion induces massive dorsal, axial mesodermal structures. When comparing the inductive capacities at various stages in *Ambystoma* they found that by a very early gastrula stage (stage 10 H.) the mesoderm-inducing capacity of the dorsal endoderm has virtually vanished, that of the lateral endoderm is already markedly reduced, and that of the ventral endoderm is slightly diminished. This renders it very likely that the entire endoderm will have lost its mesoderm-inducing capacity by stage 11: this may explain the contradictory results of Kotani and Smith, who for their grafting experiments used early and middle gastrula stages, respectively. In *Pleurodeles*, however, the germ-cell-inducing capacity of the ventral endoderm seems to persist during gastrulation.

Kotani's conclusion is corroborated by the study of Sutasurya and Nieuwkoop (1974), who found that under the inductive influence of ventral endoderm PGCs can be formed from any part of the totipotent animal-moiety of the urodele blastula, the competence being highest in the 'ectoderm' near the equator and lowest at the animal pole. In normal development PGC formation is restricted to the ventral and ventro-lateral marginal zone by an inhibiting influence from the presumptive notochord spreading dorso-ventrally through the mesodermal mantle. From these experiments Sutasurya and Nieuwkoop conclude that in the urodeles PGCs *do not develop from predetermined elements, but arise strictly epigenetically from common, totipotent cells of the animal moiety of the blastula* as part of the regional induction of the mesoderm by the vegetal yolk endoderm. When comparing PGC formation in the urodeles with that in the anurans, one is unavoidably led to the conclusion that not only do the PGCs originate from two different sites in the two groups, but that there are moreover two fundamentally different mechanisms at work. (This conclusion is clearly in flat contradiction to a statement by Blackler (1968) to the effect that the different site of origin of the PGCs in anurans and urodeles is only a minor variation and has no development significance).

In the anurans all the PGCs originate from the endodermal moiety of the egg in the vicinity of the vegetal Pole, whereas in the urodeles they arise from the animal 'ectodermal' moiety more particularly the presumptive lateral plate mesoderm in the ventral to ventro-lateral equatorial region. In the anurans all the descriptive and experimental evidence pleads in favour of the predetermined nature of the pPGCs, based on the presence of a germ-cell-specific cytoplasmic component, the 'germinal plasm', which is present in the embryo from the very beginning of development.

In contrast, in the urodeles the PGCs develop strictly epigenetically from common, totipotent cells of the animal moiety under the inductive influence of the ventral yolk endoderm. Anuran and urodele PGCs seem to have one characteristic in common, i.e., their irreplaceability. Removal of the region from which they originate leads to sterility of the developing larvae in both groups. Regulatory formation of new PGCs from adjacent regions apparently does not occur. This could lead to the conclusion that the PGCs are already determined or at least partially determined at the stage of operation. The most striking difference in the mode of origin of the PGCs between anuran and urodele amphibians seems to be the role of the germinal plasm in germ cell determination and differentiation.

It was shown that the presence of germinal plasm is also a characteristic of urodele PGCs. Although ultrastructurally it looks identical to that of the anuran PGCs, it appears only at a late stage of development. i.e., when the cellular differentiation of the PGCs has already begun and thus long after their actual determination must have taken place. In the urodeles the germinal plasm cannot therefore play the role of germ cell determinant. It is very difficult to conceive that in one animal group a certain organelle acts as a determinant of a special cell type whereas in another group it constitutes only an attendant structure, being a differentiation product of that same cell type. This incongruity rises the question of whether in the anurans the characteristic structures of the germinal plasm actually do act as germ cell-determinant, or whether another, non-structural component of the cell is responsible for germ cell determination

Coecilia

There is only one reference, Marcus (1938.), on the possible site of origin of the PGCs in the Coecilia, the most primitive group of recent amphibians. In histological sections of rather late embryonic stags of *Hypogeophis* Marcus identified the PGCs in the endoderm in the vicinity of the blastopore. At later stages he found them in the dorsal mesentery and concluded that they had arrived there either direct from the dorsal endoderm or by passing through the lateral plate mesoderm. They are finally located in the genital ridges ventral to the Wolffian ducts. Marcus states that the PGCs have an 'endodermal' appearance. The evidence is inconclusive, however, the more so since it is known that the PGCs may migrate from one tissue into another at relatively early stages of development and may engulf yolk platelets from cytolysing cells.

The Fishes

As regards the fishes a clear distinction must be made between the development of the four classes: the Agnatha or cyclostomes, the Chondrichthyes or elasmobranchs, the Osteichthyes, and the Teleostomi or teleosts. The early development of the eggs of the Agnatha and Osteichthyes, which both represent ancient groups of fishes, is rather alike and shows close parallels with the early development of the amphibians. The development of the yolk-rich egg of the Chondrichthyes is rather different from that of the former groups, while the Teleostomi constitute a highly specialized groups, while the Teleostomi constitute a highly specialized group with a very aberrant development.

Agnatha

The only data available at present on the origin of the PGCs in the Agnatha are those of Okkelberg (1921) on the brook lamprey. He found PGCs first at an early tail-bud stage, in the most caudal portion of the trunk where they are scattered in the lateral plate mesoderm. Later they accumulate in the vicinity of the Wolffian ducts. This descriptive evidence suggests a mesodermal origin of the PGCs but is inconclusive.

Osteichthyes

On the origin of the PGCs in the Osteichthyes only descriptive data from a few species are available. In *Amia* some PGCs are found in the peripheral region of the endoderm at an early stage, but the majority the present in adjacent lateral plate mesoderm. In *Lepisosteus* Allen (1911) found them in the ventral and lateral portions of the single-layered gut endoderm. Reinvestigating the original material of *Polypterus* collected by Kerr. De Smet (1970)

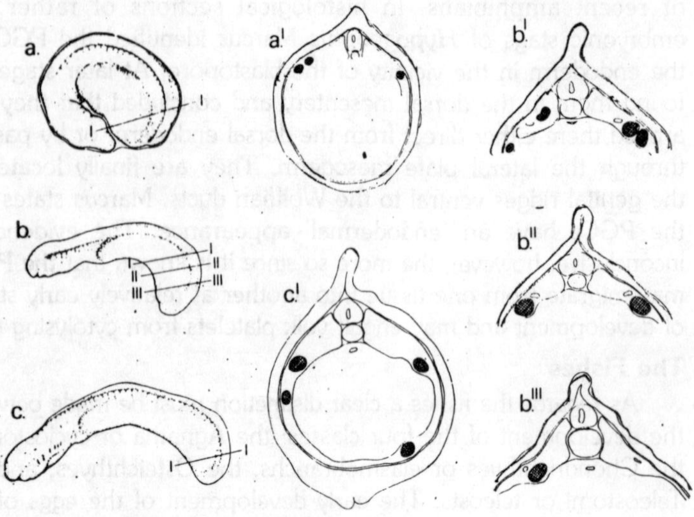

Fig. 2.2. Location of PGCs (hatched areas) in the brook lamprey embryo at three successive tail-bud stages (a-c), with cross sections through the caudal trunk (a^{I}, b^{I}–b^{III} and c^{I}).

found PGCs were first visible in an early larval stage among the yolk-laden cells of the roof of the gut, at the boundary between the gut endoderm and the lateral plate mesoderm and in the lateral plate mesoderm itself.

Finally, in *Acipenser* Maschkowzeff (1934) was first able to detect PGCs in the intestinal endoderm. On the basis of the yolk content of the PGCs the various authors concluded that in all the four species studied they are of endodermal origin. They are supposed to migrate-from their endodermal site of origin towards the lateral plate mesoderm and then to the dorsal midline, where they accumulate near the nephrostomes. It must however be said that the available descriptive evidence suggesting an endodermal origin of the PGCs in the Osteichthyes is very scanty and inconclusive.

Chondrichthyes

Data on the origin of the PGCs in the elasmobranchs are similarly scanty and inconclusive. In *Raja batis* Beard (1900, 1902a, b) found them for the first time at an open neural plate stage. The majority were located in the peripheral dorso-lateral endoderm, some in the adjacent lateral plate mesoderm, and even a single one in the overlying ectoderm. In older embryos they were found in the splanchnic mesoderm or between this and the intestinal epithelium. Beard (1920a) found a similar location of the PGCs in *Raja radiata*, *Pristiurus melanostomus*, *Scyllion vulgaris*, *Torpedo ocellata* and *Acanthias vulgaris*. This descriptive evidence points towards a relatively early and endodermal origin for the PGCs in the Chondrichthyes, but is again far from conclusive.

Teleostomi

Eigenmann (1891) concluded from the size of the PGCs in *Micrometrus* (= *Cymetogaster*) that they must have segregated from the somatic cells at an early cleavage stage. In *Findulus* Richards and Thompson (1921) were first able to identify the PGCs in the peripheral endoderm in 46 to 53-hour embryos (early tail-bud stages). They stressed the close association of the PGCs with the syncytial periblast, an observation which was confirmed by Hann (1927) and Wolf (1931). In *Scardinius* Reinhard (1924) described their possible segregation from the syncytial periblast. In the salmonid *Micropterus* Johnston (1951) identified the PGCs first in an excrescence of the *periblast* protruding the PGCs first in an excrescence of the periblast protruding over the caudal edge of the blastodisc (erroneously called the dorsal blastoporal lip). At a slightly later stage they were found in the caudal (erroneously called ventral) periblast, from where they moved into the caudal extension of the yolk sac. From there they wedged their way between the splanchnic mesoderm and the gut endoderm towards the mid-dorsal region, and subsequently towards the genital ridges. In *Oryzias* PGCs were first found in the unsegregated mesendoderm, occasionally also in the ectoderm, at an 'early gastrula stage'.

After the segregation of mesoderm and endoderm they were found in all three germ layers. Since their number increased in the absence of detectable mitotic divisions. Gamo assumed that they segregated during the entire early development. In early stages they were closely associated with the syncytial periblast. After myotome formation they were found predominantly in the peripheral endoderm, where they seemed closely associated with yolk protrusions. Gamo also noticed the capacity of the PGCs to absorb yolk. Only those PGCs which are incorporated into the gonadal anlagen become definitive germ cells, while the ectopic germ cells degenerate. According to Pala (1970), in *Gambusia* PGCs are first discernible among the deeper cells of the embryonic node in the caudal half of the blastodisc at an 'early gastrula stage'. During the further segregation of the embryonic anlage they are found in the mesodermal portion of the mesendoderm.

Nedelea and Steopoe (1970) localized the PGCs in the mesodermal layer of the embryonic shield of *Cyprinus* shortly before the completion of the overgrowth of the yolk. The only experimental evidence available seems to argue for a mesodermal origin of the PGCs. Randomly reassociated fragments of posterior thirds of embryonic shields of 'middle gastrulae' (about half way in the overgrowth of the yolk), grafted into the extra-embryonic area of a host embryo, formed gonads with PGCs in the absence of any endodermal structures. Summarizing, it must be said that there is no concordant evidence for a particular site of origin of the PGCs in the teleosts. The often-observed close association of the PGCs with the syncytial periblast may well be of a secondary nature in view of the tendency of the PGCs to associate themselves with a nutritional source. Nothing can at present be said about the mode of origin of the PGCs.

The Lower Chordates

Tunicata

Except for the Larvacea, which are sexually mature tadpole-like organisms, all the other tunicates undergo metamorphosis, during which the axial structures (notochord and somites) so characteristic of the chordates are resorbed. As far as is known PGCs are only found in metamorphosed, sessile individuals. Berrill (1975) states that among the tunicates various forms of asexual reproduction are found, during which a small fragment of the parental body is set apart for the formation of a new zooid.

In several forms sexual and asexual generations alternate and in these forms no germ line is therefore traceable. The germ cells in the hermaphroditic gonads—the ovotestes seem to arise from small masses of undifferentiated mesodermal cells. These cells are called haemoblasts, suggesting an origin from the vascular system. In summary it may be said that in the tunicates the PGCs probably originate in the mesoderm, although their mode of origin is unknown.

Cephalochordata

According to Boveri (1892) the PGCs in *Branchiostoma* larvae are found in clusters at the ventro-cranial extremities of the eleventh to thirty-sixth of the 61 somites. The '*Grenzzelle*' which Hatschek (1888) found at the ventral extremity of each somite in a younger larva may represent the segmental stem cell of the PGCs. The strictly segmental origin of the PGCs pleads in favour of their differentiation *in situ* from the intermediate mesoderm.

THE AMNIOTA

Aves

Swift (1914) was the first to locate the PGCs at early somite stages, in an anterior, extra-embryonic crescent-shaped area at the boundary between the are pellucide and the area opaca, which he called the 'germinal crescent'. In the next decade this observation was both confirmed and contradicted, but it gradually found wider approval. Willier (1937) found PGCs in chorioallantoic grafts of portions of the chick blastoderm containing part of the germinal crescent, whereas grafts without this region were sterile. Simon (1960) observed complete sterility or a very marked reduction in the number of PGCs after removal of the germinal crescent. Reynaud (1969) destroyed all the PGCs by UV-irradiation of the germinal crescent, and was able subsequently to restore a normal PGC population in the irradiated embryo by intravenous injection of a cell suspension made from the germinal crescent of another embryo.

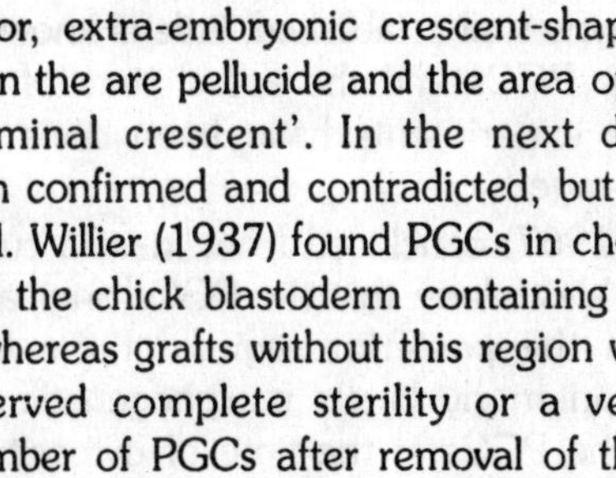

a

b

Fig. 2.3. Lateral view of part of a row of somites (s) in larva of Branchiostoma lanceolatum. (a) Segmental location of presumed stem cells of PGCs (black) in early larva. (b) Similar arrangement of clusters of PGCs in slightly older larva.

The Spatial extension of the germinal crescent apparently varies markedly in different species and breeds. In the chick and turkey it usually surrounds only the anterior half of the embryonic anlage, but in the quail it extends around the anterior two-thirds, while in a chick hydrid and in the duck its extension is even more variable and may reach the posterior side of the embryonic anlage. The PGCs are initially located in the endodermal hypoblast layer of the germinal crescent. They then segregate from the hypoblast and accumulate at the junction of the ectoderm and endoderm. The begin to enter the extra-embryonic blood islands as soon as the invaginated mesoderm reaches the germinal crescent area. Although the extra-embryonic location of the PGCs at early somite stages is by now well established, little is known about the *primary origin* of the PGCs and the actual time of their determination. Quite a few experiments have been done at the unincubated blastoderm stage.

Dubois (1967) concluded from in-vitro culture of parts of the unincubated blastoderm that the PGCs originate from the primary hypoblast in the posterior region of the blastoderm and are transported anteriorly by its morphogenetic movements. Dubois later found the PGCs at the time of oviposition to be distributed in a transverse band in the posterior half of the embryonic anlage. This seems to be at variance with the findings of Fargeix (1969), who obtained a 'normal' number of PGCs (i.e., equal to that of a abnormal embryo) from anterior halves of unincubated blastoderms, while posterior halves formed either a 'normal' or a deficient number of PGCs. Lateral halves always formed a 'normal' number of PGCs. The number of PGCs decreased in cultured posterior halves as they were isolated at later stages.

Rogulska (1968) found that duck blastoderms transected *in vivo* formed double the 'normal' number of PGCs. In those cases where the embryonic anlagen were oriented parallel to each other, the PGCs were predominantly located in the zone between them. Working with freshly laid chicken 'winter' eggs, which did not yet show hypoblast formation. Eyal-Giladi, Kochav and Menashi (1976) found that both anterior and posterior halves formed the 'normal' number of PGCs. Fargeix (1975) showed that X-irradiation of the entire blastoderm can lead ultimately to the sterility of the embryo. In such embryos the number of PGCs is unaffected at an early somite stage but decreases later, whereas unirradiated embryos the number of PGCs increases steadily.

Fargeix found that in the unincubated blastoderm irradiation of either the anterior or the posterior half led to the same reduction in PGC number. However, irradiation of the posterior half of the blastoderm incubated for 10 hours led to only a slight reduction in the number of PGCs, whereas a marked reduction was obtained after the same treatment of the anterior half. From these experiments he concluded that the pPGCs are evenly distributed throughout the unincubated blastoderm, but accumulate in the anterior half during the first hours of incubation. This is in accordance with the segregation *in situ* of the primary hypoblast from the epiblast and its subsequent anterior displacement as a result of the expansion of the secondary hypoblast from Koller's sickle. The formation of a 'normal' number of PGCs from anterior and posterior halves of unincubated blastoderms seems to be at variance with the assumption the predetermined elements are present in the primary hypoblast–unless a mechanism exist which leads to a doubling of the number of PGCs after transaction. In this connection two contradictory observations are of interest.

Fargeix (1969) states the transected blastoderms are always retarded in development, so that there would seem to be enough time for an extra clonal division of the PGCs. In contrast Swartz and Domm (1972), who studied mitoses in PGCs by colchicine treatment, observed that the clonal divisions of the PGCs, start late, i.e., not before the migration period. A possible alternative explanation for the 'normal' number of PGCs in half embryos, first suggested by Fargeix (1969) and again by Eyal—Giladi *et al.* (1976), is that predetermined elements in the sense of the anuran pPGCs do not exist in the early development of birds, but that (in both half and normal embryos) the PGCs are induced (in 'normal' number) at a later stage of development. If an early segregation of pPGCs does occur in birds, it nevertheless seems to differ from that in the anuran amphibians, since no germinal plasm has hitherto been demonstrated in birds.

The prime characteristic of avian PGCs, i.e., their high glycogen content, cannot be regarded as a germ cells determinant since it is not specific to PGCs and appears relatively late in development. It probably only represents a condition for their migratory activity. In birds the role of germ cells determinant cannot at present be attributed to any particular ultrastructural or physiological characteristic of the PGCs. Reynaud (1968) observed accelerated multiplication of the germ cell population re-established in previously sterilized chick hosts by injection of a small number of turkey PGCs. Lutz and Lutz-Ostertag (1972) called this phenomenon auto-regulation.

Reynaud (1971a) studied 'normal' germ cell multiplication in chick, quail and turkey. Auto-regulation was also observed by Fargeix (1976) X-irradiated anterior and posterior halves of unincubated duck blastoderms. Summarizing, it may be said that the early segregation of the avian PGCs from the primary hypoblast shows certain similarities to the history of the germ cells in the anuran amphibians. However, the avian PGCs are not characterized by the presence of germinal plasm or any other germ-cell-specific structure. The suggestion by Fargeix (1969) and by Eyal-Giladi *et al.* (1976) that avian PGCs may be formed epigenetically during embryogenesis (as they are in the urodeles) cannot be excluded, but does not seem to be easily reconciled with the endodermal origin of the PGCs and with the results of X-irradiation at early blastoderm stages. It must therefore be concluded that the actual mode of origin of the PGCs in birds is still unknown.

Mammalia

The data relevant to the site and mode of origin of the PGCs in the mammals come predominantly from the analysis of the early development of the mouse embryo, to which we will therefore mainly refer. Brambell (1927) was the first to recognize the extra-gonadal origin of the PGCs when he detected them in the extra-embryonic yolk sac endoderm of the 8-day mouse embryo. During gut formation they are displaced towards the hindgut endoderm. Leaving the endoderm, they migrate through the dorsal mesentery towards the genital ridges. This observation was confirmed by Everett (1943), Chiquoine (1954), on the other hand, stated that in the 8-day mouse embryo the PGCs originate from the yolk sac splanchnic mesoderm and perhaps even from the caudal portion of the primitive streak, thus suggesting a mesodermal origin. In 1960 Brambell again inferred an endodermal origin for the PGCs. Mintz (1960a) observed them first in the yolk sac endoderm in the 8-day presomite mouse embryo.

Blandau, White and Rumery (1963) found the PGCs at the base of the allantoic evagination. The endodermal origin of the PGCs was again questioned by Ozdzenski (1967), who found them in the caudal region of the primitive streak as well as in the anlage of the allantois of presomite embryos. He concluded that they migrate from the primitive streak mesoderm into the endodermal layer. Bruel-Beaudenon and Hubert (1968) speak of a 'posterior germinal crescent' in the mammals. Hamilton and Mossman (1972) again have the PGCs originating from the extra-embryonic yolk sac endoderm. Merchant and Zamboni (1973) identified the PGCs in the gut epithelium of the 9-day mouse embryo. They came to the conclusion that the PGCs associate themselves with other tissues because as the result of insufficient yolk material they need an exogenous energy source, the alkaline phosphatase activity at the cell periphery denoting an active intercellular exchange of material.

The location of the PGCs in the yolk sac endoderm of the 8- to 9-day mouse embryo was confirmed by Spiegelman and Bennett (1973). Clark and Eddy (1975) agree with Merchant and Zamboni (1973) that the PGCs reside only temporarily in various somatic tissues during their life-history, and depend upon the somatic tissues for their nutrition. Although they found the PGCs for the first time in 8- to 8-day embryos in that portion of the yolk sac endoderm that later forms the hindgut, the PGCs do not resemble the surrounding endodermal cells in their ultrastructure but look much more similar to the adjacent mesodermal cells. They therefore propose that the PGCs do not in fact originate from the yolk sac endoderm but from the mesoderm. In other mammalian species the PGCs are likewise first detected in the extra-embryonic yolk sac endoderm and the allantoic endoderm and mesoderm of presomite embryos in the guinea pig; and in the human.

At slightly older stages they are found in the hindgut endoderm in the guinea pig; Chretien (1966) in the rabbit; Sawano (1959) in the human. The only disparate observation concerns a study by Vanneman (1917) in the marsupial *Armadillo*. In this polyembryonic form the PGCs were identified for the first time between the extra-embryonic ectodermal and endodermal layers, outside the primary embryonic buds. This more or less resembles the situation in birds, but the finding must be viewed with caution since it has never been confirmed. The only experimental evidence comes from Gardner and Rossant (1976), who produced climaeras for coat colour genes in the mouse by injecting embryonic ectoderm cells into host blastocysts, and found that PGCs were formed from the injected cells. The concluded that the PGCs originate from (totipotent) precursor cells of the definitive embryo rather than from the extra-embryonic yolk sac endoderm.

The totipotent embryonic 'endoderm' would yield both PGCs and somatic cells, the segregation of which would occur only later. This situation rather resembles that in the urodele amphibians where the lateral plate mesoderm, from which the PGCs originate, arises in the process of mesoderm formation from the totipotent animal 'ectodermal' moiety of the blastula. Recently Falconer and Avery (1978) concluded that in the mouse the variability observed in chimaeras produced by aggregation of cleavage stages and in mosaics resulting from X chromosome inactivation provides additional evidence that the PGCs originate in the primary ectoderm and not in the yolk sac endoderm.

Reptilia

The origin of the PGCs in the reptiles has been traced back no further than the early somite stages. In all reptiles studied the PGCs were found in the extra-embryonic area of the blastoderm. As to the more precise location of the PGCs the reptiles fall into two categories: (a) those with an anterior germinal crescent, which may extend around the entire embryonic anlage, and (b) those with a posterior germinal crescent. The former localization more or less resembles the situation in birds, the latter that in mammals. A strictly anterior germinal crescent was found in the snake *Vipera* by Hubert (1969), while an anterior localization with extensions to the lateral and posterior sides seems to exist in the Lacertilia *Mabuya* and *Chamaeleo* and the slow-worm *Anguis*. Both anterior and posterior germinal crescents were described in *Sphenodon* by Tribe and Brambell (1932). A strictly posterior germinal crescent was found in the chelonians as well as in several *Lacerta* species.

It is evident that the two types of localization do not coincide with particular taxonomic groups, since the Lacertilia show both types. According to Hubert (1976), in the forms having an anterior germinal crescent (e.g. Mabuya, *Chamaeleo*) the PGCs are displaced anteriorly together with the primary hypoblast by the gastrulation movements, whereas in those having a posterior germinal crescent (e.g., the chelonians and *Lacerta*) the localization of the PGCs is not affected by the gastrulation process, which occurs more anteriorly. At the stages examined the PGCs were found in the endodermal layer before the mesoderm has spread between the ectoderm and endoderm.

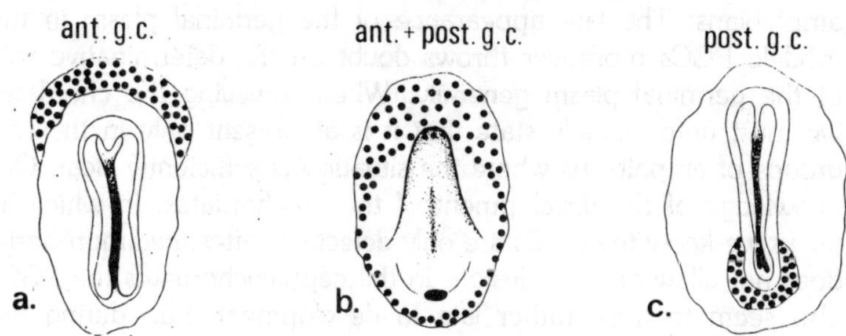

Fig. 2.4. Different extra-embryonic locations of PGCs in reptiles. (a) Anterior germinal crescent (ant.g.c.) in several Lacertilia and Ophidia. (b) Both anterior and posterior germinal crescent (ant.+post.g.c.) in Sphenodon. (c) Posterior germinal crescent in Chelonia and some Lacertilia.

According to Hubert (1976) the PGCs originate either in the primary hypoblast or in the definitive hypoblast formed during gastrulation. The only experimental evidence available, which was obtained by the extirpation of the posterior germinal crescent in *Lacerta vivipara* either before or after the migration of the PGCs towards the genital ridges, confirms the descriptive data. Summarizing, it must be concluded that little is known about the actual site of origin of the extra-embryonic PGCs in the reptiles. The observed localization in the anterior or posterior germinal crescents probably represents a secondary situation. About the mode of origin of the PGCs in the reptiles we are entirely in the dark.

General Conclusions

Surveying the entire phylum of the Chordata with respect to the site and mode of origin of the PGCs, it is evident that at least two essentially different mechanisms are involved in individual groups, viz. (a) a very early segregation of the germ line from the somatic cell lines, and its determination by a special cytoplasmic component, and (b) a strictly epigenetic development of the germ cells from totipotent embryonic cells under an inductive influence. These two alternative mechanisms correspond to an endodermal and a mesodermal origin of the PGCs, respectively. The existence of the two mechanisms has been deduced from the experimental analysis performed in anuran and urodele amphibians.

The existence of the two mechanisms has been deduced from the experimental analysis performed in anuran and urodele amphibians. The late appearance of the germinal plasm in the urodele PGCs moreover throws doubt on the determinative role of the germinal plasm generally. When surveying the chordates we must unfortunately state that it is at present only in the two groups of amphibians where the situation is sufficiently clear. Our knowledge of the development of the urochordates, in which as far as we know the PGCs are only detectable after metamorphosis, does not allow any conclusion. In the cephalochordates the PGCs also seem to arise rather late in development, i.e., during the final differentiation of the mesoderm. Their clearly segmental localization ventral to the somite segments pleads in favour of differentiation *in situ* within the epigenetically formed mesoderm, suggesting a 'urodele' type of germ cell formation.

In the Agnatha the PGCs are first found scattered in the lateral plate mesoderm at an early tail-bud stage, a situation which again resembles the 'urodele' type of germ cell formation. In the Osteichthyes there is some purely descriptive evidence for an endodermal localization of the PGCs, but this does not say anything about their actual site and mode of origin. The same uncertainty holds for the situation in the Chondrichthyes. The purely descriptive evidence regarding the Teleostomi is far from concordant and does not allow a decision one way or the other. No experimental evidence is available for the Coecilia. When considering the amniotes, the Reptilia are almost unknown terrain, apart from some evidence that the situation in the Chelonia differs from that in the Squamata. In contrast much experimental evidence is available on the birds. The PGCs seem to originate in the extra-embryonic primary hypoblast during early development. The lack of germinal plasm in avian PGCs possibly means their actual mode of origin is different from that in the anuran amphibians. It also calls into question once more the determinative role of the germinal plasm. In the mammals the still scanty experimental evidence suggests an epigenetic mode of origin of the PGCs, in line with the 'urodele' type of germ cell formation.

3

Extra-Gonadal Origin of Germ Cells

Although the origin of the primordial germ cells is studied in details, still it has several controversies. Two controversial issues with respect to the origin of the PGCs in vertebrates have dominated the literature for many decades: (a) that of the gonadal versus extra-gonadal origin of the PGCs, and (b) that of the one-time versus repeated origin of the PGCs during the life span of the organism. We feel that these controversial issues should be dealt with before the place and mode of origin of the PGCs can be satisfactorily discussed.

Gonadal versus Extra-Gonadal Origin of the PGCs

The early literature on the origin of the PGCs in vertebrates was dominated by the controversial issue of the gonadal versus extra-gonadal origin of the PGCs. Embryologists were sharply divided into opposing camps, i.e., those supporting Waldeyer's (1870) ideas and those advocating Nussbaum's (1880) concepts. *Waldeyer's theory*, which implies an origin of the PGCs from the somatic 'germinal epithelium' of the gonadal anlage, was primarily based on a study of the higher vertebrates, where the PGCs were not clearly distinguishable before they were inside the gonadal anlagen.

The concept was later extended to all the vertebrates and even to a number of invertebrate groups, Nussbaum's notion, which implies an early segregation of the PGCs from the somatic cells of the embryo long before the formation of the gonadal anlagen and topographically separated from them was chiefly based of studies in the anuran amphibians and teleosts. The issue remained controversial until the fourth decade of this century, as a result of the rather unspecific staining methods available.

During the nineteen-thirties the balance between the two points of view began to shift towards extra-gonadal origin, particularly in the lower vertebrates. When Bounoure's book appeared in 1939 the advocates of Waldeyer's concept had lost nearly all their ground as regards the lower vertebrates, but not as regards the higher vertebrates, especially the mammals, where the PGCs could only be clearly distinguished inside the gonadal anlagen. Brambell (1960) already stated: It is now recognized by a large majority of embryologists that (in the vertebrates) the PGCs arise exceedingly early (in development), long before the rudiments of the gonad are formed and at a distance from the site they will (ultimately) occupy...Their state of origin is extra-gonadal in all instances and in the amniotes and some of the fishes it is (moreover extra-embryonic). This marked shift in the general consensus was mainly due to the development of new histological and histochemical techniques which allowed identification of the PGCs a much earlier stage of development that was possible until then. Recent studies have fully confirmed Brambell's conclusions e.g., in the fishes.

One-time Versus Repeated Origin of the Germ Cells

Towards the end of the nineteenth and the first half of the twentieth century the concept of the origin of the PGCs was obscured by the controversial issue of whether there exist only a single source or multiple sources of germ cells during the life span of an individual. The concept of the continuity of the germ line as postulated in Weismann's *Keimplasma* theory implies a segregation of the germ cells from the somatic cells-preferably during early development–as well as the persistence of the former throughout fertile life. The adversaries of this concept essentially claimed the existence of several phases of asexual and sexual life. In his 1945 review Everett distinguished four different categories among those working on the origin of the germ cells in vertebrates: (1) those who deny the early segregation of the PGCs and advocate only a late, somatic origin from the 'germinal epithelium', (2) those who believe in the early segregation of germ cells but postulate the subsequent degeneration of this first generation and the ultimate formation of a second generation of definitive germ cells from the 'germinal epithelium', (3) those who agree to the early segregation of PGCs , and the subsequent development into definitive gametes, but in addition postulate a supplementary formation of germ cells from the 'germinal epithelium', and (4) those who defend a one-time, early segregation of the PGCs, which constitute the only source of the definitive gametes.

The issue is further complicated by the assumption either of a periodic or of a continuous formation of germ cells from the 'germinal epithelium' throughout the reproductive period. We have seen in the preceding section that in the vertebrates the PGCs, are formed outside the gonadal anlagen, to which they are subsequently transferred. The problem therefore boils down to the question of whether these PGCs represent the sole source of

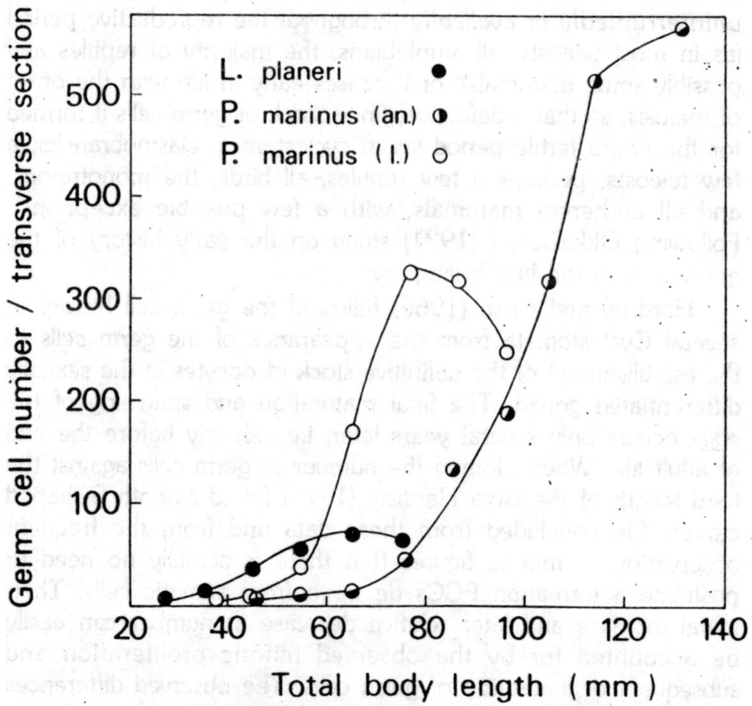

Fig. 3.1. Increase in average germ cell number per transverse body section in relation to total body length, in Lampetra planeri and Petromyzon marinus of the anadromous (an.) and landlocked (l.) races.

the definitive gametes or whether they are supplemented with or replaced by one or more generations of germ cells formed *de novo* in the gonadal anlagen during later phases. This uncertainty was mainly due to the fact that particularly in the higher vertebrates the PGCs lose most of their cytological characteristics during multiplication in the gonadal anlagen. The problem can therefore only be definitely solved by experimental analysis. Nevertheless, the evidence for the existence of a single and uninterrupted germ line in the life cycle of vertebrates is both descriptive and experimental. The descriptive evidence consists of quantitative data on germ cell history in normal animals and sterile mutants, while the experimental evidence mainly relates to defect and grafting experiments.

Quantitative Data

In the vertebrates *oogenesis* follows a uniform pattern with two main variants. Either germ cell multiplication continues *uninterruptedly* or *cyclically* throughout the reproductive period (as in most teleosts, all amphibians, the majority of reptiles and possibly some mammals), or it ceases early in life with the onset of meiosis, so that a definitive, finite stock of germ cells if formed for the entire fertile period (as in cyclostomes, elasmobranchs, a few teleosts, perhaps a few reptiles, all birds, the monotremes, and all eutherian mammals, with a few possible exceptions). Following Okkelberg's (1921) study on the early history of the germ cells in the brook lamprey.

Hardisty and Cosh (1966) followed the germ cell history in several Cyclostomata from the appearance of the germ cells to the establishment of the definitive stock of oocytes in the sexually differentiated gonad. The final maturation and spawning of the eggs occurs only several years later, i.e., shortly before the end of adult life. When plotting the number of germ cells against the total length of the larva Hardisty (1971) found smooth S-shaped curves. He concluded from these data and from the frequent observation of mitotic figures that there is actually no need to postulate a formation PGCs *de novo* from somatic cells. Their initial increase and later relative decrease in number can easily be accounted for by the observed mitotic proliferation and subsequent degeneration of germ cells. The observed differences in fecundity of the various species of lamprey must be due to differences in proliferation, since the initial average number of PGCs is nearly the same in the different species.

The difference between the actual fecundity of the adult female (number of eggs produced) and the potential fecundity of the ammocoete larva (maximum number of germ cells) can be accounted for by the observed extensive degeneration of germ cells during sexual differentiation. In the chick Hughes (1963) studied germ cell number in the left ovary from day 9 of incubation till the first day after hatching. The formation of oocytes ceases at about the time of hatching. Oogonial mitoses are responsible for the nearly 25- fold increase in germ cell number between day 9 and day 17, while the subsequent decrease by about 40% between day 17 the first day after hatching is due to the high incidence of degenerating germ cells.

Hughes concludes that also in the chick ovary there is no need to postulate a second source of germ cells during oogenesis. A similar study, on oogenesis in the rat, performed between day 14 *post coitum* (p.c.) and day 2 *post partum* (p.p.) shows that oogonia are mitotically active until day 17 p.c., after which a sharp decline in mitotic activity occurs at the onset of the leptotene stage of meiosis. The total population increases sixfold between days 14 and 17 reaches a slightly higher maximum value at day18, and is reduced by about one third, between day 18 p.c. and day 2 p.p During this period four different waves of germ cell degeneration were distinguished. In the rabbit ovary Chretien (1966) found two main periods of germ cell multiplication, a first one between days 10 and 12 p.c., leading to a fourfold increase, and a second one between days 16 and 18 p.c. resulting in an approximately eightfold increase. The germ cell number reaches a maximum at day 26 p.c., after which a marked decrease takes place by date 28 as a consequence of extensive degeneration.

Since in the rabbit PGC migration starts at about day 10 and lasts at least until day 16 to 18, it is evident the PGCs actively multiply during migration. A similar germ cell history was found in the guinea pig by Ioannou (1964), in the monkey by Baker (1966), and in the human foetus by Baker (1963). The most dramatic increase in germ sell number occurs in the ovary of the human foetus, viz. from 700-1300 during migration to approximately 600000 in the second month of gestation, and further to a maximum of around 7000000 in the fifth month. After several waves of degeneration the number falls to about 2000000 at birth and further to approximately 300000 at the age of seven.

Since only about 500 eggs ovulate during the entire fertile period, there is a tremendous overproduction of female germ cells in human development. Although in the human foetus oogonial multiplication reaches a maximum in the fifth month of gestation.

Gailard (1950) observed extensive germ cell formation in explants of ovarian cortex of 14-week to 36-week human foetuses and concluded that the potency for germ cell multiplication is not restricted to the early foetal period but certainly persists until birth. His explants first showed complete degeneration of parenchyma and young oocytes, after which new cord-like structures were formed from the 'germinal epithelium', which surrounds the explant. In these sex cords new oogonia appeared in large numbers. He left open the question of whether these oogonia originated from persisting but unrecognizable germ cells or were formed *de novo* from the somatic cells of the 'germinal epithelium'. In the mouse Rudkin and Griech (1962) found that [^{3}H] thymidine is incorporated into oocyte nuclei when administered half-way through the gestation period. The label was still found in the oocytes of female offspring at 6 weeks p.p., suggesting that the PGCs present at the time of labelling had actually formed mature oocytes in the adult female.

Borum (1966) found up to 100% labelling in the oocytes of female offspring after [^{3}H] thymidine injections into pregnant mice at days, 12 to 15 of gestation but no incorporation into oocytes upon injection at different times after birth. He concluded that all oocytes present in the adult mouse have persisted from foetal life, so that at birth the mouse is furnished with a definitive population of oocytes from which all mature ova will later be derived. Peters and Crone (1967) showed that DNA synthesis occurs in oogonia in premeiotic interphase, the rate of synthesis falling during the foetal period in the mouse but within the neonatal period in the rabbit. Once synthesized the DNA persists in the growing and maturing oocytes.

Kennelly and Foote (1966) concluded from [^{3}H] methyl-thymidine injections into female rabbits on the day-of birth and at 4 weeks p.p. that most, if not all definitive ova have already been formed at birth and that oocytogenesis *de novo* does not occur in the post-pubertal rabbit. Unfortunately little is known about PGC numbers in forms which show a continuous or cyclic multiplication of the germ cells during the reproductive period, all the above forms belonging to the second variant of germ cell multiplication. *Spermatogenesis* is usually characterized by either continuous or cyclic renewal of the germ cell population. In mammals it has been studied in recent years by a number of workers. In these studies emphasis was placed on renewal of the spermatogonial stem cells and on the phenomenon of germ cell degeneration. Hilscher and Hilscher (1979) state that when female and male gametogenesis in mammals are compared the 'gonia' stage of the female germ cells shows only a single proliferation wave, whereas that of the male germ cells shows a first proliferation wave which is comparable with that of the oogonia, followed by a second wave after a preparatory interphase. The second proliferation wave is characterized by both stem cell renewal and differentiation into spermatocytes.

According to Clermont (1972) the seminiferous epithelium of the mammalian testis is composed of five to six generations of germ cells. The spermatogonial population is renewed by a number of successive mitoses (three in man and up to six in the rat). Two types of spermatogonial stem cells can be distinguished, the type A 'reserve' and type A 'renewing' stem cells. The former type is not involved in the production of spermatocytes, while the latter type renews itself and simultaneously gives rise to differentiating spermatogonia at each mitotic cycle. However, the 'key' division between stem cell and differentiating spermatogonium does not have the character of a differential in the morphological sense. The mechanism which regulates this choice is still unknown. The general conclusion from the quantitative analysis of both oogenesis and spermatogenesis in the vertebrates is that there actually is no need to postulate any formation of germ cells *de novo* from somatic cells during any part of development or fertile life. It is therefore very likely that all the definitive male and female gametes descend from the initial population of germ cells, which has arisen extra-gonadally during early embryonic development.

Experimental Evidence

Several experiments have been performed related to the surgical removing of primordial germ cells. Elimination of the PGCs has been achieved surgically by removing that portion of the embryo in which the PGCs are located at the time of operation. It is a well-established fact that complete removal of the gonads leads to permanent sterility, but incomplete removal may lead to regeneration of the gonad and recovery of fertility. In amphibians removal of the extra-gonadal source of the PGCs prior to their migration to the gonadal anlagen leads to the formation of gonads without PGCs and to permanent sterility.

In neurulae of *Ambystoma mexicanum* Nieuwkoop (1947) removed the presumptive lateral plate mesoderm-the source of the PGCs in the urodeles-and obtained larvae which had more or less normal genital ridges but lacked PGCs. Vivien (1964) obtained similar results in *Lebistes* and *Xiphophorus* upon ^{32}P administration. Blackler (1962, 1965a) replaced the fertile region of the endoderm of *Xenopus laevis* neurulae—the source of PGCs in the anurans—by a similar, more anterior region of the endoderm of which does not contain PGCs, and obtained completely normal-looking tadpoles with normal gonadal anlagen but without PGCs. When reared to maturity the animals remained sterile. In the anurans similar results were obtained by the elimination of the so-called 'germinal plasm', which is assumed to act as a germ cell determinant.

Pricking of the vegetal pole of UV-irradiation of the vegetal hemisphere performed at the 1- to 14-cell stage can lead to sterility of the gonadal anlagen of the larva and to subsequent permanent sterility of the gonadal anlagen of the larva and to subsequent permanent sterility. It may therefore be concluded that destruction of the PGCs or their putative 'determinants' leads to permanent sterility. In other words, no formation of germ cells *de novo* occurs in animals, which have no PGCs from an early stage of development. Although this seems a strong argument in favour of the continuity of the germ line, it is essentially negative and therefore inconclusive, for formation of germ cells *de novo* from somatic cells of the 'germinal' epithelium could depend upon a stimulating influence from existing viable germ cells. This rather unlikely postulate can only be disproved by heteroplastic or xenoplastic recombinations of presumptive germ cells and somatic gonadal tissue or by using specific nuclear markers. Such experiments were performed by Blackler and Fischberg (1961), who exchanged the fertile endoderm region between two strains of *Xenopus laevis* of which one contained the Oxford nuclear marker.

Blackler (1962) made a similar exchange between two subspecies of *X. laevis* while Blackler and Gecking (1972a, b) performed the same experiment between *X. laevis* and *X. mulleri*, with subsequent intraspecific and interspecific matings. In all these experiments the germ cells showed the characteristics of the donor type the furnished the fertile endoderm region. Similar experiments were carried out in urodeles by Smith (1964), who exchanged the ventro-lateral marginal zone (presumptive lateral plate mesoderm, representing the source of PGCs in the urodeles) between gastrulae of the white and the black axolotl.

The successful cases in which the exchange was complete showed only progeny of the donor type, as tested by mating as well as histological examination. Because in Blackler's experiments the overlying ectoderm and mesoderm were also exchanged, while in Smith's experiments the graft not only furnished the source of the PGCs but also presumptive lateral plate mesoderm, it is possible, though rather unlikely, that the gonadal anlage, which is normally formed from the intermediate mesoderm, partly developed from graft tissue. Disregarding this minor objection, these experiments demonstrate that the presence of PGCs does not lead to a formation of germ cells *de novo* from the somatic components of the ground. Moreover, no recent publications have led to any claim contradictory to the conclusions stated above.

The most elegant proof of the continuity of a single germ line was furnished in birds. First the extra-gonadal and extra-embryonic location of the PGCs in the so-called anterior germinal crescent of the blastoderm at early somite stages was demonstrated by surgical extirpation, cauterization. UV-irradiation, γ-irradiation and X-irradiation. In birds the PGCs are transported by the blood stream from the extra-embryonic germinal crescent to the gonadal-anlagen. Reynaud (1969) was above to obtain repopulation of the gonadal anlagen after intravenous injection of a PGC suspension made from germinal crescent endoderm into a host embryo which had previously been sterilized by UV-irradiation of its germinal crescent. When he made xenoplastic recombinates of turkey PGCs with chick hosts or vice versa, he found that the gametes formed in the F1 generation were all of donor type.

The colonization of the host gonadal anlagen by foreign germ cells constitutes a crucial experiment, demonstrating that in birds the PGCs formed during early development are the sole precursors of the definitive gametes. In mammals the evidence for the existence of a single and uninterrupted germ line is not as conclusive as in amphibians and birds, but is nevertheless fairly strong. X-irradiation (more than 168r) of mouse gonadal anlagen after their colonization with PGCs leads to the development of sterile gonads, which nevertheless who more or less normal proliferation of the gonadal epithelium.

Mouse gonadal epithelium—prior to colonization with PGCs—when grated into the kidney capsule of host embryos remains sterile, whereas similar grafts made after colonization with PGCs form typical testicular or ovarian tissues containing germ cells. These results plead strongly against any formation of germ cells *de novo*. Further evidence comes from genetical studies. Several workers have described several alleles of a mutation in mice called *dominant white spotting* (W), which in homozygous condition leads to sterility at birth. In the mutants the PGCs are formed at 8 days of gestation in the yolk sac endoderm in the normal number, which ranges from 10 to 100.

The cells behave normally and migrate towards the gonadal anlagen. However, in normal mice the number of PGCs increases exponentially and reaches a value of approximately 5000 at the end of the migration period, whereas in the mutants the number of PGCs does not increase during migration. They subsequently degenerate, leading to total and permanent sterility of the gonadal anlagen. Twenty-five percent of the offspring of heterozygous parents of the W series show the defect and are sterile at birth. This germ cell deficiency, which begins to manifest itself on the ninth day of embryonic development and is fully expressed at birth, furnishes a strong argument against any formation of germ cells *de novo* for the gonadal epithelium. This conclusion is further supported by culture *in vitro* of sterile mutant half-gonads fused with younger normal fertile half-gonads.

In these chimaeric explants epithelial cells of the mutant gonad are in close contact with normal PGCs but no proliferation of mutant germ cells was ever observed. A similar germ cell deficiency seems to exist in the steel mutant (*Sl*) described by Bennett (1956). Radiation-induced damage (400 r) likewise results in depletion of the PGC stock leading to permanent sterility. At lower doses the gonads may become repopulated, but all evidence points towards repopulation by remaining germ cells and not from somatic cells of the gonadal epithelium.

Conclusions

Summarizing, it may be concluded that in the vertebrates the PGCs have an extra-gonadal origin during early embryonic development. The concept of a continuous, single germ line origination in early embryonic development has become highly probable for amphibians and birds and likely for mammals, while there is circumstantial evidence for it in agnathan fishes. Moreover, in recent years no compelling evidence against the general validity of this concept has been presented. Although further analysis is highly desirable, particularly in fishes and reptiles, it may be assumed for the present that in the vertebrates germ cell history is characterized by the one-time origin of PGCs during early development and that the resulting PGC population gives rise to all the definitive gametes of the adult. Waldeyer (1870) introduced the term 'germinal epithelium' for the putative germ cell producing outer layer of the gonadal anlage. The now convincingly demonstrated extra-gonadal origin of PGCs makes this term, which is unfortunately currently used in the literature, a very confusing one (see also Roosen-Runge. 1977).

It would be better to speak of 'gonadal epithelium' when describing the proliferative layer of the somatic gonadal anlage, and this is the term we shall use in the following chapters. In the lower chordales the problem of the origin of PGCs is still unsolved. Moreover, gonadogenesis differs markedly from that in the vertebrates. Since the further history of the germ cells has not been studied, the conclusions drawn above for the vertebrates cannot yet be extended to the entire phylum of the chordates.

4

FATE OF THE PRIMORDIAL GERM CELLS

In all the vertebrate groups a displacement of the PGCs occurs from their extra-gonadal (in the amniotes extra-embryonic) site of initial appearance to their ultimate destination in the gonadal anlagen: Cephalochordates are the exceptions where the primordial germ cells arise *in situ* in their definite location and no migration is therefore required. The migration of the PGCs from their site origin towards the gonadal anlagen has been most thoroughly studied in birds, and reasonably well studied in the anuran amphibians. Since the migration route is quite different in the two groups, and since as far as we know these groups essentially typify the two main routes along which PGCs may reach the gonadal anlagen, we will first focus our attention on the anuran amphibians, deal only briefly with the urodeles and fishes, subsequently discuss extensively the situation in the birds, and finish with some additional data on the mammals and reptiles.

It seems desirable to distinguish between two different kinds of movement of the PGCs during their migration, i.e. (a) 'passive' displacement by morphogenetic movements, and (b) 'active' displacement by their own amoeboid motility. During their active displacement the PGCs may make use of the circulatory system of the embryo. This is often called 'passive transport' but should be clearly distinguished from the passive displacement by morphogenetic movements, during which the PGCs do not show any motility. The term 'vascular transfer' therefore seems more appropriate.

THE ANAMNIA

Amphibia

Anura

The displacements of the presumptive and true PGCs in the anuran amphibians have already been described in the proceeding chapters. It is now generally agreed that the displacement of the germinal plasm from the vegetal surface of the egg towards a more internal position during cleavage is due to the morphogenetic movements involved in the cleavage process. The subsequent displacements of blastomeres containing *germinal plasm* (pPGCs) towards the center of the yolk mass and even as far as the floor of the blastocoel can likewise be satisfactorily explained by the morphogenetic pregastrulation movements, which were first described by Schechtmann (1934) in *Triturus* and analyzed further by Gipouloux (1962) in *Discoglossus*.

The displacements of the PGCs during gastrulation and neurulation also seem to be due to morphogenetic movements affecting the endodermal yolk mass. Since up to this stage of development no active movements of the PGCs are required, these displacements of the PGCs and their forerunners may be designated as *passive displacement*. At the end of this first period of the development the PGCs are located in the center of the yolk endoderm in the caudal half of the embryo. Kamimura *et al.* (1976) found that in *Xenopus* the PGCs subsequently move to a peripheral position in the yolk mass (stage 31, N & F.). They then move to the dorsal endoderm during stages 33 to 36, accumulate in the form of a dorsal endodermal crest or ridge around stage 40, and separate from the endoderm at stage 41. Subsequently they move to the dorsal root of the mesentery, and finally laterally towards the genital ridges.

Wylie et. al. (1976) observed that in *Xenopus* the PGCs actually accumulate underneath the gonadal epithelium of the genital ridges. Vannini and Giorgi (1969) and Giorgi (1974) hold the opinion that the displacements of the PGCs from the dorsal endodermal crest towards the dorsal root of the mesentery and subsequently towards the genital ridges are mainly caused by growth processes in the surrounding tissues (passive displacement). There is, however, rather convincing evidence that the displacements of the PGCs during this phase of development are for the greater part to their active movements. The PGCs show a

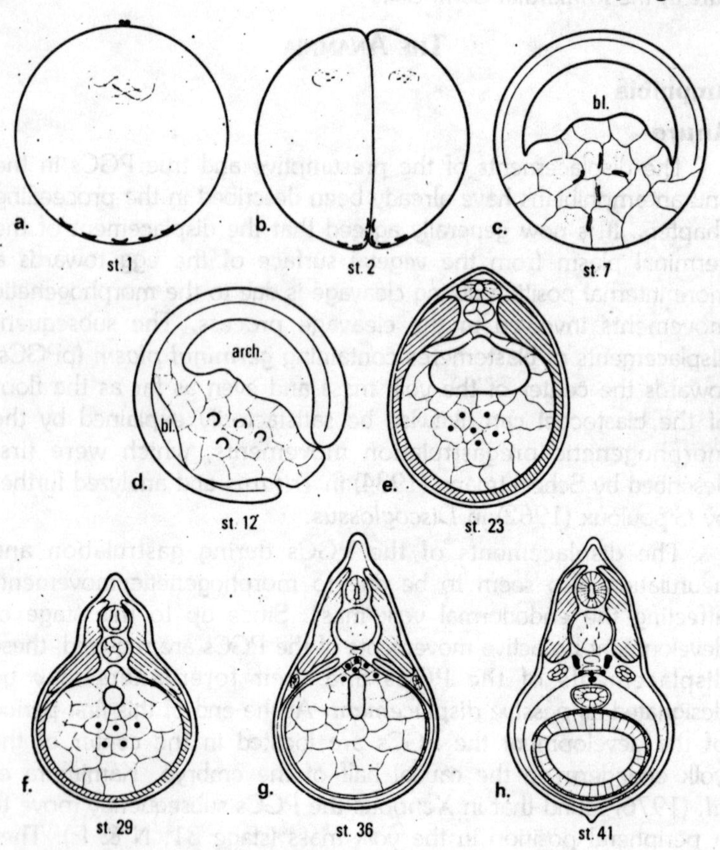

Fig. 4.1. Fate of germinal plasm and migration of PGCs in the anuran amphibians. (a) Localization of germinal plasm in the form of subcortical cytoplasmic patches near the vegetal pole of the fertilized, uncleaved egg. (b) As for (a), but at the 2-cell stage. Beginning of ascent along cleavage furrow. (c) Internal (passive) displacement of endodermal blastomeres containing germinal plasm by pregastrulation movements at early blastula stage. (d) Further (passive) displacement of endodermal blastomeres by gastrulation movements, and intra-cellular displacement of germinal plasm from a peripheral towards a juxta-nuclear position. (e) and (f) Nearly unchanged localization of PGCs at tail-bud stages. (g) Active dorsal migration of PGCs inside endoderm at early larval stage and accumulation in dorsal endodermal ridge. (h) Further active migration of PGCs through dorsal mesentery towards genital ridges. arch., archenteron; bl., blastocoel.

distinct cell boundary in the light microscope. When studied with the electron microscope this is found to be due to wide intercellular spaces between the PGCs and the surrounding cells. Cambar et al., (1970) described the ultrastructure of the PGCs in *Rana* during their migration and after they have settled in the genital ridges.

The PGCs show cytoplasmic processes extending into the intercellular spaces. With time-lapse cinematography Wylie and Roos (1976) actually observed active amoeboid movements *in vitro* of PGCs isolated during the migration period. Gipouloux (1967) carried out an experimental analysis of the displacements of the PGCs in *Discoglossus* and other anurans. After dorsoventral transposition of the endodermal mass the PGCs migrate through the endoderm in a direction opposite to normal, though in reduced numbers. In non-rotated endoderm there was normal dorsal migration (PGCs in dorsal root of mesentery) when either notochord plus somites plus Wolffian ducts, notochord plus somites, or somites plus Wolffian ducts were present.

Partial migration (PGCs partially in dorsal endoderm and partially in mesentery) was observed when somites alone were present, and still more restricted migration (PGCs in dorsal endoderm only) in the presence of part of the somites or of the Wolffian ducts only. Implanted notochord or Wolffian duct exerted a weak attraction on only some of the host PGCs. The effect was stronger with implanted somites, stronger still with implanted notochord plus Wolffian ducts or with notochord plus somites, and strongest with implanted notochord plus somites plus Wolffian ducts.

Gipouloux concludes from these experiments that the dorsal mesodermal organs exert an attracting influence upon the PGCs, leading to the establishment of a dorsoventral gradient of a chemotactic agent inside the embryo. This primary attraction of PGCs is a property of the caudal, but not of the cephalic mesoderm. In embryos brought into ventral parabiosis, so that two axial systems are situated in opposite sides of a single endodermal mass, the PGCs are partially immobilized by the opposite attracting influences, leading to a markedly reduced migration. That the attractant is a chemical agent is rendered likely by the observation that PGCs are also attracted by an implanted piece of agar previously immerse in an extract of dorsal mesodermal structures (Gipouloux, 1967). However, Gipouloux's study of the diffusibility of the attractant through Millipore filters and of its heat and enzymatic degradation is in our opinion too preliminary to allow any definite conclusion regarding the possible chemical nature of the attractant. Gipouloux (1967) also found that suppression of genital ridge formation by removal of the axial structures or by mercuric chloride treatment leads to a 'permanent' accumulation of the PGCs in the dorsal root of the mesentery.

Nieuwkoop (1947) had observed and attraction of host PGCs by an extra genital ridge formed in urodele larvae after implantation of either notochord or Wolffian duct anlagen. This was confirmed by Gipouloux (1967) for the anuran embryo. Moreover, there is the interesting observation by Martin (1959) that in *Alytes obstetricans*, which has L and R genital ridges of unequal lengths, the PGCs are also distributed unequally between the two gonadal anlagen. These data plead strongly in favour of a chemotactic action exerted by the genital ridges.

Urodela

Nieuwkoop (1947) concluded for his xenoplastic transportations of ventrolateral marginal zone that in the urodeles the PGCs are

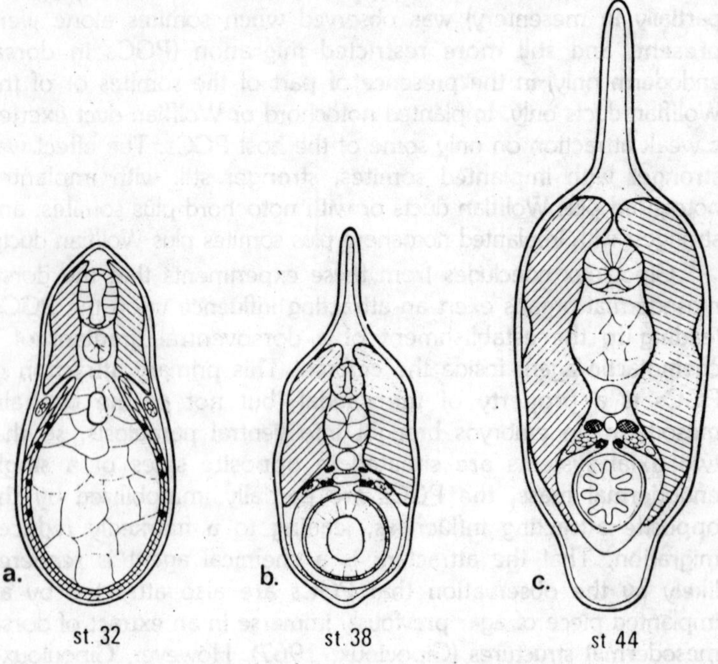

Fig. 4.2. Migration of PGCs (black) in the urodelen amphibians. (a) Successive steps in the migration of PGCs along the mesoendodermal interspace towards the dorsal midline, and (b and c) their subsequent migration towards the genital ridges.

initially more or less evenly distributed throughout the presumptive lateral plate mesoderm. This was confirmed by Maufroid and Capuron (1973) by means of grafts of ventral and ventrolateral marginal zone into the blastocoelic cavity of host embryos. Maufroid and Capuron (1972) studied the dorsal displacements of the PGCs in *Pleurodeles* during stages 16 to 27 by means of unilateral ablation of various dorsoventral regions of the lateral plate mesoderm. Ikenishi and Nieuwkoop (1978) found cilia on some of the PGCs in *Ambystoma* during the migration period. However, nothing is known about the mechanisms upon which the displacement of the PGCs through the lateral plate mesoderm and their ultimate accumulation in the genital ridges are based.

Coecilia

The migration of the PGCs in the Coecilia has not been studied.

Fishes

The only evidence available on the migration of the PGCs in the fishes is of a descriptive nature. In the lamprey (Agnatha), where according to Okkelberg (1921) the PGCs seem to originate in the lateral plate mesoderm, the migration route resembles that in the urodele amphibians. Among the Osteichthyes Allen (1911) described the migration route in *Amia* and *Lepisosteus*.

In *Amia* the PGCs migrate from the peripheral endoderm into the lateral plate mesoderm, then accumulate in its mediodorsal region, and migrate from there to the genital ridges. This situation seems more or less intermediate between that in the anuran and that in the urodele amphibians. In *Lepisosteus* the PGCs first migrate towards the dorsal portion of the gut and subsequently through the dorsal mesentery to the genital ridges, a route which resembles that in the anuran amphibians. No relevant data are available on the migration of the PGCs in the Chondrichthyes. In the Telestomi the available data are scanty and controversial. The authors who assume an endodermal origin for the PGCs taken them to migrate from the yolk sac endoderm—or even earlier from the periblast–through the splanchnic mesoderm towards the dorsal midline, and subsequently through the dorsal mesentery towards the genital ridges in *Fundulus*; *Micropterus*; *Channa*; and in *Oryzias*. The authors who defend a mesodermal origin of the PGCs assume them to arise in the unsegregated mesendoderm.

After segregation of the two layers they are found in the mesodermal layer or distributed throughout all three germ layers, with a preference for the mesodermal layer. Although the PGCs may show a temporal association with the peripheral endoderm, they subsequently migrate through the splanchnic towards the somatic mesoderm and further towards the genital ridges (*Gumbusia*; and in *Cyprinus*). Several authors have observed pseudopod formation on the surface of the PGCs, which they took to be an indication of active migratory movements, but other authors were more inclined to ascribe the movements of the PGCs entirely (Richards and Thompson, 1921, in *Fundulus*) or at least partially to passive displacement by the surrounding tissues. Stolk (1958), who observed many highly ectopic PGCs in *Abramites* and *Cirrhina*, even assumed vascular transfer to be involved, but this has never been confirmed.

THE AMNIOTA

Aves

The presumptive PGCs are very probably passively displaced by the morphogenetic movements occurring during the formation of the primary and secondary hypoblast and the subsequent formation of the primitive streak and the invagination of the tertiary hypoblast or embryonic endoderm. The subsequent, active displacements of the PGCs from the germinal crescent towards the genital ridges may be divided into four phases. The first phase is their segregation from the endodermal layer and their accumulation between the endoderm and ectoderm during stages 4 to 8 (H & H). In the second phase the PGCs being to penetrate the vascular network. This occurs after the invagination of the extra embryonic mesoderm. Its subsequent extension between the ectoderm and endoderm, and the start of the differentiation of the blood islands (stage 10, H. & H.).

In the next phase the PGCs are found successively in the extra-embryonic blood vessels at stage 12, and in the embryo proper with the onset of cardiac pulsation and blood circulation at stage 13. By 2 days of incubation (last phase) they start to leave the visceral branches of the aorta—the gonadal anlagen at that time being situated close to the omphalomesenteric artery—and begin to penetrate the gonadal epithelium. The majority of the PGCs have settled in the gonadal epithelium by 3 days of incubation. During the formation of the indifferent gonads the PGCs are scattered in the cortical layer but also in the medullary tissue. In the young ovary they accumulate in the cortex formed by the

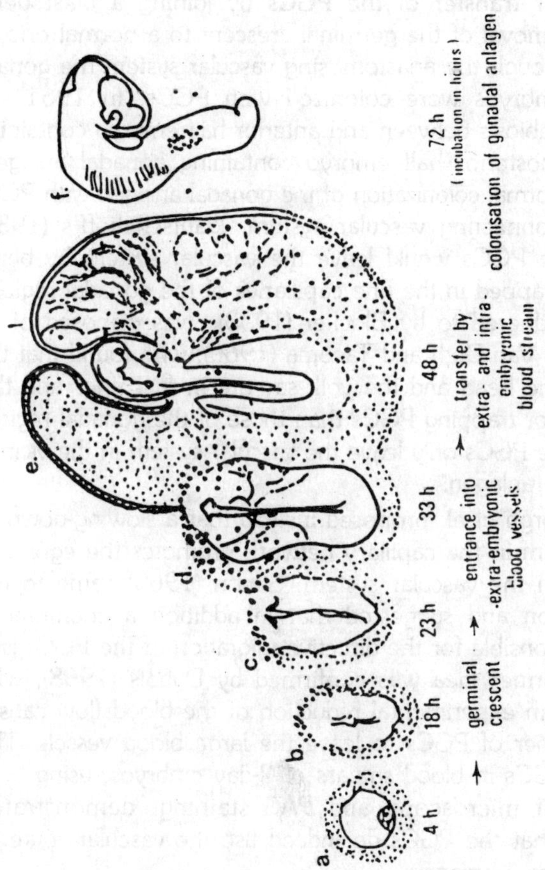

Fig. 4.3. Migration of PGCs in the avian embryo.

the anterior half of the blastoderm containing the germinal crescent laterally with respect to the posterior half containing the presumptive gonadal anlagen, and in this way showed that the PGCs actually reach the gonadal anlagen by the only plausible route, i.e., the vascular system.

In 1960 Simon provided an even more elegant demonstration of the vascular transfer of the PGCs by joining a blastoderm sterilized by removal of the germinal crescent to a normal one by parabiosis. Through the anastomosing vascular system the gonads of the two embryos were colonized with PGCs. In 1961 she performed parabiosis between and anterior half embryo containing PGCs and a posterior half embryo containing gonadal anlagen, and obtained normal colonization of the gonadal anlagen with PGCs through the connecting vascular system. Dantschakoff's (1936) notion that the PGCs would leave the vascular system by being mechanically trapped in the fine capillaries of the gonadal region– an idea also adhered to by Blocker (1933)– was disposed of by Van Limborgh, Van Deth and Tacoma (1960), who found that the capillaries of the head and the yolk sac are in fact finer and this more suitable for trapping PGCs than those of the gonadal region; nevertheless the PGCs only leave the vascular system in the vicinity of the gonadal anlagen.

Van Limborgh et al. proposed instead that a slowing-down of the blood stream in the capillary network promotes the egress of the PGCs from the vascular system. Simon (1960) came to the same conclusion and suggested that in addition a chemotactic process is responsible for the directive migration of the PGCs (see below). The former idea was confirmed by Dubois (1968), who observed that an experimental reduction of the blood flow causes a greater number of PGCs to leave the large blood vessels. The discovery of PGCs in blood smears of 2-day embryos, using both phase contrast microscopy and PAS staining, demonstrated unequivocally that the PGCs do indeed use the vascular route to reach the gonadal anlagen.

Further support for this route is given by Reynaud (1969), who was able to repopulate the gonads of embryos previously sterilized by UV-irradiation of the germinal crescent by intravenous injection of a cell suspension made from germinal crescent endoderm of another embryo. The insertion of a piece of shell membrane between the hypoblast and the presumptive vascular mesoderm does not prevent the PGCs from leaving the hypoblast (although it of course prevents their entering the vascular system), and an isolated hypoblast cultured *in vitro* shows a spontaneous but random migration of the PGCs. Thus, the first step in the migration of the PGCs, i.e., their movement from the endodermal layer of the germinal crescent into the vascular system, is an active process. PGCs not only have amoeboid motility but actively invade various tissues under in-vitro conditions.

The entry of the PGCs into the vascular system apparently does not require a specific attraction by the vascular mesoderm. As already suggested by Simon (1960) the PGCs leave the vascular system under the influence of a chemotactic attraction exerted by the gonadal epithelium. *In vitro* the PGCs may leave the germinal crescent when it is associated with the gonadal region of another embryo, and may colonize the gonadal epithelium in the absence of a vascular system. The attraction exerted by young gonadal epithelium can also act upon the PGCs in a colonized but still undifferentiated gonad. It turned out that even spermatogonia still possess migratory capacity, sensitivity to the attractive influence, and the ability to colonize gonadal epithelium at least up to the 12th day of incubation.

The majority of the oogonia lose their migratory capacity on the 8th day. Since the attraction can be exerted across a permeable barrier such as the shell membrane, the most satisfactory explanation is that of positive, selective chemotaxis. By delayed injection of a PGC suspension made from germinal crescent endoderm, Reynaud (1969) was able to show that a sterile host gonad obtained by UV-irradiation of the germinal crescent can still trap PGCs at 5 days of incubation. Thus the competence of the PGCs to react to the attracting influence lasts much longer than the normal period of PGC immigration and only disappears with the onset of gametogenesis, while the attracting capacity of the gonad disappears during the formation of a primary sex cords.

Rogulska (1969) demonstrated a direct transfer of PGCs from an intra-coelomic graft of germinal crescent to an adjacent host genital ridge at 3$^{3/4}$ days of incubation, i.e., long after normal migration is completed. By means of intra-coelomic grafts of mouse hindgut to the gonadal region of a chick embryo, Rogulska, Ozdzenski and Komar (1971) found that chick gonadal epithelium also attracts mouse PGCs, demonstrating that the attractant is not species or class specific. Let us now consider the properties of the PGCs and the gonadal epithelium more closely in order to understand better the chemotactic process. The PGCs in birds undergo a measure of cytogenesis.

Clawson and Domm (1963a, b) described the PGCs of the germinal crescent as containing much yolk and only little glycogen. The migrating PGCs, however, contain far less yolk but much more glycogen. The glycogen content diminishes during migration, so that after arrival in the gonadal ridges the PGCs again contain little glycogen. This was confirmed by Fujimoto *et al.* (1975). Dubois and Cuminge (1968) described more or less parallel changes in the amount of osmophilic lipid droplets, an observation confirmed by Ukeshima and Fujimoto (1975). In addition, Fujimoto *et al.* (1976a) described a structural change in the glycogen and its unipolar intracellular localization during the settling of the PGCs in the gonadal epithelium. The glycogen and the lipids are considered to play an important role as energy sources during the migration of the PGCs. The PGCs show pseudopodial extensions during their migration. Microfilaments and microtubules, however, have not been demonstrated.

According to Cuminge and Dubois (1971) the PGCs show an intense lytic activity and actively penetrate the gonadal epithelium. There they take up soluble substances and cellular debris by pinocytosis and phagocytosis, respectively. The gonadal epithelium has a syncytial and spongy character and shows large intercellular spaces. The same features are found in sterile gonadal epithelia, so they are not due to the lytic activity of the PGCs, Autoradiographic electron-microscopic analysis has shown that the gonadal epithelium synthesis exportable proteins or protein complexes, which are set free by a merocrine excretory process. Although in the nucleus only a single slow type of protein replacement is observed, two classes of proteins are synthesized in the cytoplasm, i.e., slowly replaced structural proteins and rapidly replaced exportable glycoproteins. However, it has not been conclusively shown that the attractive factor produced by the gonadal epithelium is in fact a protein.

Swartz (1975) obtained some evidence that the chemical attractant may be steroid in nature, since injection of androgens or oestrogens interferes with PGC migration. In the synthesis of the exportable proteins the Golgi complex plays an important role. The fact that concanavalin A, which binds to glycoproteins, inhibits the migration of PGCs placed under the attracting influence of young gonadal epithelium, suggests that glycoproteins play a decisive role in the chemotactic migratory capacity of the PGCs. In contrast to those of the gonadal epithelium, the glycoproteins synthesized by the PGCs show only a slow turnover and are not excreted. They are found preferentially in the lysosomal apparatus and on the plasma membrane, where they are concentrated on microvilli, pseudopodia and at sites of pinocytosis and phagocytosis.

Very little is known about possible attractive influences that play a role in the ultimate distribution of the PGCs between the cortical and medullary tissues of the differentiating gonad. Summarizing, it may be said that the initial step in the migration of the PGCs, i.e., their movement from the hypoblast of the germinal crescent into the vascular system, is an active but non-directional process. Impressive evidence argues for the involvement of a chemotactic mechanism in the migration of the PGCs from the vascular system into the gonadal epithelium. Very little is known about the mechanisms involved in the ultimate displacements of the germ cells inside the differentiating gonad.

Mammalia

Leaving aside their possible primary site of origin, the PGCs are clearly recognizable for the first time in presomite and early somite stages in the yolk sac endoderm and adjacent splanchnic mesoderm near the allantoic evagination, in the marsupial *Tatusia*; man; mouse, the guinea pig; and in the rabbit. Slightly later they are found in the endoderm and adjacent mesoderm of the hindgut. They then leave the gut and migrate through the splanchnic mesoderm towards the dorsal mesentery, and from there towards the genital ridges *Tatusia*, the guinea pig; the mouse; rabbit; and in man. It is still unclear whether the PGCs migrate through the endodermal epithelium of the yolk sac and hindgut, or through the adjacent splanchnic mesoderm, associating themselves temporarily with the endodermal cells for nutritive purposes only. The great majority of authors the opinion that the initial displacements of the PGCs during gut formation are mainly or entirely passive, i.e., they result from the morphogenetic movements associated with gut formation.

Opinions differ as to the mode of their subsequent translocation from the hindgut to the genital ridges. Some authors have observed large intercellular spaces around the PGCs, and the formation of

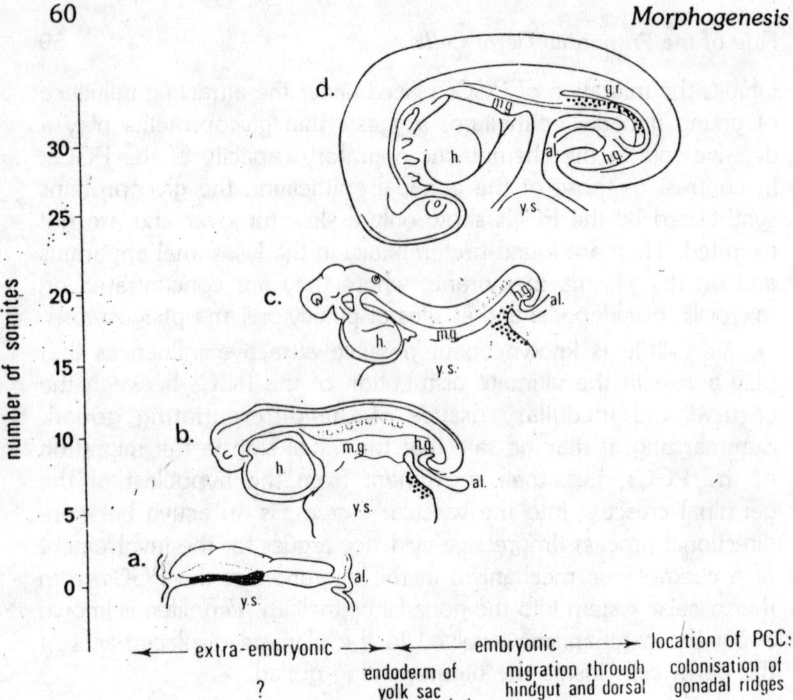

Fig. 4.4. Migration of PGCs in the mammalian embryo. (a) Absence of identifiable germ cells at the neural-plate stage. (b) Appearance of PGCs in the wall of the yolk sac. (c) Passive displacement of the PGCs during hindgut formation. (d) Colonization of the gonadal anlagen by actively migrating PGCs. all., allantois; g.r., gonadal ridge; h., heart; h.g., hindgut; m.g., midgut; y.s., yolk sac.

pseudopodia containing microtubules, which would suggest active migration. On the other hand, Jeon and Kennedy (1973) found tight junctions between PGCs and adjacent somatic cells, which pleads against active migration, but Spiegelman and Bennett (1973) deny the presence of intercellular junctions. Nearly all authors who favour active migration of the PGCs in the mammals assume interstitial migration through the various tissues. Semenova –Tian-Shanskaya (1969) in addition advocates partial transfer by way of the vascular system.

Witschi postulated as early as 1948 a chemotactic mechanism for PGC migration, but relevant experimental evidence is lacking. McKay et al. (1953), Chiquoine (1954), Mulnard (1955) and McAlpine (1955) found a positive alkaline phosphates reaction not only in the PGCs but also in the gonadal epithelium. Gondos and Hobel (1971) found indications of active migration of the germ cells towards the periphery of the seminiferous tubules in the male gonad, in the form of pseudopodial extensions and the presence of microtubules. Blandau *et al.* (1963) observed undulating surface movements of oogonia in ovarial squash preparations. In summary, it may be said that in the mammals most of the evidence pleads in favour of active, interstitial migration of the PGCs from the hindgut towards the genital ridges, whereas the earlier displacements from the yolk sac to the hindgut are probably mainly passive. The route taken by the PGCs during their active migration may be influenced by their tendency to associate with somatic cells for nutritional purposes.

Reptilia

The reptiles are an interesting group exhibiting different types of PGC migration. Risley (1933) found the PGCs to be localized in a posterior crescent in the turtle *Sternotherus*. He described what he called active interstitial migration of the PGCs from the site towards the gonadal anlagen through the splanchnic mesoderm and the dorsal mesentery, in much the same way as described above for the mammals. A similar situation was found by Hubert (1976) in various *Lacerta* species. Hubert (1970b) described the formation of pseudopodia containing microfilaments on the surface of the PGCs. In *Sphenodon* Tribe and Brambell (1932) described an anterior as well as a posterior germinal crescent. They postulated that the PGCs of the posterior crescent would reach the gonadal anlagen by interstitial migration through the splanchnic mesoderm and dorsal mesentery, but those of the anterior crescent would enter the vitelline veins and reach the gonadal ridges by way of the vascular system- i.e., passing through the vitelline veins, the heart, and aorta and the omphalo-mesenteric artery.

The presence of an anterior germinal crescent and the subsequent vascular transfer of the PGCs was also described in *Mabuya* and *Chamaeleo* by Pasteels (1953), Hubert (1969) found in strictly anterior germinal crescent in *Vipera*, but in *Anguis* the PGCs were found around the entire embryonic enlarge, though markedly predominating anteriorly. He observed vascular transfer of the PGCs towards the gonadal anlagen from both the anterior and the posterior extra-embryonic regions. Summarizing, it may be said that the Chelonia the migration of the PGCs from the posterior germinal crescent towards the gonadal anlagen seems to be exclusive interstitial. Interstitia migration as well as vascular transfer is found in the Squamata. PGCs from a posterior site generally take the interstitual migration route, whereas those from an anterior site over the much longer instance to the gonadal ridges by means of vascular transfer.

Conclusions

Surveying the displacements of the PGCs in the chordates, it is evident that in the lower chordates, where the gonadal anlagen-arise in the vicinity of the PGCs, no transport mechanism is required. In the vertebrates, where the PGCs have an extra-gonadal site of origin, they are both passively and actively displaced towards the gonadal anlagen. The early displacements occurring during embryogenesis are mainly passive and result from morphogenetic movements of the surrounding tissues. Although morphogenetic processes may still play role in germ cell displacement during later phases of development, and even during gonadogenesis, the considerable distance between the location of the germ cells in the early embryo and the gonadal anlagen is bruged in the main by active migration, in which chemotactic processes play a leading role.

The chemotactic attraction originates first from the dorsal mesodermal organs, leading to an accumulation of the PGCs near the dorsal root of the mesentery, and later from the epithelium of the genital ridges. The former phase has been extensively studied in the amphibians, the latter in the birds. Depending upon the absolute distance to be covered, the PGCs may make use of the blood circulation to reach the neighbourhood of the genital ridges. Vascular transfer occurs mainly in forms with meroblastic eggs, but is also related to the time of onset of blood circulation—early in the Amniota and late in the Anamnia. The use of the vascular route seems to be correlated with the nature of the passive displacements of the PGCs during early embryogenesis, i.e., whether they lead to an anterior or posterior localization of the PGCs. Posterior localization is usually associated with interstitial migration, whereas vascular transfer occurs in the case of anterior localization. A careful analysis of the displacements of the PGCs during early embryogenesis seems of crucial importance. The reptiles, which show example of both types of migration, are a very interesting group in this respect and should be studied much more extensively.

5

Sex Organs

All the living creatures develop from the pre-existing organisms and performs all the activities for their growth and maturity. But as none of the living creatures is immortal, so the individuals do something for the continuity of their race. And the activity for the continuation of the race is called *reproduction*. In all vertebrates, special organs are developed for the purpose, which are called the *reproductive organs* or *sex organs*. In a large number of sexually reproducing animals there are two distinctly different kinds of individuals, males and females. The males produce sperms and the females produce eggs. The term egg also refer to the familiar hen's egg which contain yolk, albumen and a shell in addition to the female gamete, the real egg.

Sometimes the same individuals possesses the capacity to function as both male and female. Such an animal is termed *hermophrodite* or *bisexual*. Hermophroditic animals are rare among vertebrates. A common example is hagfish. All the vertebrates reproduce sexually, by the fusion of sperm cells with egg cells. Not only does this union of gametes bring together in a single fertilized egg the hereditary factors form both parents, but it induces the egg, otherwise inert, to start division and the formation of an embryo. The reproductive system of vertebrates consists of the *reproductive organs* or *gonads*, in which the germ cells become differentiated, and the reproductive ducts, through which the germ cells leave the body.

During the development of the individual, the reproductive and urinary system arise in close association so that structurally they form what is known as the *urino-genital* system. The functions of the two parts of this system are entirely unrelated, and we shall be concerned here with only the reproductive or genital system. The gonads of vertebrates produce gametes and, in most cases, steroid hormones. These hormones serve various functions, such as inducing or facilitating sexual behaviour and/or parental behaviour, preparing the reproductive tract for receiving the gametes, caring for the zygote, and other functions. Here we present a general outline of the anatomy of the male and female reproductive systems of representatives of the different classes of the vertebrates.

Male Reproductive System

In most vertebrates, the testes are paired, but in some, e.g., the cyclostomes, the left and right testes are fused, and in some teleost species only one testis develops as in *Notopterus notopterus*. In most classes of vertebrates the sperm produced in the testes are conducted to the outside by a duct system: in cyclostomes and the Salmonidae, however, such a duct system is lacking. In vertebrate species that do have a sperm duct system, this system may be separate from the urinary Wolffian duct system, as is the case of teleosts. In all other vertebrates, sperm are transported in tubules than originate as pronephric and mesonephric tubules.

Cyclostomes

In lampreys, the one testis consists of two gonads fused in the middle. It is located on the midline and is attached to the body wall by a mesentery. The testis consists of lobules, each containing several ampullae lined with germinal epithelium. When the lamprey becomes sexually mature and starts spawning, these ampullae break down, and the sperm are released into the coelom. There is no duct system, and the sperm leave the body through an opening secondarily connecting the coelom to the cloaca, as in *Lamperta* or through a median opening between the anus and the urinary opening, as in *Myxine*.

Elasmobranchs

The testes are paired and suspended from the body wall by mesorchia. Close to each testis lies the epigonal organ, which consists of lymphoid or hemapoietic tissue. "the testes consist of ampullae, which are arranged in zones in the spiny dog fish,

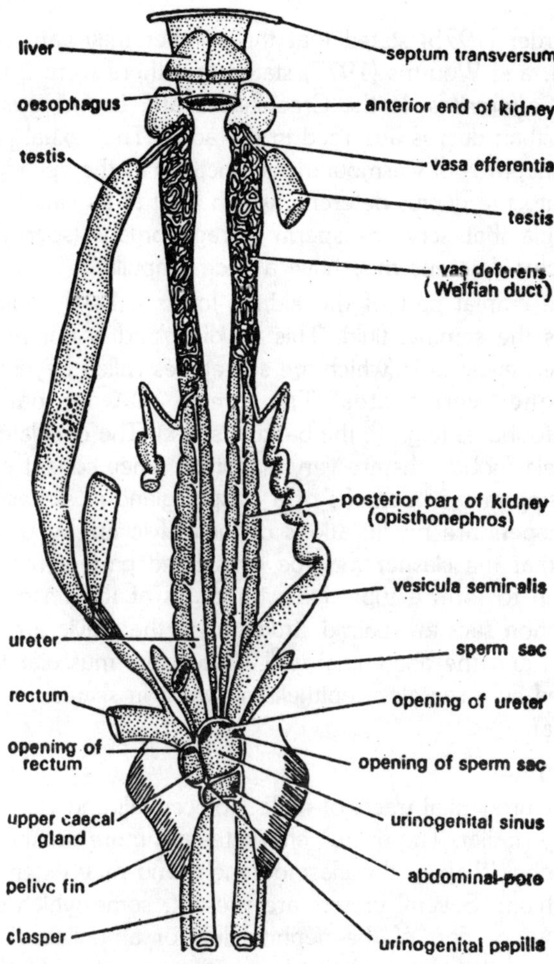

Fig. 5.1. Scoliodon. Male urinogenital organs.

Scyliorhinus canicula. In the basking shark, *Cetorhinus maximus*, the testes consist of many lobules separated by connective tissue, each lobule corresponding to the entire testis of the spiny dog fish. Only the cranial portion of the opisthonephros communicates with the testis. The number of efferent ducts that lead to the ductus deferens varies according to the species: it is 8 in *Centrina*, 5 in *Scyliorhinus canicula*, 2-3 in *Squalus*, 1 or 2 in *Mustelus vulgaris*, and 1 or 2 in *Galeus*. In the rays, there is one efferent duct.

Harder (1975) stated that the number may vary between 1 and 18, and Wourms (1977) stated that there were 2-6 of these *ductuli efferentes*. In the Greenland shark, *Laemargus borealis*, the Wolffian duct is absorbed in the adult. The sexual part of the opisthonephros of elasmobranchs functions as the epididymis. This leads into the ductus deferens, which may at its caudal end have diverticula that serve as sperm storage organs (sperm sacs), or the ductus deferens may have a wide ampulla.

The cranial part of the kidney forms a Leydig gland which secretes the seminal fluid. This Leydig gland is not homologous with the Leydig cells (which are sometimes called Leydig gland) of the higher vertebrates. The sperm may be packed into spermatophores (e.g., in the basking shark). The copulatory organs of the elasmobranchs are very elaborate. They consist of claspers and a siphon sac in sharks or a clasper gland in skates and rays. The claspers are modifications of the pelvic fin. Wourms (1977) stated that the clasper may be considered part of the pelvic fin rolled up to form a tube and with edges of the fin overlapping. The siphon sacs are paired structures in the pelvic area between the skin and the abdominal wall. They have muscular tissue and are lined by a secretory epithelium, which in skates, may be very glandular.

Holocephali

The urogenital tracts of male Holocephali and elasmobranchs are very similar. The ductuli efferentes are more numerous in the Holocephali than in the elasmobranchs and may extend into the mesorchium. Several ureters are present, some which open into the posterior part of the nephric duct or ampullae. The ductus deferens of the Holocephali has a complex system of chambered ampullae with epithelium differentiated into regions. The chimeras of Holocephali have one pair of pelvic fin claspers homologous with the claspers of the elasmobranchs and a single clasper in front of the dorsal fin. In the Holocephali, both claspers are covered with dermal teeth, whereas in the elasmobranchs the claspers are smooth.

Dipnoi

In the Australian lung fish (*Neoceratodus forsteri*), some renal tubules throughout the length of the kidney from an efferent duct system. In the South American and African lung fishes, the

Leptosirenidae, however, only some posterior renal tubules are used. No copulatory organs have been found in the Dipnoi.

Chondrostei

The renal-testicular tubule relationship in *Sturgeons*, *Acipenser*, and gars, *Lepisosteus*, are similar to those in the Australian lung fishes. In the Polypteridae, the duct systems are separated, and the sperm duct opens in the urinary duct. Amoult (1964) and Holden (1971) have presented evidence that the anal fin is used as a copulatory chamber in *Polypterus senegulus*. Holden (1971) speculates that this copulatory mechanism is required to conserve sperm, because the testes of these fish are quite small, even during the breeding season. The testes of sturgeons are lobed and are suspended by mesorchia. The male sturgeon has a well-developed Mullerian duct, but its function has not been determined.

Teleosts

There is no connection between testes and kidney in teleosts. In some species, the sperm duct has its own opening to the exterior, whereas in others there is no sperm duct, e.g., in the Salmonidae and Anguillidae, and the sperm are released into peritoneal cavity and then leave via peritoneal canals. The testes of most species of teleosts are paired, but sometimes only one testis is present; e.g., *Notopterus notopterus* has only a left testis. The testes are often oval or cylindrical, although in *Perca* and *Cyprinus* they are Y-shaped, whereas in the Ictaluridae. *Glyptosternum* and *Trachycorystes* and testes are branched, fingerlike organs. In this type of gonad, spermatogenesis takes place only in the anterior part. The posterior lobes secrete a fluid, which, in *Trachycorystes*, seals off the vagina after mating, in the Ictaluridae and *Glypstosternum*, the function of this secretion is not known. The testes are suspended by mesorchia, but when the testes fuse, the mesorchium becomes attached to the mesentery of the intestine.

Acinar and Tabular Testes

Teleost testes have been classified on the basis of either the presence of acini (lobules) or tubules or on the basis of the distribution of spermatogonia in the tubules. The testes may be of acinar type which is reportedly found in the Clupeidae. Cyprinidae, Salmonidae, and Esocidae; the tubular type is found in Perciformers, Grier *et al*. (1980), after a critical review of the literature, concluded that the criteria for classifying teleost testes as either lobular or tubular were not well established and that the classification of Leydig cells, Sertoli cells, and lobule boundary cells had led to misconceptions when this dual classification was used. They proposed a classification based on the intratubular distribution of spermatogonia.

Intratubular Distribution of Spermatogonia

In the Salmoniformes and Perciformes, tubules form either anastomosing or branching networks, and spermatogonia are present along the entire length of each tubule. The spermatogonia and their associated Sertoli cells from nearly solid cords within the tubules when the testes are regressed. When spermatogenesis begins, the Sertoli cells from the border of the cysts within which spermatogenesis occurs. In the Atheriniformes, the spermatogonia are present only the distal end of each tubule.

The Sertoli cells from cysts by the time spermatogonia are transformed into primary spermatocystes. After spermiation, the Sertoli cyst cells become efferent duct cells. The identity and homologies of Leydig, Sertoli, and lobule boundary cells has caused controversies. These have been reviewed by Grier et al. (1980) and Grier (1981). It appears from the available microscopic and ultrastructural evidence that the Leydig cells may be interstitial cells or they may be distributed around the efferent ducts at the periphery of the testes.

The Sertoli cells are tubular cells, which have several functions, i.e., nourishment of germ cells, production of steroids, phagocytosis of the residual bodies of spermatids, and in some cases, participation in the formation of efferent duct cells. The Sertoli cells have erroneously been called lobule boundary cells in earlier investigations. The true lobule boundary cells of teleost testes lie outside the basement membrane and they may be myoid in nature, as they are in mammalian testes. The ultrastructural evidence argues against the steroidal capacity for these cells. The posterior lobe of some fishes, e.g., *Clarias* spp. and *Heteropneustes* spp., both genera belonging to the Siluridae, secrete materials that participate in the formation of spermatophores (sperm packets).

In Atheriniformes, which have internal fertilization and in which naked sperm bundles (spermatozeugmata) are produced, the Sertoli cell-spermatid or Sertoli cell-sperm evolves in slightly different ways in different species. In the Poeciludae, sperm nuclei become embedded within the cytoplasmic recesses of the Sertoli cells, whereas in the Goodeidae, the flagella of the spermatid become associated with the Sertoli cyst cells. Spermatophore formation, at least in *Horaichthys setnai*, involves the Sertoli cells in the sense that their secretory product coalesce around the mature sperm. Grier (1981) has pointed out the coevolution of testes and anal fin in the various groups of teleosts, which has resulted in modifications of those into different structures for the transfer of sperm into the female reproductive tract.

Generally, species that have a gonopodium have unmodified testes and do not produce either spermatozeugmata or spermatophores, although in the most primitive representative of the Anablepidae, *Anableps dowi*, partial spermatozeugmata are formed Grier (1981) has speculated that the ancestral "type" from which the Anablepidae have evolved may have had a testicular structure in which spermatozeugmata were formed. The sperm duct, the *ductus deferens*, does not produce secretions, but there are sometime accessory sexual organs, such as the seminal vesicles of the Indian catfish. *Heteropneustes fossilis*, which do produce secretions.

The term *seminal vesicles* has been used for different structures. For instance in the frillfin goby. *Bathygobius soporator*, the seminal vesicle do contain sperm and thus are truly seminal vesicle. So-called seminal vesicles, which are separate glandular structures, are found in the following teleost fish: the toad fish. Opsanus tau; the gobies, *Gobius niger*, *G. minutus*, *G. paganellus*; the loach, *Cobitis forsilis*: the goatfish, *Mullus barbatus*; the northern pike *Esox lucius*; the mudsucker, *Gillichthys mirabilis*; the Indian catfish. *Heteropneustes fossili. Ictalurus furcatus*, *I. catus*, *I. nebulosus*, and the South American catfish. *Trachycorystes striatulus*. In the Ictaluridae mentioned and the South American catfish, the seminal vesicles are part of, or are derived from, the testes. The secretions of the seminal vesicles reach the exterior via the sperm duct and genital papilla. That the seminal vesicles of the mudsucker do not respond to androgens suggests that these seminal vesicles are not derivatives of the Wolffian duct system.

We cannot, however, use such a failure to respond to androgens as definitive evidence. For instance, in the musk shrew (*Suncus murinus*), the uterus does not respond to estrogen, although it is clearly a Mullerian derivative. The sperm ducts of the Salmonidae and Asguillidae have secondarily disappeared. The copulatory organ of teleosts varies from being absent in many groups, to the bizarre priapus of the Phallostethidea (phallus= penis, stethus=chest) and Neosthethidae, in which vestiges of the pelvic fin, the pelvic girdle, parts of the pectoral girdle, and the first pair of ribs from the copulatory organ. The copulatory organ may thus be variously formed. It may be formed by the genital papilla, which may be elongated e.g., in *Zenarchopterus*, some Cottoidei, and in some species of the Blennoidei. Some brotulid fish, e.g., of the genus *Typhlias*, *Stygicola*, *Dinematichthys*, *Brosmophycis*, and the species *Lucifuga subterranean* and *Dipulus caecus*, have a penislike structure. The morphology of this structure was studied by Turner (1946), who concluded that it was of an elaborately modified genital sinus. The anal fin may be modified to form a copulatory organ called a gonopodium. In the clinid fish, *Starksia*, the copulatory organ is formed by the anal fin, which is attached to the genital papilla.

The copulatory organ of the Jenynsiidae and Anablepidae is peculiar in that in an individual, it can move in one direction only, left in some individuals, right in others. (The females correspondingly have either a left or a right genital opening). Turner (1948) found that the copulatory organ of *Jenynsia lineata* is formed by elongation of anal fin rays 3,4,6,7 and 8 (1,2 and 5 are resorbed) and a thickening with bilateral asymmetry of rays 6 and 7, so that the fine becomes converted into a tube into which the sperm duct opens. The anal fin of the Embiotocidae is enlarged and fleshy, so that it looks tubercular. In the cottid fish, *Oligocottus snyderi*, the first two rays of the anal fin are separate, and the first ray, which is much larger than the others, is prehensile and can be bent around the female, so that it can act as a clasper as well as in intromittent organ.

In *Tomeurus gracilis*, the gonopodium, which extends beyond the tip of the head, is separated from the anal fin and is displaced under the pectoral fin. It is, however, not an intromittent organ, for the spermatophores are deposited near but no in the female's genital opening. In the scorpaenid fish, *Sebastiscus marmorathus*, the mesonephric duct and urethra put out behind the anus and forms the copulatory organ. In the Phallostethidae and Neosthetidae, the pectoral girdle, first pair of ribs, vestiges of the pelvic fins and of the pelvic girdle, together help from the copulatory organ. It is not clear what structures are involved in the formation of the "copulatory organ" of *Trachycorystes striatulus*.

Such an intromittent organ occurs also in the genera *Asterophysis*, *Pseudoauchenipterus*, *Auchenipterichthys*, and *Tatia*, and probably in *Ceratocheilus*. Mohsen (1961) described a pear-shaped pseudopenis for *Skiffia lermae* (Goodeidae), into which the ducti deferentes and the urinary duct drain. It is not clear what the anatomical origin of this pseudopenis is and whether it is an intromittent organ. In certain species of the genus *Corynopoma*, spoon-shaped appendages are found on the left side of the body. These organs are not intromittent organs, although they may function in establishing sexual contact. In *C. landonia,* these appendages develop from scales, whereas in *C. aliata* and *C. riisei*, they develop from the edge of the gill cover.

Amphibians

The testes are paired, and the sperm reach the exterior via the ductus deferens, which opens in the cloaca. Near the testes lie the fat bodies, which are essential for normal spermatogenesis, since their removal leads to a decrease in testicular weight in *Rana esculens* and *Triturus viridescens*. It has been seen that the connections between the kidney and testes are similar to those found in elasmobranchs. A more critical examination, however, shows that in most Anura, the Wolffian duct serves as the urinary duct and the semen duct, except that in the genus *Alytes*, the two duct systems are completely separate.

In the urodeles, the urinary and gonadal systems are not separate in *Megalobatrachus*, *Triturus* (*Diemictylus*), and *Hynobius*, the mesonephric tubules drain directly into the cloaca and not into the Wolffian duct. The ductus deferens of some anurans have a bottle-shaped dialation (ampulla) at the caudal portion, which functions as a sperm storage organ. This has been found, e.g., in the toad (*Bufo bufy japonicus*), in the tree frog (*Hyla arborea japonica*), in *Rana nigramaculata*, in the bullfrog (*Rana catesbeiana*), in *Rhacophorus buergeri* and *Rh. japonicus*. In other species, e.g., *Rana japonica*, *R. ornati ventris*, *R. tagoi*, *R. tsushimensis*, and *okinavana*, there is knoblike outgrowth of the outer wall of the Wolffian duct, which function as a true seminal vesicle for storing sperm.

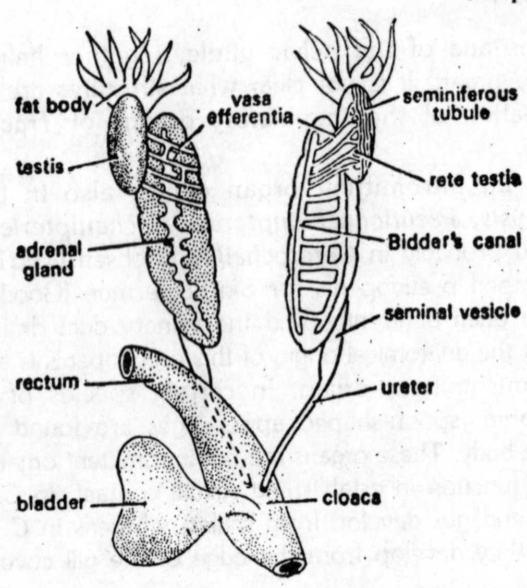

Fig. 5.2. Frog. Male urinogenital organs.

In still other frogs, e.g., *Rhacophorus arboreus*, the part of the Wolffian duct lateral to the kidney is coiled and serves as a sperm storage organ. The ductus deferens opens into the cloaca. Copulatory organs are rare in amphibia. In the primitive limbless Caecilia or Gymnophiona, fertilization is internal and the males have a modified cloaca, which they evert and introduce into the cloaca of the female. Among the Anura, fertilization is internal in the "tailed" frog, *Ascaphus truei*, which has a permanently everted cloaca that serves as a copulatory organ, and in the viviparous toad of East Africa (*Nectophrynoides vivipara*), in which no copulatory organ has been found. In the urodeles, fertilization is internal, but the sperm are not introduced into the cloaca by the male; they are deposited instead on the ground in spermatophores, which are subsequently picked up by the female by means of the cloacal lips. The formation of the spermatophore of the salamander requires two of the three sets of cloacal glands, e.g., the pelvic gland located in the roof of the cloaca and the cloacal glands that cover the cloacal wall. The third set of glands, the abdominal glands, which in some salamanders extend forward into the abdominal cavity, do not take part in spermatophore formation.

The testes of the Anura are generally compact, and consist of convoluted seminiferous tubules. The interstitial tissue, which contains the Leydig cells, lies among the tubules. Spermatogenesis is uniform in all the tubules. The testes of the urodeles differ from those of the Anura; they consist of lobules connected by narrow bridges of tissue, each of which drains into a short duct that joins the ductus deferens. These lobules of the urodeles are transient, and are replaced, in contrast to the seminiferous tubules of the Anura, which are permanent structures. The similarity of the testes of urodeles with the acinar testes of some teleosts striking. The lobules are held together by connective tissues, and the entire testis is surrounded by a fibrous coal.

Reptiles

In the reptiles, the testes are paired organs suspended by mesorchia. In lizards, fat bodies are present, but these do not appear essential for spermatogenesis if the animal receives sufficient food. The testes contain seminiferous tubules and Leydig

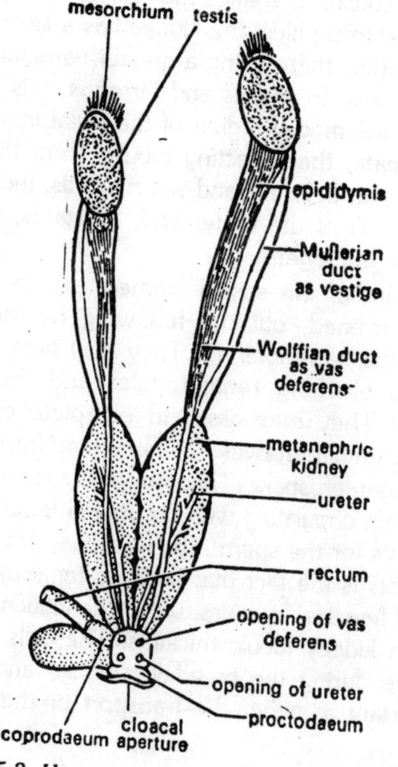

Fig. 5.3. Uromastix. Male urinogenital system.

cells, which are usually located among the tubules. However, in all teiid lizards investigated (17 species of *Cnemidophorus* and one of *Ameiva*), the Leydig cells form a circumtesticular tunic. As in all amniotes, the Wolffian duct has lost its connection with the mesonephros and has become the ductus deferens. The genital region of the mesonephros functions as the epididymis. The arrangement of the ductuli efferentes varies among reptiles. In snakes, several seminiferous tubules drain into one ductile, and the ductuli are arranged along the length of the testes; in lizards (Sauria), they are reduced to a single marginal canal.

In the Anguidae, they are intermediate; several of the ductuli efferentes anastomose and drain into the ductuli epididymides. In the Cholomidae, there are many efferent ductuli that anastomose and form a network the *rete testis*, which differs from the mammalian *rete testis* in being located outside the testis. The epididymis is large in lizards (Sauria), worm lizards (Anguidae), and turtles (Chelonia); in snakes the size of the epididymis varies with the species. In reptiles, the kidney has a sexual segment, the size of which varies, there being a greater variation among lizards than among snakes. In snakes and *Varanus*, this sexual segment corresponds to the medial region of the distal convoluted tubules; in other Squamata, the collecting tubules form the main part of the sexual segment. In lizards and worm lizards, the sexual segment includes part of all of the ureter and, in lizards, increases under the influence of androgens.

The functions of this sexual segment and its secretions have not been determined, but the following functions have been proposed, for the secretions: (1) They may help separate semen from urine by blocking renal tubules and the ureter during copulation. (2) They may assist in complete emptying of the ampullae and seminal grooves. (3) They may form the copulatory plug, which in garden snakes (*Thamnophis sirtalis*) prevents other conspecifics from copulating with the mated female. (4) They may contain nutrients for the sperm. (5) They may act as pheromones; evidence for this is the fact that male garter snakes do not court recently mated females. In males, under the influence of androgens, a part of each kidney tubule thickens. The cells of these tubules are very active during the breeding season, and their secretion may be important in either the transport or the nourishment of the sperm.

Male lizards and snakes have a paired hemipenis, the two arms of which are caudal extensions of the cloaca, each of which can be everted independently of the other into the cloaca of the female. During this copulation, the sperm run along an external groove, and the spermatic, sulcus of the hemipenis into the female's cloaca. Spines of ridges help to maintain intromission. The penis of chelonians and crocodiles is a modification of the floor of the cloaca. It consists of a median groove with ridges of erectile tissue along each side. The caudal end of the penis is a raised gland, which acts as an entering wedge prior to the filling of the erectile blood sinuses; consequently intromission in turtles also involves an eversion of the cloaca. After testosterone propionate treatment of young crocodiles (*Crocodylus palustris*), there is no change in the sexual segment of the kidney tubules, a slight response of the Wolffian duct, but an enormous development of the penis.

Birds

The testes of birds, which can be very large (e.g. 360 mg in a 30 g grass parakeet, *Melopsittacus undulatus*) are suspended by mesorchia. There are no tubuli recti, but the rete testis connects the seminiferous tubules to the ductuli efferentes. The epididymis is small compared with the epididymis of mammals. It is homologous with the caput epididymidis of mammals, and different

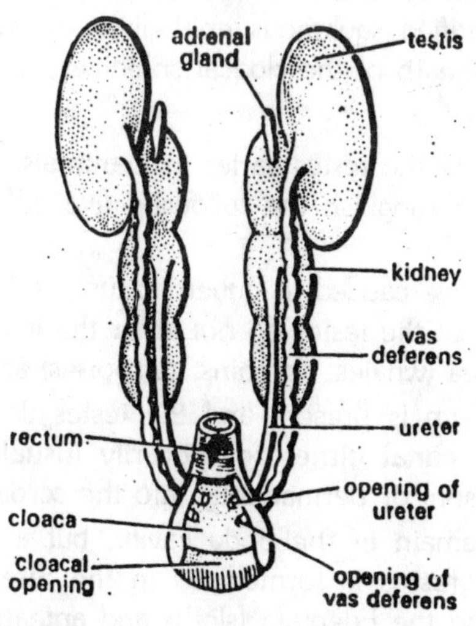

Fig. 5.4. Pigeon. Male urinogenital system.

regions of the ductus deferens are homologous with the corpus and cauda epididymidis of mammals. The ductus deferens, under the influence of either endogenous or exogenous androgens, becomes coiled. It has no glandular areas that might correspond to the accessory glands of the mammalian reproductive system.

In some species there is an enlargement of the ductus deferens, called the "seminal vesicle" or "seminal glomus," which is also highly coiled and in which sperm are stored. Wolfson (1954) demonstrated that in a number of passerine species (*Junco hyemalis*, *Zonotrichia albicollis*, *Z. leucophrys*, and *Melospiza melodia*), the temperature of the seminal vesicle was 0.8-4.70 C below the deep body temperature. He proposed that such low temperatures might be beneficial for the sperm and might correspond to the effect of the scrotum on mammalian testes.

The copulatory organs of many birds are small, and they are not intromittent organs. During mating, the cloaca of the two sexes are brought into brief contact, the so-called cloacal kiss. An example of such a small copulatory organ is that of the rooster. Greese, swans, ducks, the Antidae and ostriches, cassowaries, emus and rheas have an intromittent organ, which is called a penis but which is actually a pseudopenis. The male Japanese quail has foam glands, which produce a foam that presumable aids in the deposition or transport of sperm. These glands, which respond to androgen treatment by an increase in secretory activity, are located between the stratified squamous epithelium of the proctodeum and the fibrous sheath of the cloacal sphincter muscle.

Mammals

The location of the testes varies in mammals. Giersberg and Rietsehel (1968) distinguish the following five different types of arrangement.

1. The testes move cauded, a gubernaculum and inguinal canal are formed, but the testes do not enter the inguinal canal, as in the Cetacea (whales, dolphins, porpoises) and armadillos.
2. A gubernaculum is present and the testes descend through the inguinal canal either temporarily (usually during the breeding season) or permanently into the scrotum.
3. The testes remain in the body cavity, but a gubernaculum that later regresses is formed, as in the Sirenia (seacows), elephants, and the Edentata (sloths and anteaters).

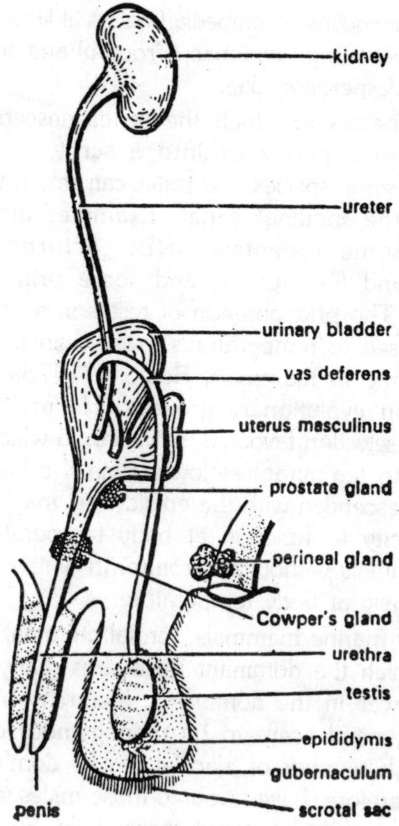

Fig. 5.5. Rabbit. Male urinogenital organs (Lateral view).

4. The testes remain in the body cavity, and there is no gubernaculum, as in the monotremes (duck-billed platypus, *Ornithorhynchus anatinus*, and the spiny anteaters. *Tachyglossus* sp., *Zaglossus* sp.), some primitive insectivora (the Macroscelididae or elephant shrews, and *Tenrec ecaudatus*), and the Hyraxes.
5. There is a gubernaculum, and the testes descend temporarily or permanently into a pouch of the cremaster muscle, but there is no true scrotum. The testes descend temporarily in *Solenodon*, in ,moles, hedgehogs, shrews, aardvark (*Orycteropus afer*), and many rodents, and in the lagomorphs (rabbits). The testes descend permanently in Phasocolomidae (wombats), Tapiridae, Hippopotamidae, Rhinocerotidae,

Manidae (pangolins), Pinnipedia (seals seal lions, walruses), some land carnivora (e.g., hyaenas, *Crocuta*) and some Chiroptera, e.g., the Vespertilion dae.

In most species in which the testes descend either into a cremaster muscle pouch or into a scrotum, the descent is permanent. In some species, the testes can return to the abdominal cavity or to the inguinal canal. Examples are: *Orycteropus* (aardvarks), some rodents (*Rattus*, *Sciurus*, *Tamias*, i.e., *chipmunks*) and Chiroptera, and some primates (*Loris* and *Perodicticus*). The phenomenon of testes in a scrotum, in which they are exposed to temperatures lower than the body, appears for the first time in mammals. Bedford (1978a, b) has tried to explain this on evolutionary grounds. He makes a convincing argument that selection favoured individuals in which the epididymis was exposed to temperatures lower than the body and that the testis, which descended with the epididymis, may secondarily have lost the capacity to function at body temperature. There are, however, mammals without a scrotum, in which the testes cleary function normally at body temperature.

Except for marine mammals, scrotal mammals are polygynous species, in which the dominant male sires many more offspring than males lower in the dominance hierarchy. Ascrotal species, however, in general seem to be monogamous or live in family units. The high number of ejaculations by dominant polygynous males may, therefore, have favoured those males with epididymides in which sperm could be stored successfully. Such storage would probably be more successful at lower temperatures because metabolic rates would be lower; thus eventually, individuals with epididymides (and as a consequence testes) a temperatures lower than the body would have a selective advantage. The apparent contradiction of polygynous ascrotal marine mammals may, in fact, be only an apparent contradiction. In the southern elephant seal (*Mirounga leonina*), the temperature of the testis is 6°C below body temperature, apparently as a result of blood flow from the rear flippers. In some seals, e.g., *Callorhinus* and *Zalophus*, the testes are located in a scrotal pouch. Many mammalian species have one or more well-developed accessory reproductive glands. These glands are:

1. *Prostate gland*. This originates as part of the urogenital sinus and has simple or pseudostratified columnar epithelium. The

secretions contain enzymes, such as diastase, glucoronidase, fibrinolysin, acid phosphatase as well as citric acid. The gland has ducts that drain into the urethra.

2. *Proprostate glands*. These are found in the rabbit, and appear to be part of the prostate; however, they are separated from the prostate by a connective tissue system. These glands originate, as does the prostate, from endoderm of the urogenital sinus. Holtz (1972) emphasizes that some names given to these glands are not appropriate, e.g., coagulating gland (because the glands are not the source of the gel mass in the ejaculate), and prostate and glandula vesicularis (because they are not Wolffian duct derivatives). The proprostate is an acinar gland, rich in smooth muscles and with a columnar epithelium. It secretes a white granular substance, which gives the gland a white colour and has two large excretory ducts that lead to small urethral diverticula.
3. *Ampullary glands*. These are derived from the Wolffian duct system and have a simple cuboidal, or columnar epithelium; their secretions are high in ergothioneine.
4. *Bulbar glands*. Also called the *interior bulbourethral glands*, these glands seem to be present in some of the Sciuridae. They are glandular developments of the bulbourethral glands inside the corpus spongiosum. The secretions drain via a glandular duct (ductus penis) at the ventral side of the urethra.
5. *Seminal vesicles*. More properly called vesicular glands, these are also derived from the Wolffian duct system. They have a simple or pseudostratified epithelium and secrete fructose, citric acid, proteins and ascorbic acid.
6. *Inguinal glands*. These are sebaceous glands and sweat glands found on the surface of the inguinal canal and in skin folds of the genital region. Holtz (1972) suggests the name *perineal glands* for these in the rabbit.
7. *Paraurethral glands*. Also called glands of Littre, these are probably synonymous with the glands named paraprostate glands by McKeever (1970) and Holtz (1972). These glands originate from the epithelium of the urethra and are thus part of the urogenital sinus. They are usually small structures, the ducts of which drain into the urethra. According to Holtz (1972), there are two different types of paraprostate glands in the rabbit: (a) glands consisting of small clusters of fluid-

filled vesicles, which, under microscopic examination, resemble the bulbourethral glands, and (b) small, compact, whitish glands, which are histologically identical with the prostate.

8. *Prostatic utricle*. Also called *uterus masculinus*, this is usually vestigial or even absent but can, according to Prasad (1974), be well developed. It is derived from the Mullerian duct, and the ducts drain into the urethra.

   Since this is a derivative of the Mullerian duct system, it might be expected to be sensitive to estrogens. As estrogens have been suspected of inducing cancer of derivatives of the Mullerian duct system in women, it may be prudent to consider seriously the wisdom of the use of estrogens in treating cancer of the prostate.

9. *Preputial glands*. These are modified sebaceous glands with a stratified, squamous epithelium and with ducts draining at the surface of the glands. They are enormously developed in the beaver.

10. *Urethral gland*. This gland is found in some Chiroptera between the prostate and the openings of the ducts at the bulbourethral glands. It lies around the urethra, is covered by the muscles of the urethra, and its secretions drain via short ducts into the urethra.

11. *Bulbourethral glands* (Cowper's glands). These glands are also part of the urogenital sinus. They have a simple columnar epithelium which secretes a viscous fluid containing sialoprotein, and have ducts that drain into the urethra.

The presence of these various glands in some mammalian species. The effect of the presence of these accessory sex glands on fertility has been studied in the boar (*Sus scrofa*), which produces a large amount of fluid in the ejaculate (about 200 cc) 26 per cent of which is contributed by the seminal vesicles, 56 percent by the prostate, and 19 per cent by the bulbourethral glands. These large quantities of secretions make this species a very suitable animal for studying these glands.

Removal of the seminal vesicles, or the bulbourethral glands, or both, had no effect on fertility. The effect of removal of the prostate was not reported for the boar, but apparently removal of the prostate in humans does not affect fertility. In rats, removal of the seminal vesicles and the prostate and coagulating gland reduces fertility severely, but removal of the following glands or combination of glands has no effect on fertility; bulbourethral glands, seminal vesicles, prostate and coagulating gland, bulbourethral glands and seminal vesicles, bulbourethral gland and coagulating gland, seminal vesicles and prostate.

Prototheria

The penis of the Prototheria, like that of the reptiles and birds, is located on the floor of the cloaca, and lies in a preputial fold of the ventral wall of the cloaca. The penis conducts the semen but not the urine, to the outside, because the semen-urine duct gives off a branch to the cloaca before it enters the penis. The ductus deferens drains into the urogenital sinus, dorsal to the ureters, not ventral, as in the Metatheria and Eutheria. The penis contains fibrovascular tissue which is homologous with the corpus cavernosum penis of the Metatheria and Eutheria. The glans of the duct-billed platypus is bifurcated.

Metatheria

The penis is located in a cloacal fold, which is surrounded, together with the anus, by a cloacal sphincter muscle, so that it is not visible generally except during erection. However, *Phascogale* (marsupial pocket mice) and *Dasyurus* (Australian "native cats") the penis is free and not in a cloacal fold. The penis of the opossum *Didelphis virginiana* is bifurcated, as is the penis of *Dasyurus* and *Phascolomis*, although in these last two genera, the bifurcation is not as marked. The penis of the Macropodidae and Tarsipes is definitely not forked. In *Peamelidae* (bandicoots), the urogenital sinus conducts only the semen to the outside; the urine is voided via the cloaca. In *Didelphis* and *Caenolestes* (bat opossum), the distal opening of the urinary duct lies at the base of glans and the ductus deferens ends in the tip of the penis. In *Myrmecobius* (numbats). *Thylacinus* (pouched wolf) and *Dasyurus*, the urethra divides into two branches, one branch going to one tip of the bifurcated glans, the other to the other tip.

Eutheria

The perineum divides the cloaca into the dorsocaudal rectum and the ventrocranial urogenital sinus, so that the penis lies on the ventral side of the anus. The distance between anus and penis can be small, as in beavers (*Castor* sp.) and hares (*Lepus europeaus*). Prasad (1974) states that in beavers, the urogenital openings and the anus open into a cloaca, a situation also found in the Prototheria, Metatheria, Edentata, and in the Insectivora, and *Neomys* sp. In other species, the penis lies, with its preputial fold, near the umbilicus. The double skil fold, which surrounds the penis, partly or completely forms the prepuce, which in the horse consists of two circular preputial folds which are telescoped so that the penis can be pulied inside the folds. The intrapreputialis penis corresponds to the glans penis in humans. Several types of penis have been distinguished:

1. *Indifferent type* (edentates and rodents). The penis is short, and the corpus spongiosum penis is not covered by a tunica. The corpus cavernosum can be either fibrous tissue or trabecular erectile tissue.
2. *Vascular type* (Perissodactyla, Carnivora, Primates, and some Insectivora). The penis, if not in erection, is long and flexible. The corpus cavemosum penis and corpus spongiosum urethra are surrounded by a fibrous capsule, the tunica albuginea. Erection results from the corpora cavernosa and corpora spongiosa filling with blood.
3. *Fibro-elastic type* (ruminants and whales). Usually the penis is long and fibrous and has a sigmoid flexure in the resting stage. There is little erectile tissue in the corpus cavernosum, the erection consisting of the straightening out of the sigmoid flexure. The corpus spongiosum is surrounded by a well-developed tunica.
4. *Intermediary type* (elephants, and the sirenia). This type combines features of type 2 and 3.

According to Slipper, as cited by Prasad (1974), there is a general correlation between the type of penis and the length of coitus. Animals with a fibro-elastic type penis have a short coitus, the ones with the indifferent type have a relatively short coitus, the ones with the intermediary type have a longer coitus than the other two types, but shorter than animals with the vascular type in which coitus may be fairly long. However, coitus in swine which have an intermediate type penis, may last as long as 30 minutes! In most species of Chiroptera, Rodentia, Carnivora, Insectivora, and many primates (but not humans) there is a penilea bone or baculum present. The male reproductive systems of some mammalian species are illustrated, *in situ*.

Female Reproductive System

In most vertebrates, the female has two ovaries, but there are exceptions to this, and among birds, most species have only one functional ovary, although some species have a high incidence of two functional ovaries. The female genital duct system of vertebrates parallels that of the males to the extent that in cyclostomes there is no genital duct system, in the Salmonidae the funnel is a secondary formation, and in the teleosts, the oviduct is not a Mullerian duct system but is formed from part of the ovary and mesentery.

Cyclostomes

The lampreys have one very large ovary, which originates from the fusion of the two gonad primordia. The ovary contains many follicles, (24,000-236,000), each of which consists of a fibrous theca externa, a glandular theca interna, and a granulosa layer. The follicles are all in the same stage of development at any particular time, and as Dodd (1972) has pointed out, not oogonial nests remain at the time of metamorphosis. The unique feature may be related to the fact that lampreys spawn and die shortly thereafter. The follicles form neither corpora atretica nor corpora lutea, although atresia does occur in *Petromyzon marinus*. Atresia consists of reabsorption of the yolk and follicle cells, so that only stroma cells are left.

Table 5.1. Vertebrates with one Functional Ovary

Species	*Experience for one-ovary condition*
Lampreys	Fusion of two gonad
Hagfishes	One gonad fails to develop
Perches, *Perca*	Fusion of two gonads
Pike perch, *Lucia-Stizostedion sp.*	Fusion of two gonads
Stone loach, *Noemacheilus sp.*	Fusion of two gonads
European bittering, *Rhodeus amarus*	Fusion of two gonads
Japanese ricefish, *Oryzias latipes*	One gonad fails to develop
Guppy, *Pocilia reticulata*	One gonad fails to develop
Sharks	
Scyliorhinus	Left ovary becomes atrophic
Pristiophorus	Left ovary becomes atrophic
Carcharhinus	Left ovary becomes atrophic
Galeus	Left ovary becomes atrophic
Mustelus	Left ovary becomes atrophic
Sphyrna	Left ovary becomes atrophic

Viviparous ray, *Urolophus*	Left ovary functional
Ovoviviparous ray, *Dasyatis*	Right ovary absent
Blind worm snakes, *Typhlopidae*	Left ovary and oviduct absent reason unknown
Birds	Left ovary functional in most species; right ovary regresses embryos
Duckbill platypus, *Omithorhynchus anatinus*	Left ovary functional
Bats	
Miniopterus nata ensis	Left ovary functional
Minopterus schrelbersis	Right ovary functional
Rninolophus	Right ovary functional
Tadarida cyancephala	Right ovary functional
Molossus ater	Right ovary functional
Mountain viscacha Lagidium ceruanumi	Right ovary functional, but after its removal left ovary becomes functional
Water buck, *Kobus defassa*	Left ovary functional

The single large ovary of the Myxiniidae (hagfishes) results from the failure of development of one of the two gonad primordia. There are few oocytes (1-21), and these are arranged in a row along the ovary. Each oocyte is enclosed in a tough shell which is secreted by the follicle. One can distinguish four layers in the follicular wall, an inner, simple, follicular epithelium: two layers of connective tissue; and an outer squamous layer. In the hagfishes there are corpora atretica and "corpora lutea," which originate either from follicular cells or from fluid-filled cysts. There are no gonoducts in either lampreys or hagfishes, and the gametes are shed in the body cavity and leave via an orifice secondarily connecting the coelom with the cloaca, in *Lampetra*, and through a single median orifice between anus and urinary opening in *Myxine*.

Elasmobranchs

The ovaries originally are paired in all species, but during development they may become asymmetric. The left ovary atrophies, for instance, in the sharks (*Scyliohinus*, *Pristiophorus*, *Corcharinus*, *Galeus*, *Mustelus* and *Sphyrna*), although both oviducts are present. Among the rays in the viviparous genus *Urolophus*, the left ovary is functional and the right is non functional, but each has an oviduct. However, in the ovoviviparous ray, *Dasyatis*, both the right ovary and the right oviduct are absent. In the skates, which are oviparous, both ovaries and both oviducts are functional.

Table 5.2. Avian Species in Which 50 Percent or More of the Individuals Examined have two Functional Ovaries

Species	*Number examined*	*Incidence of two Ovaries*
Apteryx australis (kiwi)	2	2/2
Podiceps cristatus (great crested grebe)	4	2/4
Fulmarus glacialis (furmar)	2	2/2
Cathartes aura (turkey vulture)	5	2/5
Circus aeruginosus (marsh harrier)	6	4/6
C. approximans (swamp harrier)	17	16/17
C.pygarus	5	4/5
C. cyaneus cyaneus	13	13/13
C. cyaneus hudsonius (marsh hawk)	31	22/31
C. macrourus	7	5/7
Accipiter gentiles atricapillus	14	13/14
A cooperi (Cooper's hawk)	3	3/3
A fasciatus	5	5/5
A. nisus	30	20/30
Leracidea berigora	2	2/2
Buteo buteo (common buzzard)	11	6/11
E.jamacensis (red-tailed hawk)	15	12/15
B.platypterus	5	4/5
Falco columbarius (pigeon hawk)	4	2/4
F. vespertinus	9	6/9
F. tinnuculus (European kestrel)	20	12/20
Eclectus roratus	4	2/4
Bubo bubu (eagle owl)	7	7/7
Strix aluco (tawney owl)	4	4/4

The ovaries are suspended by mesovaria. They are usually naked, i.e., *gymnovaria*, and the follicles develop from the germinal epithelium covering the ovary. However, the hollow ovary of the basking shark, *Cetorlhinus maximus*, which weights about 12 kg (the animal weighs about 3-4 tons), is invested by a fibrous coat. In this shark the germinal epithelium forms a network of tubules which opens into a pocket into which the oocytes are released. They then travel through the peritonium to the oviduct. This ovary is exceptional and should, according to Wourms (1977), not be considered as typical for the viviparous condition in sharks. The ovary contains: (1) follicles, which in the oviparous *Scyliorhinus*

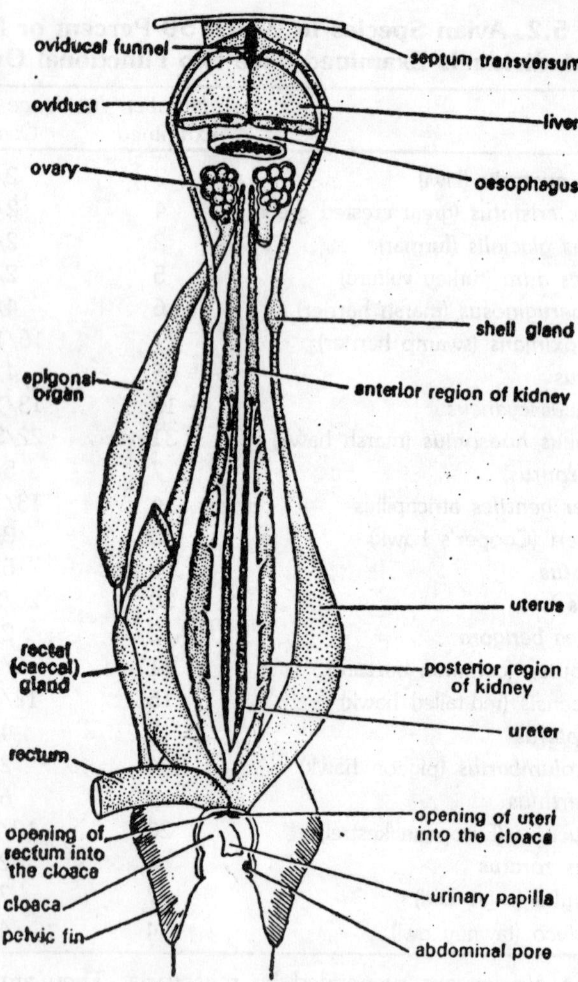

Fig. 5.6. Scoliodon. Female urinogenital organs.

canicula are quite large, but in the viviparous basking shark are not more than 5mm in diameter at most; (2) corpora atretica, which are derived from egg-containing follicles that have degenerated; (3) "corpora lutea" which are reptured follicles that have reorganized into structures that resemble mammalian corpora lutea; (4) corpora lutea found, e.g., in the ovoviviparous *Squalus acanthias* in which the granulosa shows evidence of 38-HSD activity; and (5) "functional corpora atretica", a name here used for lack of a better one, which are derived from atretic follicles but show luteinization and the presence of steroids. Functional corpora aretica are found in ovoviviparous electric rays. *Torpedo marmorata* and *T. ocellata*. The reptured follicles, "corpora lutea," do not show evidence of steroidogenesis in these rays. The (Mullerian) oviduct has four regions:

1. *Uterus*. This may play a role in the nutrition of the embryos. The left and right uterus may fuse posteriorly and open into a common vagina, or they may open separately into the cloaca. The common vagina, if present, also opens into the cloaca. In some cases a hymen is present at the posterior end of the oviduct.
2. *Isthmus*. This connects the shell gland to the uterus.
3. *Funnel* (anterior ostium tubae). This collects the ovulated oocytes. It is formed either by fusion of the anterior end of the left and right oviduct, as in the basking shark, or by asymmetric development of one primitive funnel, as in *Scyliorhinus*.
4. *Shell gland* or *nidamental gland*. This structure is well developed in oviparous and ovoviviparous species, but may be vestigial in viviparous ones. It is a tubular gland which secrets albumin, mucus, and in those species producing an egg case, egg case proteins. In *Scyliorhinus*, sperm are stored in this part of the oviduct.

Holocephali

In the chimaera, *Hydrolagus colliei*, two ovaries and two oviducts are present. In general, the reproductive system resembles that of the elasmobranchs, except that in the Holocephali, the two oviducts always have one common funnel, and each oviduct has a separate opening into the urogenital sinus lateral to the excretory duct, whereas in the elasmobranchs the oviducts open via either a common vagina or separate gavinas into the cloaca. The uterus contributes to the formation of the egg case in *H. colliei*, but does not do so in elasmobranchs.

Chondrostei

The Acipenseridae (sturgeqns) have a vestigial connection between ovary and kidney. The ovary starts out as a ribbonlike structure, as it does in teleosts, but it rolls up laterally and dorsally, so that it forms a hollow organ. The Mullerian Oviduct is continuous with the ovary and opens into the cloaca next to the Wolffian duct (Baer, 1964: Harder, 1975). The oocytes, after release from the follicle, pass through the Mullerian oviduct to the cloaca.

Teleosts

The ovaria start as ribbonlike structures, but they roll up in a lateral and dorsal direction, so that the side at which ovulations occur is turned inward and surrounds a tube, the oviduct. The oviduct is called *entovarial*, when the free side of the ovary is against ovarian tissue, but *parovarial*, when the free side of the ovary is against the wall of the peritoneal cavity. The oviducts are thus extensions of these peritoneal folds, which enclose the ovary during its development. The ovaries are usually paired, hollow organs suspended by mesovaria. The left ovary of the smelt (*Osmerus eperlanus*) is located anteriorly and the right ovary posteriorly. Only one gonad primordium develops in the Japanese rice fish, (*Oryzias latipes*) and the guppy (*Poecilia reticulata*). In the perches (*Perca* spp.), the ovaries are fused completely, whereas in the pike perch (*Lucia*, formerly *Stizostedion*, *perca*), they are fused only caudally.

The ovaries are fused medially in the stone loach (*Noemacheilus* sp.), and in the European bitterling (*Rhodens amarus*). Oocytes and embryos develop in the wall of the ovary. The teleost ovary may contain follicles, corpora atertica, pre- and postovulatory "corpora lutea" and pre- and postovulatory corpora lutea. The follicular wall may, in addition, from (1) a calyx from the theca closing the ovarian wall after rupture of the follicle, (2) a calyx nutricius from the follicular wall that secretes nutritious material (embryotrophe) for the embryo either into the ovary or into the follicle containing the embryo. The oviducts, which, as noted, are not Mullerian ducts but peritoneal folds that have grown backward to form oviducts, are absent in some species.

In the Salmonidae (salmons and trouts), the anterior part of the oviduct has disappeared, and the oocytes fall into the body cavity and then reach the funnel like remnant of the posterior part of the oviduct. In the Galaxiidae, Glyodontiidae, Notopteridae, Osteoglossidae, and Anguillidae, and in *Misgurnus*, the oviducts have completely disappeared, and the gametes leave the body cavity via a genital pore, which is not, however, homologous with the abdominal pore found in many sharks. The genital opening may be in various locations. The midgullar position of the female's genital opening in the Phallosthide is adapted to the specialized penis of the male.

Teleosts have some specialized structures and associated behaviours, e.g., the ovipositor of the bitterling (*Rhodeus amarus*), a skinlike structure derived from the genital papilla. In this species, the growth of the oviposterior and the nuptial colour of the male require the presence of a fresh water mussel, *Unio pictorum* or *Anodonta intermedia* (Banarescu, 1973). Males select the female with the largest oviposterior. At spawning, the females thrust their ovipositor into the *excurrent* siphon of the mussel and thus have to push their oocytes upstream. Breder and Rosen (1966) state that, exceptionally, the female lays her oocytes in the current siphon. The male subsequently spawns near the *incurrent* siphon, or the sperm would never enter the mussel.

The oocytes are fertilized inside the mussel and they develop in its gill chambers. After about a month the embryos hatch and the young bitterlings leave the mussel. The larvae of the mussel attach themselves to the gills and fins of the young fish and live their as parasites for a time. The Jenynsiidae and Anblepidae have their genital opening either on the left or right, corresponding to the males having a left or right gonopodium. Among the Jenynsiidae, three-fifths of the males are dextral and two-fifths sinistral, whereas for the females the ratio is just the reverse.

Amphibians

The ovaries are paired, lobed, hollow organs consisting mainly of a cortex. The outside of the ovary is covered with germinal epithelium; the cavity is lined with cells of medullary origin.

Anura

The ovaries of the Anura, which are largely oviparous, have follicles and scars from ruptured follicles, corpora atretica and "functional corpora atretica," that is, structures formed from non-ovulated follicles, but resembling corpora lutea. In the viviparous toad *Nectophrynoides occidentalis*, Xavier (1974) identified corpora lutea formed from ruptured follicles. These corpora lutea contained cholesterol, showed a positive reaction for 3β-HSD, and were capable of synthesizing progesterone from pregnenolone. The quantity of progesterone synthesized paralleled the size of the corpora lutea, and both decreased as pseudopregnancy (ovulation without mating) progressed beyond four months.

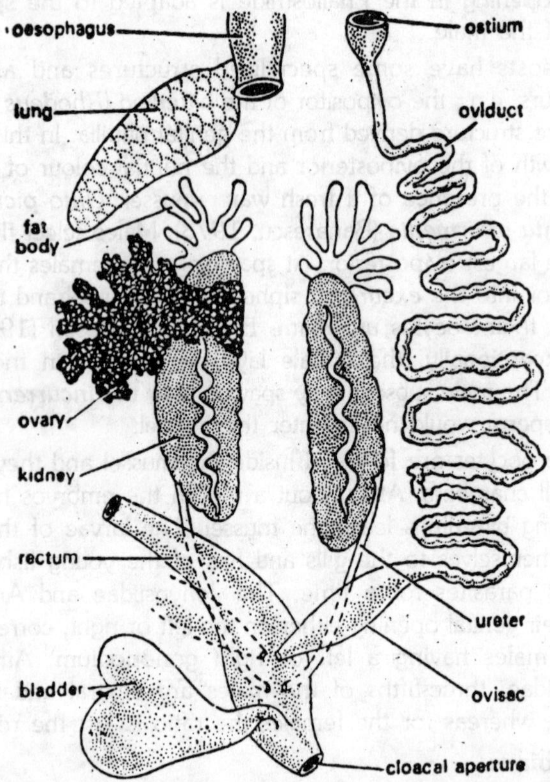

Fig. 5.7. Frog. Female urinogenital organs.

Ruptured follicles appear to have a steroid secretory function in *Rana esculenta* and *R. cyanophlyctis*, which are oviparous, and in some ovoviviparous species. The females of some species carry the fertilized eggs either in a pouch, e.g., *Gastrotheca marsupiata* and *G. pygmaea*, or in separate chambers, e.g., *Pipa pipa* and *Cryptobatrachus evansi*. The developing embryos probably derive nutrients from the mother.

Urodeles

The ovaries resemble those of the Anura, with follicles, corpora lutea, and corpora atretica, which originate from follicles in which the yolk has been resorbed. According to Lofts (1974), there is little or no evidence that these structures secrete steroids. Corpora lutea, which result from a reorganization of the ruptured follicle into a progestin-secreting structure, are found in *Triturus cristatus* and *Salamandra salamandra*. The (Mullerian) oviducts of amphibia have three regions: the pars recta, the pars convoluta, and the pars uterine. The oviducts end in the cloaca, which in the urodeles, shows some specialization. In salamanders, three sets of glands are associated with the oviduct: (1) *pelvic glands*, which are located in the roof of the cloaca, and serve as reservoirs for the storage of spermatozoa; (2) *cloacal glands*, present in all ambystomids, salamandrids, and the primitive plethodontids, which may play a role in formation of the egg case; and (3) *abdominal glands*, present in *Ambystome*, *Necturus*, and *Eurycea*, which are apparently rudimentary. In the urodeles with internal fertilization, the female cloaca may have well-developed cloacal lips, with which the spermatophores can be picked up from the ground.

Reptiles

The ovaries are paired, hollow organs, suspended by mesorchia and with squamous epithelium lining the ovarian cavity. The follicles project from short stalks, and the walls, which surround the large yolk, consist of a granulosa layer, a theca interna, and a theca externa. Corpora lutea are formed after ovulation as the result of the luteinization and proliferation and/or hypertrophy of the granulosa cells and supporting connective tissue from the thecal layers. In oviparous species, the corpora lutea regress soon after oviposition. In the lizard, *Anolis carolinensis*, follicles that have become atretic form corpora atretica. The corpora atretica seem to determine the response of the ovarian follicles to exogenous and probably endogenous gonadotrophins. After ovulation, the oocytes may according to Bellairs (1970), wander through the body cavity and enter either of the two oviducts. Abdominal fat bodies have been found in all temperate zone reptiles that have been examined and in some anoline lizards of warmer regions.

The lipids from these bodies pass to the liver and are incorporated into yolk precursons. In the lizard, *Uta stansburiana*, removal of the fat bodies inhibits follicular development or, when the follicles are developed, causes atresia of the developed follicles. There appears to be a reciprocal relationship between ovary and fat bodies, because ovariectomy prevents rapid fat mobilization from the fat bodies. The oviduct, which the Mullerian ducts, are convoluted. They consists of (1) an infundibulum or funnel: (2) a convoluted segment with a layer of longitudinal and a layer of circular muscle, and a mucosa with many alveolar glands and

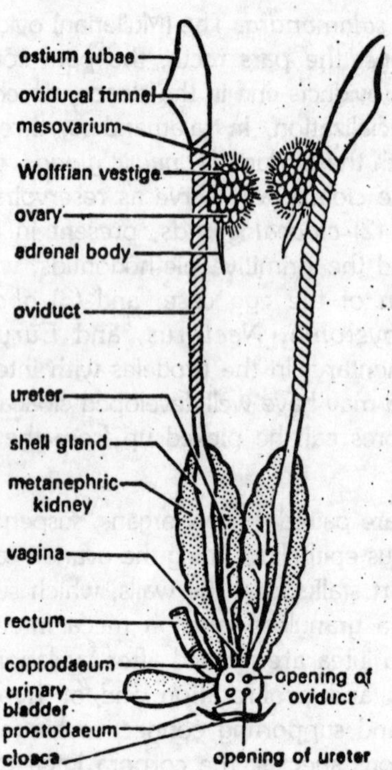

Fig. 5.8. Uromastix. Female urinogenital organs.

ciliated epithelium; (3) an isthmus; (4) a uterine segment, which has a strong musculature and is lined with ciliated and non-ciliated columnar epithelium; and (5) a vagina which has mucous glands, and traverses the cloaca. The cloacal openings for the paired oviducts are separate in *Lacerta viridis*, *Sceloporus* sp., *Hemidacytylus* sp., *Phrynosoma* sp., but united in *Gerrhonotus* sp. and *Cnemidophorus* sp. In the blind, limbless lizards (*Anniella*) both ovaries are functioned and of the same size, but only the right oviduct is functional. The left oviduct is smaller than the right oviduct and has numerous glands, the secretions of which probably pass out via the cloaca. In blind worm snakes (Typhlopidae), the left oviduct is absent.

In some reptiles the sperm retain their fertilizing capacity for a considerable length of time after insemination. It is not surprising, therefore, that there are structures for storing sperm in the oviduct. In snakes, turtles and lizards, sperm are retained in "seminal vesicles," which are specialized alveolar glands at the base of the infundibulum. From these glands, ducts lead into the oviductal lumen. In the snake, *Thamnophis elegans terrestris*, the sperm are retained in the posterior two-thirds of the oviduct , and the sperm do not reach the "seminal vesicles" until February-March, if the snake mates in the fall.

If insemination occurs in the spring, however, the sperm reach the "seminal vesicles" in April. In the adder, *Vipera aspis*, the sperm remain in the vagina after fall mating and proceed to the "seminal vesicles" in the spring just before they fertilize the oocytes. Iguanids have "seminal vesicles" in the anterior segment of the vagina, whereas gekkonids have these structures between uterus and infundibulum. The seminal receptacles, found in the caudal part of the albuminous region of the oviduct of the box turtle (*Terrapene carolina*), are morphologically unspecialized glands.

Birds

Most avian species have only a left ovary and a left oviduct, but some species have been found to have two functional ovaries, and in most of these species, it is the left oviduct that is functional. Oocytes ovulated from the right ovary will, in the absence of a

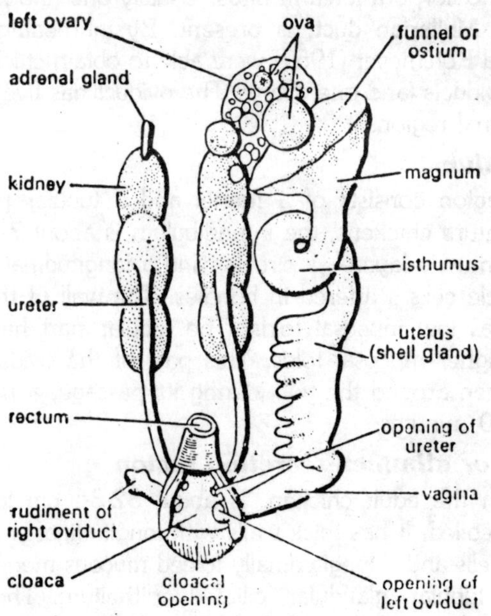

Fig. 5.9. Pigeon. Female urinogenital system.

right oviduct, drop into the body cavity and will be absorbed, since a mesentery separates the right ovary from the left oviduct, so that an oocyte from the right ovary cannot be picked up by the left oviduct. The ovary is attached to the body wall by a mesovarium. It has many follicles embedded in a sparse stroma of connective tissue. During the laying season, there is a series of large follicles, which show a hierarchy of size.

The follicular wall consists of a granulosa layer, a theca interna, and a theca externa. The large follicles are filled with yellow yolk; some of the smaller follicles are filled with either yellow or white yolk, depending on the stage of development. On the wall of the large follicles is a clear band, the stigma, which macroscopically is devoid of blood vessels and along which the follicle ruptures. In addition to the large and developing follicles, one finds atretic and ruptured follicles. In atretic follicles, the yolk is reabsorbed, and the granulosa cells become dissociated from the follicular wall and move into the follicles.

Atretic follicles are not always present in laying chickens, and ruptured follicles rapidly disappear after ovulation. In laying chickens, they are difficult to find five days after their rupture. There are no corpora lutea in birds. Usually one (the left) oviduct, which is a Mullerian duct, is present. By inbreeding, however, Morgan and Kohlmeyer (1957) were able to obtain chickens strains with two oviducts (and one ovary). The oviduct has five anatomical and functional regions.

Infundibulum

This region consists of a funnel and a tubular part. In the sexually mature chickens, the infundibulum is about 7-9 cm long. It has two muscle layers—a circular and a longitudinal layer—and a few muscle cells scattered in bundles. The wall of the funnel is thin and has low mucosal folds. The tubular part has a thicker wall and higher mucosal folds. This part of the oviduct secretes thick albumen around the yolk during its passage, a passage that takes 15-30 minutes.

Magnum or albumen-secreting region

This, in the adult chicken, is about 32-34 cm long, white, and highly coiled. It has thick walls with longitudinal circular layers of muscle cells and a longitudinally folded mucous membrane lined with high, tubular, glandular, ciliated epithelium. The glandular epithelium gives this region of the wall its thickness. Thick albumen is secreted in this region around the yolk as it transverses the magnum (in about 2-3 hours).

Isthmus

This region is about 10 cm long in the adult chicken and is separated from the magnum by a clear translucent band of non-glandular tissue. The isthmus has a thicker circular muscle layer than the infundibulum and the magnum, and its glandular development is less marked than that of magnum. The glands are tubular and secrete the proteinaceous egg membranes around the yolk and its surrounding thick albumen, as the egg is transported through the isthmus (in about 1 hour).

Shell gland or uterus

This, in the adult chicken, is about 11 cm long and has very thick muscular walls. It has a prominent longitudinal muscle layer and is lined with tubular and unicellular goblet-type cells. The cells of this lining secretes the thin albumen, the shell, and any pigments that may colour the shell. In the chicken, the egg remains in this region 20-26 hours.

Vagina

This is about 10 cm long; it is separated from the shell gland by a sphincter muscle and ends in the cloaca. It has a longitudinal muscle layer that is easily distinguished and a circular muscle layer that is well developed. The mucous membrane has low narrow folds and is lined with ciliated and non-ciliated cells. There are no glands and no secretion in the vagina proper; the egg is "finished" in the shell gland. In addition to the glans just mentioned in connection with egg formation, there are two areas where spermatozoa are stored and can retain their fertilizing capacity (about 7-14 days in chickens and 40-50 days in turkeys). These two regions are at the extreme ends of the oviduct, i.e., at the infundibulum and at the utero-vaginal junction. The ovary receives its blood from the gonadrenal artery, which is a branch of the dorsal aorta that sends a branch, the ovarian-oviductal artery, to the ovary. The blood is drained by a cranial ovarian-oviductal and a caudal ovarian vein, each of which drain into the vena cava. The venous system of the ovary is very well-developed.

The innervation of the ovary is complex; adrenergic and cholinergic fibres have been found. The adrenergic supply originates from the 5th, 6th, 7th, sometimes the 4th thoracic, and the 1st and

2nd lumbar ganglia of the sympathetic chain. The origin of the parasympathetic fibres is not known.

Mammals

Ovary

The ovaries are usually paired, but in some species (Table) one may be inactive; that is, either it does not ovulate or it ovulates only after the active ovary has been removed. For example, if the right ovary of the mountain viscacha (*Lagidium peruanum*) is removed, the left one becomes functional. The right ovary is also the functional one in the following bats: *Miniopterus*, *Schreibersi*, *Rhinolophus* sp., *Tadarida cyanophala* and *Molossus ater*. The left ovary is dominant in the bat, *Miniopterus natalensis*, and in the waterbuck, *Kobus defassa*. Each ovary is suspended from the dorsal body wall by the mesovarium or secondarily by the broad ligament.

In many species, the ovary lies in a membranous sac called the *ovarian bursa*, which is a fold of the mesosalpinx, the

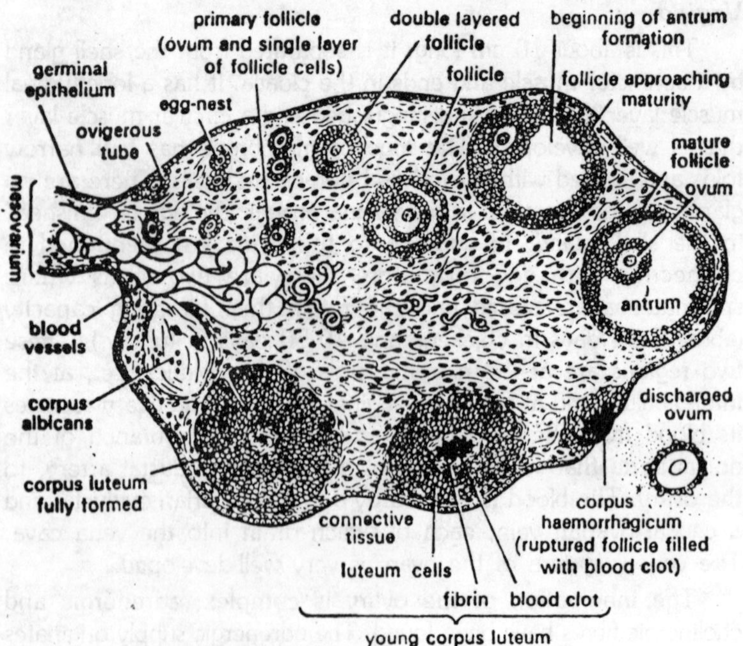

Fig. 5.10. Rabbit. T.S. ovary.

mesentery of the oviduct. The function of bursa has not been determined. The ovary consists of the stroma, the germinal epithelium, the germ cells and their associated structures, interstitial cells, and various vestigial structures. The germinal epithelium, which covers the ovary, may be either directly on the ovarian stroma (e.g., in shrews) or it may be on a fibrous ovarian tunica albuginea. It may consist of squamous or low columnar cells. The germ cells and their associated structures include oocytes, follicles in different stages of development, connective tissue, corpora lutea, old corpora lutea called *corpora albicantia*, and interstitial tissue.

Follicles

These can be characterized as small, medium and large follicles, according to Pedersen and Peters (1968). The main distinctions, are made on the basis of oocyte size, number of layers of follicular cells, and the presence of follicular fluid. The follicular wall has a granulosa layer, a theca externa, and a theca interna. The oocyte is surrounded by an acellular layer, the zone pellucida, which separates it from the surrounding follicular cells. In the large follicles, the oocyte is surrounded by granulosa cells, which form a structure, the cumulus oophorus, which connects it to the follicular wall. Immediately surrounding the zona pellucida is a layer of columnar or pseudostratified columnar cells, which is called the *corona radiata*.

Table 5.3. Occurrence of ovarian bursa in mammals

	Anatomy of bursa				
Animals	*1*	*2*	*3*	*4*	*5*
Prototheria					
Tachyglossus	+				
Ornithorhynchus				+	
Mernatheria					
Phalangeridae				+	
Phascolomidae		+			
Wacropodidae				+*	
Eurneria					
Insectivora					
Tenrecinae	+				
Erinaceinae	+				
Soricinae	+				
Talpidae	+				

Chiropters	+				
Primates					
Lemuridae			+		
Hominidae			+	+	+
Edentata					
Bradypodidae	+				
Dasypodidae		+			
Lagomorpha					
Lepus europaeus			+		
Oryctolagus cuniculus		+			
Rodentia					
Sciuridae		+			
Castoridae		+			
Circetidae	+				
Muridae	+				
Gliridae	+				
Dipodidae					+
Hystricidae		+			
Cavidae			+		
Dasyproctidae					+
Cetacea					+
Carnivora					
Canidae	+				
Ursidae	+				
Procyonidae	+				
Mustelidae	+				
Viverridae	+	+			
Hyaenidae		+			
Felidae		+	+		+
Pinnipedia		+			
Hyracoidea					
Procaviidae		+			
Perissodactyla					
Equidae				+	+
Artiodactyla					
Suidae		+			
Camelidae	+				
Cervidae			+	+	
Bovidae		+			

Atretic follicles

These are degenerating follicles. The first sign of atresia is the pykonosis of the epithelial follicular cells, which are adjacent to the lumen of the large follicles and in layers just external to the corona radiata of secondary follicles. Subsequently the cumulus oophorous shrinks, the granulosa cells disappear, and fibroblasts and leukocytes invade the atretic follicle.

Corpora lutea

These originate from ruptured follicles and are formed by enlargement and multiplication of the granulosa cells. The corpora lutea consist of a solid mass of polyhedral cells. Mossman and Duke (1973a) proposed the term "luteal gland" in preference to corpus luteum, but the use of the term *corpus luteum* is such long standing that substitution of a new term seems to have little chance of adoption. The corpora lutea of the Madagascar hedgehog (*Setifer setosus*) develop from a follicle in which no antrum or cavity is formed. The oocytes is fertilized inside the follicle, and the zygote is extruded by contractions of the follicular sheath and a shift in position of the granulosa cells, so that the zygote becomes exposed on the surface of the follicle.

Accessory corpora lutea

These structures can have two different origins: (1) granulosa cells of atretic follicles, as in the North America porcupine (*Erithizon dorsatum*) and in the primates, *Tupaia* sp., and *Urogale* sp., (2) the epithelium of medullary cords, as in juvenile easter chipmunks (*Tamias striatus*), in which medullary follicles ovulate and form large corpora lutea.

Corpora albicantia

These are small masses of connective tissue which are the remains of degenerated corpora lutea.

Thecal glands

These consist of a layer or zone of glandular cells adjacent to the granulosa cells of mature follicles.

Interstitial glands

These may originate from the ovarian stroma, medullary cord cells, and the theca interna of atretic follicles. Mossman and Duke (1973b) stress the inaccuracy of calling these structures "luteinized follicles."

Some structures in the ovary are vestigial. The *rete ovarii* plays a role in oogenesis. It is a regular network of spaces lined with a simple low columnar epithelium. It is well developed in the tarsier, *Tarsius* sp. (a primate) and in pangolins, *Manisa* sp. The *epoophoron* consists of vestiges of efferent ductules and in homologous with the epididymis of the male. *Gartner's duct* or the duct of the epoophoron is a vestige of the ductus deferens. The *paraoophoron* consists of the more rostral and the more caudal mesonephric tubules—or ductuli aberrantes. It is homologous with the (also vestigial) paradidymis of the male. The ovary receives blood from the ovarian artery, a branch of the dorsal aorta, and blood leaves the ovary via the ovarian vein, which drains into the caudal vena cava. Some branches of the ovarian artery and vein anastomose with ovarian branches of the uterine blood vessels. The ovary receives fibers from the vagus nerve and also adrenergic fibers. It has been seen that adrenergic mechanisms are involved in ovulation.

Oviduct

The oviduct is suspended by the mesosalpinx on the one side; the other side is attached to the mesotubarium superius, which connects the oviduct to the fimbria. The fimbria consist of a fringe of irregular processes. The relationship between oviduct and mesenteries can vary considerably among species. Beck and Boots (1974) give a comprehensive summary of the various relationships in a large number of species. Starting at the cranial end, the oviduct consists of: (1) infundibulum, with a fimbriated, funnel like opening, (2) the ampulla, (3) the isthmus, (4) the utero-tubal junction. The wall of the oviduct consists of several layers. The *tunica serosa*, the outermost layer is a thin layer of connective tissue covered by a single layer of squamous epithelium. The *tunica muscularis* consists of smooth muscle, which may be arranged in four distinct patterns, i.e., (1) a predominant circular layer of smooth muscle without distinguishable longitudinal layers, (2) thin longitudinal layers on both sides of a thicker circular layer (3) a thick circular layer with an outer longitudinal muscle layer, and (4) a circular layer with an inner longitudinal muscle layer.

The *tunica mucosa*, the innermost layer, consists of a lamina propria of connective tissue and a lamina epithelialis, which has ciliated cells and three types of non-ciliated cells: secretory cells, wedge-type cells called "peg cells", and basal cells along the base

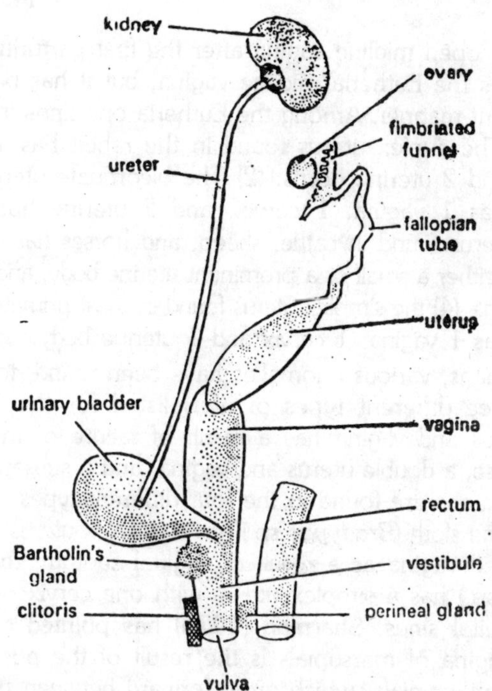

Fig. 5.11. Rabbit. Female urinogenital system (Lateral view).

of the epithelium. In species such as pigs, mice, rats, humans and rhesus monkeys, the presence of ciliated cells depends on the stage of reproductive cycle; however, the occurrence of cyclic changes in cilia is controversial. The anatomy of the utero-tubal junction varies among species. Beck and Boots (1974) described ten different morphological types of utero-tubal junctions. Type 1-3 lack mucosal projections of oviductal or uterine origin. Type 4-10 have specialized mucosal projections of oviductal or uterine origin. The oocytes or the zygotes are transported rapidly through the utero-tubal junction, which may play a role in sperm transport.

Uterus

The uterus is suspended by the mesometrium. There are different types of uteri. In the prototheria, the uteria are paired and open into the urogenital sinus; no vagina is present. In the metatheria, or marsupials, there are 2 vaginae, 2 cervices, 2 uteria. Sharman (1976) has illustrated some of the variations of this general plan and two such variations. The kangaroo (*Macropus* sp.) has an open midline vagina after the first parturition, so that it resembles the Eutheria midline vagina, but it has been formed in a different manner. Among the Eutheria one finds four kinds of uteri: (1) The duplex uterus found in the rabbit has 1 vagina, 2 cervices, and 2 uterine horns. (2) The bicornuate uterus found in the pig, has 1 vagina, 1 cervix, and 2 uterine horns. (3) the bipartite uterus found in cattle, sheep, and horses has 1 vagina, 1 cervix and either a small or a prominent uterine body, and 2 separate uterine horns, (4) the simplex uterus found in most primates, including humans, has 1 vagina, 1 cervix and 1 uterine body.

In humans, various anomalies have been found; for example, all the three different types of uteri listed for the Eutheria, a doubt uterus and vagina has a result of failure of the Mullerian ducts to fuse, a double uterus and vagina, and a subseptate uterus. Small variations are found in the four different types of uteri. For example, the sloth (*Bradypus* sp.) has a simplex uterus, one cervix, but a paired vagina as a result of sagittal septum; the armadillo (*Dasypus* sp.) has a simplex uterus with one cervix opening into the urogenital sinus. Sharman (1976) has pointed out that the peculiar vagina of marsupials is the result of the position of the ureters of marsupials, which grow forward between the Wolffian and Mullerian ducts. This prevents the fusion of the vaginae in the midline (because of the intrusion of ureters). The lateral vaginae serve for transported of spermatozoa, but the birth of the young occurs via the midline birth canal, which connects the uterine opening *directly* with the arogenital sinus.

Cervix

The cervix is a sphincter at the posterior end of the uterus. Its wall is thick, and the inner wall has ridges. It consists of a serosa, an outer longitudinal and an inner circular muscle, a submucosa, a mucosa consisting of columnar epithelium at the cranial end and squamous epithelium at the caudal end, and many goblet cells that secrete mucus. The consistency of this secretion varies during the menstrual cycle and is sometimes used to diagnose the time of ovulation in women, as in the crystallization pattern of the mucus.

Vagina

The relation of the Mullerian ducts, urogenital sinus, and the integument in the formation of the vagina. The location of vacinal and urethral openings in a few species. The vaginal wall consists of a serosa, an outer longitudinal and an inner circular muscle layer, a submucosa, and a mucosa consisting of stratified squamous epithelium. In some laboratory species, especially the rat, mouse, and hamster, the stage of the estrous cycle is correlated with differences in the histological appearance of the vaginal epithelium, so that vaginal smears can be used to diagnose the stage of the estrous cycle. During proestrus epithelial cells predominate; during estrus there are some remnants of epithelial cells and many large cornified and epithelial cells and leukocytes; during diestrus few epithelial cells remain and few to many leukocytes are found.

Prostate

Although the prostate is mostly a male secondary sexual gland, a female prostate is well developed in some species. In laboratory rats, the "Witschi rat," developed by selective breedings, has a well-developed prostate.

Table 5.4. Part of Prostate Ground in Female Mammals

Species	*Part of prostate*
Insectivora	
Hemicentetes semispinosus	Dorsal-lateral
Erinaceus europaeus	Dorsal-lateral
Talpa europaea	Ventral
Chiroptera	
Taphozous	Dorsal
Nycteris luteola	Ventral
Coleura afra	Dorsal
Primates	
Macaca mulatta	Dorsal (?)
Cercopithecus aeithiops	Dorsal (?)
Homo sapiens	Dorsal
Lagomorpha	
Oryctolagus cuniculus	Dorsal (?)
Sylvilagus floridanus	Dorsal
Rodentia	
Arvicola sapidus	Ventral (?)
Apodermus sylavaticus	Ventral
Arvicanthis cinerus	Ventral
Rattus norvegicus	Ventral
R. rattus	Ventral
Mastomys erthroleucus	Ventral
Mus musculus	Ventral

External Genitalia

The external genitalia consist of the labia majora, labia minora, and the clitoris, but only the human has true labia. Generally, the clitoris is poorly developed unless the fetus has been exposed to larger amounts of androgens than normal, e.g., in the adrenogenital syndrome, in the freemartin, or after experimental androgen treatment. However, in the hyaena (*Hyaena crocuta*), the clitoris is very well developed. This extreme development and anatomy make it almost impossible to distinguish a male from a female, especially since the females also has a false scrotum.

Sexual skin

Some primates have a so-called sexual skin, which is contiguous with the external genitalia. It consist of a highly vascularized edematous or pigmented dermis. Its colour and vascularization vary during the reproductive cycle.

6

Oogenesis

The growing oocyte can be regarded as a special form of differentiated cell, organized to carry out a unique function. It accumulates a particular collection of gene transcripts, and by the end of its growth phase has developed a complex of unusual cytological structures. We have seen that a set of genes expressed during oogenesis bears a close, largely overlapping relation to that expressed in the early embryo nuclei. Thus from the standpoint of nuclear function, *oogenesis* is the process during which the major portion of the *zygotic* pattern of gene activity is established. This pattern is also impressed on the entering male genome, and during cleavage is propagated to the blastomere nuclei. Though there are of course genes that function only in the early embryo and not during oogenesis, and *vice versa*, in quantitative terms both of these classes are minor, in respect to either the mass or the complexity of the transcripts for which they account. This chapter is focussed on gene expression during the growth phase of oogenesis, which begins following premeiotic DNA replication.

Throughout this period newly synthesized transcripts are being delivered to the oocyte cytoplasm, and until the metaphase of the initial reduction division occurs at maturation, the oocyte remains in the first meiotic prophase. Thus it contains a 4C rather than a 2C genome. In the typical case transcription of the genes utilized during oogenesis begin, early in the growth phase, before the onset of vitellogenesis and continues almost until maturation. At maturation there occurs the breakdown of the germinal vesicle, or oocyte nucleus. By this time all nuclear transcription has ceased, and a new set of biosynthetic activities is instituted. The molecular biology of oocyte maturation *per se* has been reviewed extensively and is not further considered here.

For earlier events of oogenesis, such as germ cell migration, oogonial multiplication, and the many other biological and biochemical aspects not directly relevant to germ line transcription and gene expression during oogenesis, recent complications such as those edited by Jones (1978) and Browder (1985) may be consulted. Follicle cell functions are reviewed in these works as well, and in articles by Mahowald and Kambysellis (1980) : Wallace (1983). At the molecular level perhaps the best studied follicle cell activity is synthesis of the chorion or egg shell, particular in lepidopteran. Though follicle cell function lies outside the scope of the following discussion, it is to be noted that chorion synthesis is particularly interesting in the Diptera, where it has been observed that the genes coding for the major proteins of this structure are amplified during differentiation of the ovariole.

Gene Expression in the Nurse Cell-oocyte Complex

Oogenesis is always to some extent a cooperative process, involving cells other than the oocyte itself. In most, though certainly not all forms, the growing oocyte is surrounded with follicle cells of somatic origin. Follicle cell, provide essential transport, hormonal and secretory functions, but with the possible rare exceptions mentioned below they do not directly feed transcription products to the oocyte. A natural division separates forms of oogenesis in which the *germ line* gives rise both to oocytes and to nurse cells that directly provide the oocyte with cytoplasmic organelles and macromolecules, including transcripts, from those in which the oocyte nucleus is alone responsible for the synthesis of its nonyolk macromolecular constituents. Oogenesis in which the egg cytoplasm is the cooperative product of both nurse cell and oocyte genomes is termed *meroistic oogenesis*. It is known mainly in insects, though it occurs as well in other protostomial invertebrates.

Origin of the Nurse Cell-Oocyte Complex

Meroistic oogenesis has been studied in most detail in certain holometabolous insect orders, the most prominent of which are Coleoptera. Hymenoptera, Hemiptera, Lepidoptera, and Diptera. Many other insect orders, e.g. the Orthoptera and Odonata, carry out oogenesis by the alternative route, in insects termed *panoistic* oogenesis. Nurse cells are here absent, and the oocyte growth process is autonomous, except for the assistance in yolk protein and metabolite uptake provided by the follicle cells. The chromosomes of meroistic insect oocytes typically display a condensed structure. They either do not synthesize RNA at all during oocyte growth, or do so at a very low relative rate. The RNA accumulated during the growth phase in the oocyte cytoplasm is instead transported there from the nurse cells via open cytoplasmic.

The major forms of egg chamber arrangements observed. In *polytrophic* egg chambers the oocyte is directly connected to each of several nurse cells and these to one another, by discrete intercellular channels, termed *ring canals*. In *telotrophic* egg chambers a single common duct enters each oocyte, and through this pass the combined products of a syncytial mass of nurse cell nuclei. The role of nurse cells in feeding the oocyte was remarked upon by many classical writers. Early observers noted in several species that organelles as large as mitochondria pass from nurse cells to oocyte. An interesting variation exists in turbellarian flatworms, where the nurse cells, filled with yolk, are encapsulated in a cocoon along with the oocytes after oogenesis is completed, and the nurse cell contents are used to sustain growth of the embryos just as are *intracellular* components in other eggs. This unusual course of events draws attention to the essential aspect of nurse cell function, that of providing the egg with materials it will require for development. Wherever nurse cells are found, the functional nature of the oocyte-nurse cell interaction is evident. In certain annelids, for example, one or two nurse cells with large polyploid nuclei are applied to each oocyte, and the oocyte-nurse cell complex is released into the lumen of the ovary relatively early in oogenesis. Oocyte growth then occurs at the expense of the nurse cells, which shrink progressively until they become small compared to the relatively enormous oocytes.

Cytoplasmic Transport in Polytrophic Egg Chambers

The structure of the nurse cell-oocyte complex of *Drosophila* as it appears at light microscope magnification. The five deeply staining nurse cell nuclei seen in the section are highly polyploid, while the oocyte nucleus of course contains only the 4C meiotic prophase genome. Three ring canals are shown, one connecting the oocyte with the follicle cells, and two connecting adjacent nurse cells. The growth phase of *Drosophila* oogenesis is divided into 14 stages beginning with the formation of the oocyte-nurse cell complex and its positioning at the posterior end of the germanium, Oogonial replication and the establishment of new egg chambers begin during puparial life and continue in the adult. Grell and Generoso (1982) showed that the premeiotic S-phase that marks the onset of oogenesis proper occurs in the first pupal pre-oocytes within the interval 132-162 hr after oviposition. During this period the synaptonemal complexes are extended, and this is also the time at which meiotic recombination may occur.

The entire 14 stages require 4½ days in pupae, and 6 days in the adult with the difference localized mainly to the rate at which stages 1 and 2 are traversed polyploidization of the nurse cell nuclei begins in stages 2, during which they increase their DNA content from 4C to 8C and this process continues through the previtellogenic stages 2-7. Together these stages require about 50 hr. After the 32C stage the nurse cell DNA does not replicate uniformly, in that satellite DNA, and some other sequences, including histone gene DNA are underreplicated. The amount of DNA present in the nurse cell genomes at stage 8, when vitellogenesis begins, is the result of 4-5 further replications of about 75% of the total DNA beyond the 32C stages. Two complete further replications have occurred by stage 10A, when the process is complete.

There are then four nurse cell nuclei that contain an amount of DNA equivalent to about 1500 haploid genomes and eleven that contain about half this amount of DNA Hammond and Laird (1985). The mass of oocyte constituents meanwhile increases continuously, first as the result of the flow of materials through the ring canals, and later by the pinocytotic ingestion of yolk as well. The mechanism of cytoplasmic transfer through the ring canals, and later by the pinocytotic ingestion of yolk as well. The mechanism of cytoplasmic transfer through the ring canals may be electrophoretic.

Though there are no equivalent observations in *Drosophila*, a potential gradient of >1 V/cm has been demonstrated across the open ring canals of the polytrophic *Hyalophora* egg chamber, and a similar potential gradient exists in the telotrophic ovariole of the bug *Rhodnius*. Injection of fluorescein labeled compounds into both types of meroistic ovaries shows that negatively charged molecules are transported by intracellular electrophoresis from nurse cells to oocyte, and once inserted in the oocyte these molecules are discouraged by the polarity of the potential gradient from diffusing backwards toward the nurse cells. Here the mobility within the Hyalophora egg chamber of fluorescein labaeled lysozyme, a positively charged protein, and of its negatively charged methylcarboxylated derivative, are compared, after injection into nurse cells. Just as originally summarised by classical observers using the light microscope, the canal area is seen to be packed with mitochondria, which have apparently been caught up in the cytoplasmic streaming that carries materials from nurse cell to oocyte. At this stage the flow from nurse cells to oocyte is accelerated as the result of a new process, the direct injection of nurse cell cytoplasm into the oocyte.

As the injection proceeds, the cytoplasmic volume of the nurse cells decreased and that of the oocyte increases concomitantly. This phase continues for about 5 hr in *Drosophila*, until the nurse cells have relinquished most of their cytoplasmic mass. They then degenerate. The velocity of the cytoplasmic stream of the ring canal has been measured during the injection process as about 2μm sec^{-1} and once having entered the egg it dissipates in a circular motion that effectively distributes the entering cytoplasmic constituents. At its peak the total flow carried by the four ring canals entering the oocyte is calculated to be about 1.3×10^4 μm^3 min^{-1}. A similar rate of cytoplasmic flow into the oocyte was estimated by Telfer (1975), in the polytrophic ovaries of the moth *Hyalophora*. The end result of previtellogenic growth, followed by the direct transfer of nurse cell cytoplasm and yolk accumulation, is an enormous change in oocyte volume. In *Drosophila*, for example, the volume of the oocyte increases 90,000 fold within the few days required for oogenesis.

Origin of the Nurse Cells

In most Diptera the accessory cells that perform the function of feeding RNA and other cytoplasmic constituents to the growing oocyte through ring canals derive exclusively from the germ line. However an exception is known in the dipteran family Cecidomyidae, where polyploid follicle cells that are of somatic origin fuse with a single true nurse cell to form a syncytial nurse chamber that contributes cytoplasmic RNA to the oocyte. Careful analysis has shown that in the more typical example provided by *Drosophila* and in *Lepidoptera* and *Hymenoptera*, the nurse cells of a given egg chamber, together with the oocyte, constitute a clone, descended from a single oogonial-stem cell. The development of the egg chamber was described for *Drosophila* by Koch et al., (1967), for the moth *Hyalophora cecropia* by King and Aggarwal (1965), and for the wasp *Habobracon juglandis* by Cassidy and King (1972). The structure and ontogeny of oocyte-nurse cell complexes in various insect groups are reviewed by King (1970), Telfer (1975), and King and Buning (1985).

The oocyte nurse cell complex in *Hyalophora* includes seven nurse cells and in *Drosophila* 15 nurse cells. These complexes are constructed the terminal three and four oogonial divisions, respectively. The ring canals that connect each nurse cell to other nurse cells and/or to the oocyte are highly organized membranous structures associated with actin microfilaments. The disposition of the ring canals, which originate by incomplete cytokinesis following each oogonial mitosis, indicates the order of appearance of the nurse cells and the sequence of steps by which the egg chamber is constructed. It can be seen that except for that nurse cell which is formed first the oocyte is the cell with the largest number of ring canals (four), although oocytes with only three ring canals have been reported. Both the first nurse cell and the oocyte initially form synaptonemal chromosomal complexes. The synaptonemal complexes developing in the nurse cell nucleus later disappear, whereas in the oocyte the meiotic prophase movements proceed, and the growth phase of oogenesis ensues.

Germ Line Functions Required in Egg Chamber Differentiation

Additional insight into the process of oocyte-nurse complex formation in *Drosophila* has derived from observations on female sterile mutations that affect egg chamber formation. Morphological effects of the mutations *fes* [*fs*(2)B] were described by Johnson and King (1972). In homozygous *fes* females cytokinesis is often *complete* rather than incomplete, with the result that large numbers of abnormal cell clusters containing less than 16 interconnected cells are formed. The presumptive oocytes are thus deprived of their complement of appropriately connected nurse cells. Normally the interconnected cystocytes all divide synchronously and this form of coordination is also absent in *fes* mutants. Several mutations are known that cause disorganized cystocyte divisions, resulting in masses of ovarian cells referred to as "tumors," including *narrow* (*nw*) and various alleles of *fused*, and of ovarian tumor (*otu*). An interesting example is the dominant mutations *fs*(2)D, which in heterozygous females results in inhibition of cystocyte division (heterozygous males are fertile).

Only a few egg chambers are produced, and most of these contain less than 16 cystocytes. These egg chambers contain no cells identifiable as oocytes, which is consistent with the concept that a condition for the switch to the oocyte as opposed to the nurse cell pathway of differentiation is the presence of four intercystocyte ring canals. Telfer (1975) drew attention to a branched cytoskeletal structure, the *fusome*, described in detail by classical cytologists and noticed more recently as well, which extends through the ring canals and interconnects the cystocyte complex. At division one pole of each spindle seems to be anchored in the fusomes, which may thus serve to orient each successive cleavage and ensure the appropriate geometry. The effects of the *fs*(231) mutation have been interpreted by King (1979) and King and Riley (1982), as a derangement of fusome organization. In this mutant the cystocytes often construct long linear chains, in which branches from less often than normally, and three dimensional reconstructions from serial electron micrograph sections show that fusomes are often not connected in the adjacent cells, *fs*(231) cystocytes frequently become polyploid and differentiate into pseudonurse cells.

Incomplete cytokinesis takes place about half the time, and cells with four ring canals occur at a frequency of about 10^{-7} while oocytes appear only at a frequency of 10^{-7}. Thus, while it may be necessary, the possession of four ring canals does not seem to be sufficient to trigger the differentiation of an oocyte. Another mutation isolated by Schupbach presumably affects this switch directly, as it causes the formation of egg chambers that contain 16 nurse cells, but no oocyte. The development of the oocyte-nurse cell complex can be regarded as a classical example of a clonal differentiation process. The number and orientation of the divisions by which the final 16 products are produced from the stem cell, and the selection of one of the two cystocytes that contain four ring canals as the oocyte are all seem from these examples to be subject to specific genetic controls. The implication of the genetic evidence that function of these genes occurs only in cells that are of germ line origin, is that the differentiation of this lineage involves the activation of tissue-specific genes, just as in other cell lineages.

Transcriptional Role of Nurse Cells

The structure of the meroistic egg chamber, and the observation of cytoplasmic transport from nurse cells to oocyte, suggest that the maternal transcripts present in the mature egg at fertilization were synthesized originally in the nurse cell nuclei. An indirect argument supporting this proposition was adduced by Ribbert and Bier (1969). They compared the length of time required for oogenesis in panoistic and meroistic oogenesis, and correlated this with the number of genomes putatively cooperating in the preparation of the oocyte. In the panoistic oogenesis of the Cricket *Acheta domestica*, for example, oogenesis takes more than three months, while in the meroistic dipterans, such as *Drosophila* of the blowfly *Calliphora*, oogenesis is completed within a few days.

The nurse cells each achieve a DNA content of 750-1500C in *Drosophila*, as we have seen, and of 256C in *Calliphora*. Thus there are 1-4 $\times 10^3$ more genomes putatively involved in providing oocyte transcripts in these Diptera than in the 4C oocyte nucleus of the cricket. This argument indicates the great adaptive value of the meroistic method of oogenesis. A related speculation is that perhaps nurse cells are usually of germ line origin because some prior processes of germ line differentiation are required for expression of the specific set of genes needed during the growth phase of oogenesis, irrespective of whether the transcription of these genes occurs in the germinal vesicle or the nurse cell nuclei. We now consider available molecular evidence relating directly to transcription in nurse cell nuclei and its role in the provision of maternal oocyte RNAs.

Sites of Heterogeneous RNA Synthesis in the Meroistic egg Chamber

Under normal conditions the polyploid chromosomes of most dipteran nurse cells are not *polytene*, i.e., the multiple chromatids are not aligned in register. Thus banded chromosomes such as are present in other polyploid cell types are usually not observed in nurse cells. However, the transition from polyploidy to polyteny in nurse cells seems to be fairly trivial. In some *Anopheles* species polytene chromosomes form spontaneously in nurse cells, while they are absent in close relatives. Among the unexplained effects of the *fes* mutation of *Drosophila* is the appearance of polytene chromosomes in nurse cell and giant polytene chromosomes also appear in nurse cells of some ovarian tumor (*otu*) mutants. In Calliphora polytene chromosomes can be induced in nurse cells merely by a mild cold treatment Bier, (1960). Ribbert (1979) also found that in *Calliphora* reduction of polymorphism through inbreeding results in a high incidence of polyteny in nurse cell chromosomes.

Cytogenetic studies on these Calliphora polytene chromosomes provide evidence that is this species all or most regions of the nurse cell genomes are equally replicated. Furthermore, the autoradiograph patterns obtained by Ribbert (1979) and accompanying observations on chromosomal puffing, show that transcriptional activity in nurse cell chromosomes is very widespread, and is also relatively intense. The amount of transcriptional activity was estimated as about an order of magnitude greater than in comparable polytene chromosomes of trichogen cells, and many loci were observed to remain in a puffed configuration throughout most of oogenesis. These observations imply a high complexity for nurse cell transcription, since RNA synthesis visibly occurs in a great many parts of the genome. A classic series of autoradiographic experiments carried out by Bler (1963) demonstrated that RNA synthesized in the nurse cell nuclei is transported rapidly into the oocyte. An example, from the housefly *Musca*.

Here newly synthesized RNA can be seen localized over the polytene nurse cell nuclei after a 30 min labeling period. Five hours later the labeled RNA has moved into the nurse cell cytoplasm and is apparently pouring through a ring canal into the cytoplasm of the oocyte. Note that no RNA synthesis can be observed over the oocyte at 30 min, even though the film is clearly over-exposed with respect to the amount of incorporation in the nurse cell nuclei. Similar autoradiographic results have been reported for other dipteran species and also for various coleopteran, hymenopteran and lepidopteran species a list which includes ovaries of both telotrophic and polytrophic construction and in addition for the single nurse cell-oocyte complexes of a polychaete annelid. Transport of newly synthesized pole(A) RNA from nurse cell to oocyte has been directly demonstrated by Paglia et. al. (1976a, b) in manual dissection experiments carried out on the polytrophic ovaries of the silk moths *Antheraea polyphemus* and *Actias luna*. Whole follicles were labeled *in vitro*, and ribonucleoprotein particles containing newly synthesized nurse cell poly(A) RNA were observed accumulating in the oocyte cytoplasm after several hours. Ribosomes and newly synthesized rRNA are transported as well.

When further synthesis in the nurse cells was blocked with actinomycin, their content of labeled poly(A) RNA decreased more than 10-fold within 6 hr, while that of the associated oocytes increased correspondingly. No significant incorporation of precursor into vitellogenic oocyte RNA was observed on incubation of follicles from which the nurse cells had been ablated, while isolated nurse cell complexes are transcriptionally active. Furthermore, polysomes are present only in the nurse cells, and the mRNA, tRNA and ribosomes of the growing oocyte can thus be seen clearly to be stored for use later in development. By injecting 3H-uridine into late pupae that were then permitted to enclose. Paglia et. al. (1976a) showed that poly(A) RNA synthesized during vitellogenesis is indeed sequestered in mature chorionated eggs. In some meroistic insects, particularly some species of midges, bugs and beetles, heterogeneous RNA of the oocyte cytoplasm is also contributed by the germinal vesicle, though most derives from the nurse cells. This can be inferred from autoradiographic evidence of germinal vesicle RNA synthesis and has been shown more directly by other means. *In situ* hybridization with 3H-poly(U) has been used to determine the source of the oocyte poly(A) RNA in the telotrophic ovary of the milkweed bug *Oncopeltus fasciatus*.

During the growth phase of oogenesis poly(A) RNA accumulates in the oocyte cytoplasm and the newly synthesized transcripts are evidently supplied by the nurse cells. Thus these cells and the nutritive cord are heavily labeled by the probe, while the germinal vesicle contains no detectable poly(A) RNA. However, towards the end of oogenesis the nutritive cord is interrupted by the growth of the chorion, but thereafter a further net increase in oocyte poly(A) RNA nonetheless takes place. This is apparently due to late RNA synthesis in the germinal vesicle. Transcription of poly(A) RNA also occurs in the germinal vesicles of postvitellogenic oocytes of the bug *Dysdercus*. In the gall midge *Wachtliella* elimination of 16 of the 20 chromosomes present in germ line cells occurs in all other cells early in embryogenesis. During oogenesis the four somatic (S) chromosomes remain condensed in the oocyte nucleus, surrounded by a concentric fibrous lamellar structure, while the remaining (E) chromosomes are dispersed. Autoradiography shows that the S chromosomes are transcriptionally silent in the oocyte, while the E chromosome are active.

The nucleolar organizer is located on an S chromosome and nucleolar, formation and S chromosome activity are detected only in nurse cells. If the E chromosomes are experimentally eliminated from the germ line stem cells, oogenesis does not take place. Though the heterogeneous RNAs produced in the oocyte by the E chromosomes may be required for oogenesis, the relative importance of nurse cells and germinal vesicle transcripts remains to be determined. The observations reviewed so far are qualitative, and even where no germinal vesicle activity is reported it remains possible that rare maternal transcripts could be synthesized by the oocyte genomes. These might easily have escaped observation in autoradiographic experiments focussed as they are on the overwhelming synthetic activity of the thousand-fold polyploid nurse cell nuclei. The differences between those meroistic insects that utilize germinal vesicle transcripts, and the lepidopteran and dipteran examples that apparently do not, could be merely quantitatively.

On the other hand, many of the same RNA species produced in small quantities in the germinal vesicles of meroistic oocytes may also be represented in the much greater flow of nurse cell transcripts. These uncertainties should be kept in mind in considering the locus of action of germ line mutations that display maternal effects and the origins during oogenesis of given species of transcripts, any one of which might represent only a small fraction of the total maternal RNA.

Germ Line Functions that Affect Oocyte Structure

Several germ line mutations have been described that block oogenesis after the completion of the cystocyte divisions and formation of the egg chamber. Except that the locus of action of these mutations is the nurse cells or oocyte, for the most part the primary physiological or structural defects they cause are not known, and none are yet analyzed at the molecular level. Their significance here is to remind us that there are many additional genetically controlled aspects of oogenesis besides synthesis of maternal transcripts and proteins required after fertilization. An interesting example analyzed by Waring et. al. (1983) concerns a recessive female sterile mutation, *fs*29, that maps to region 12E1-12F1 on the X chromosome. The gram line function of this gene was demonstrated by transplanting homozygous mutant pole cells into normal eggs, and mutant pole cells into normal eggs.

The visible effect of the homozygous mutation is a slightly abnormal chorion structure, particularly at the anterior dorsal side, where in *fs*29 oocytes the respiratory appendages project at the wrong angle and the adjacent anterior region of the chorion is abnormally formed. A consequence is failure of fertilization, which normally occurs via the micropyle, an anterior chorion structure. Waring et al., (1983) found a sharp decrease in the amount of yolk protein taken up by *fs*29 oocytes, although synthesis of yolk occurs normally in both fat body and follicle cells. *Drosophila* follicle cells are the source of about 35% of two of the three major yolk proteins. The deficiency of *fs*29 animals appears to lie in the mechanism by which yolk is sequestered by the oocyte, which includes an-extremely specific binding to surface receptors followed by pinocytotic internalization.

Probably the interruption follicle cell-oocyte contracts that produces the fatel structural abnormally in the chorion in a peripheral side effect of decreased turgidity in the anterior region of the oocyte, due to decreased yolk content. A second mutation, *fs*(1) K10, that affects chorion structure in the same region and also acts in the germ line, has been described by Wieschaus et al. (1978). In normal egg chambers morphogenesis of the dorsal appendages is carried out by nests of follicle cells apposed bilaterally to the anterior surface of the oocyte, and the dorsoventral polarity of the oocyte is foreshadowed by the thicker follicle cell layer on the dorsal side. Mutations at the fs(1) K10 locus dorsalize the egg chamber, in that a thick follicle cell layer occurs on the future ventral side as well, and they result in eggs with enlarged dorsal appendages.

In the few cases in which development is initiated, the embryos lack ventral structures and display a dorsalized phenotype. Function of the K10⁻ gene is necessary only in the female germ line and normal progeny can be derived from transgenic is fs1(K10) females bearing the wild type gene. Another female sterile mutation that according to pole cell transplantation tests acts in the germ line, though it affects follicle cell function, is tiny. Though it maps to the same region of the X chromosome, *tiny* complements fs29 and any of its known alleles. In homozygous (or hemizygous) *tiny egg* chambers the distal follicle cell migration occurring early in oogenesis is blocked, and an abnormally thick follicular wall that prevents expansion of the oocyte is formed. This in turn produces a thickened irregular chorion. Di Mario and Hennan (1982) suggested that the primary lesion in *tiny egg* chambers is a defect in the nurse cell oocyte surfaces over which the follicle cells are supposed to migrate.

Ribosomal RNA Synthesis

Two different mechanisms by which ribosomal RNA is supplied to the oocyte have been observed in meroistic insects. In most forms the ribosomal RNA derives wholly from the nurse cells, along with other transcript species. As this is of course the bulk form of RNA in the oocyte its transfer is easily detected. In *Drosophila* and *Calliphora* egg chambers this occurs mainly when the nurse cell cytoplasm is injected into the oocyte while in the *Hyalophora* ovariole most of the tRNA is transferred to the oocyte prior to the terminal injection. Early vitellogenic *Hyalophora* oocytes already contain about 1µg of total RNA, most of which is undoubtedly ribosomal, and this is also the RNA content of the seven nurse cells taken together. During vitellogenesis the oocyte doubles its total RNA content, and this is increased 3µg by the terminal injection process. Ribosome transfer from nurse cells to oocyte was demonstrated in the polytrophic *Antheraea* egg chamber Hughes and Berry (1970), and in the telotrophic egg chamber of *Oncorpeltus* by Davenport (1976). In these experiments the ovariole was exposed to isotopic precursor, and the subsequent appearance of labeled ribosomes in the oocyte was monitored. Entry of newly synthesized rRNA could be interrupted in *Oncopeltus* by ligation of the nutritive cord in *Antheraea* did not occur after removal of the nurse cell cap.

In many organisms the rRNA genes are amplified during oogenesis, a special mechanism required to meet the enormous demand for preformed ribosomes early in embryonic development. However, significant ribosomal gene amplification, i.e. relative to the remainder of the genome, has been reported not to occur in the polyploid nurse cells of a number of meroistic insect species. In *Calliphora* rRNA synthesis takes place in extrachromosomal nucleoli, but even here the fraction of ovariole DNA that is ribosomal is only 1.3 times that measured in diploid brain cells Renkawitz and Kunz, (1975). In four dipteran species *Drosophila hydei*, *Drosophila virilis*, *Sarcophaga barbata* and *Rynchociara angelae* the ribosomal DNA is actually underreplicated by about a factor of two during nurse cell polytenization.

Nor is there preferential accumulation of ribosomal DNA during oogenesis in *Drosophila melanogaster* : in the silk moths *Antheraea pernyi* and *Bombyx mori* on in *Oncopeltus*, where the nurse cell nuclei of the telotrophic egg chambers are polyploid only to the extent of 32-62C. These measurements imply that the total number of ribosomal genes present per egg chamber, i.e., the total product of nurse cell ploidy rRNA genesper haploid genome, and the number of nurse cells, is sufficient to provide the oocyte with the requisite quantity of ribosomes in the time available, so that ribosomal gene amplification is unnecessary. This can be seen explicitly for *Drosophila* oocyte, where the transcription rate and other necessary parameters can be estimated. There are about 250 rRNA genes not haploid *Drosophila* genome, and so assuming neither under-nor over DNA replication, at stage 9 when nurse cell polyploidization is complete, the number of rRNA genes per egg chamber is about 3.8×10^6. At a polymerase translocation rate of 10 nt sec-1 and assuming a minimum polymerase spacing of about 100 nt it would require only about 30 hr for this number of genes to produce the approximately 4.5×10^{10} ribosomal RNA per egg. Mermod et al. (1977) showed that earlier in oogenesis the rate of egg chamber rRNA synthesis is directly proportional to the degree of nurse cell polyploidization and the total number of rRNA genes.

Ribosomal RNA synthesis could be a rate limiting process in Drosophila oogenesis. Thus in bobbed mutants in which various portions of the ribosomal gene cluster are deleted, the length of time required for oogenesis depends on the number of ribosomal genes remaining (Mohan, 1971). For example, in mutant egg chambers containing only 1/3rd the normal complement of rDNA, oogenesis required 206 hr compared to 75 hr in controls. It is interesting that as a result of this compensatory mechanism the mature eggs of *bobbed* females contain a normal quantity of rRNA. Ribosomal DNA amplification does occur in the *oocytes* of some insect species, including water beetles such as *Dytiscus* and *Colymbetes* ; the dipteran *Tipula*; and the neuropteran

Chrysoma. In these examples a single DNA body that contains rRNA genes is found in the definitive oocyte nucleus, but not in nurse cell nuclei. Autoradiographic observations and *in situ* hybridizations with ribosomal sequence probes indicate that rDNA amplification in these bodies begins during the cystocyte divisions, and continues, in the early stages of previtellogenic oogenesis.

The best studied example is the water beetle *Dytiscus*. Gall and Rochaix (1974) found that the DNA body in this species contains about 3×10^6 rRNA genes, arranged in small circles, each containing one to five genes. This number is remarkably close to that present in the nurse cell chromosomes of *Drosophila*. In midoogenesis the DNA body disperses and intense rRNA transcription begins. No other transcriptional activity can be detected in the oocyte nuclei. The chromosomes are present in a condensed mass, the karyosphere, and as usual in meroistic oocytes, they appear silent in autoradiograph experiments. The meroistic form of oogenesis is usefully considered from a logistic point of view. A great many copies of genes are put to work cooperatively in the meroistic egg chamber, all producing sequences required by the oocyte. The obvious adaptive value of this elaborate strategy is that it accelerates by a large factor the overall rate of oogenesis.

The Transcriptional Role of Oocyte Lampbrush Chromosomes

The oocytes of many animals contain spectacular meiotic prophase chromosomes that are distinguished by the presence of thousands of lateral loops. Such chromosomes were first observed by Flemming (1882) in sections taken through urodele oocytes, and they were the subject of a detailed study by Ruckert (1892), carried out on isolated germinal vesicles of shark oocytes. Ruckert recognized that they are paired chromosomes from which the numerous lateral loops project, and this aspect of their structure suggested to him the designation *lampbrush chromosomes*. These chromosomes are never found in the germinal vesicles of meroistic oocytes, which as we have seen are generally quiescent in regard to transcription, nor do they occur in the oocytes of many other species in which the maternal transcripts derive solely from the germinal vesicle. Though there remain some mysteries, and their patterns of transcription are in some respects unusual, we shall conclude from molecular and cytological analyses that their basic function is probably the synthesis and maintenance of the large pool of maternal transcripts resident in the oocyte cytoplasm.

Lampbrush Chromosome Structure : Transcription Units and Chromomeres

The modern era of lampbrush chromosome cytology began with the establishment of two basic structural tenets. First, lampbrush chromosomes were shown to be bivalent meiotic prophase structures in which each loop of an apposing pair contains a single DNA duplex, while the central axis contains two such duplexes. The meiotic homologues are typically united by several chiasmata. Second, the loops were perceived to be the sites of chromosomal RNA synthesis. Thus, an active locus is represented by four loops, two deriving from each chromosomal axis. The axes of lampbrush chromosomes from which the loops project were recognized classically to consist of a linear array of compacted beads, referred to as *chromomeres* Gall (1954) showed that DNA is concentrated in the chromomeres, as these are the only structures in lampbrush chromosomes that can be visualized by the Feulgen reaction.

The paired structure of the chromomeric axis was demonstrated by Callan (1955), by stretching local regions with microneedles to the point where the chromomeres would separate transversely into two distinct chromatin strands. From measurement of the kinetics of lampbrush chromosome breakage by DNase 1, Gall (1963) deduced the presence of two DNA duplexes in the axis and of single duplexes in the loops. Important evidence also derived from the maps constructed for newt lampbrush chromosomes by Callan and Llyod (1960). These studies showed that distinctive loops can be recognized at invariant locations in the chromosomes. Particular loop morphologies are the property of species, subspecies, or individuals. Heterozygotes or hybrids between related species, generally produce heterozygous sets of lampbrush chromosome which display the allelic alternatives characteristic of each parent, and the frequencies at which these alternatives appear are distributed in wild populations as predicted by a Hardy-Weinberg calculation.

A general conclusion is thus that the loop structure is determined by the genetic locus, i.e., the DNA sequence which it contains. Lampbrush chromosome maps have now been assembled for a number of different species, including several urodeles the anuran *Xenopus* and some amphisbaenian reptiles.

Transcription Units of Lampbrush Chromosomes

Early autoradiographic studies carried out with the light microscope showed that intense RNA synthesis occurs throughout the length of most loops, though there are a few giant loops that display unusual labeling patterns. The loops contain newly synthesized proteins as well as RNA. Furthermore, most loops appear to have a polarity, in that the loop matrix is thicker at one end than at the other. Electron microscopy carried out by the Miller spreading technique has now provided a convincing molecular interpretation of both the widespread transcriptional activity and the polarity observed at light microscope resolution. The transcription units of lampbrush chromosomes are packed densely with nascent RNA molecules.

Observations on both *Xenopus* and *Triturus* lampbrush chromosomes show that the polymerase molecules by which the nascent transcripts are anchored to the DNA fibril of the loop are typically only 100-200 nt apart. Though some variation is observed both within and among transcription units, the structural feature that immediately distinguishes lampbrush chromosomes is that almost all the transcription units functioning in lampbrush chromosomes are synthesizing closely packed because initiation is occurring as frequently as permitted by the polymerase translocation rate. By contrast, when the same methods of visualization are applied to active somatic cell types, e.g., sea urchin or *Drosophila* embryonic cells, only about 10% of the transcription units are densely packed with nascent transcripts, just as expected from consideration of the RNA prevalence distributions characteristic of such cells. The lengths of the transcription units observed in lampbrush chromosomes vary from about 2 kb to at least 70 kb of extended chromosomal DNA in *Xenopus*, and from about 10 to 100 kb in several urodeles.

The enormous size or transcription matrices such as those accounts easily for the polarized form of the prominent loops observed by classical methods, since the nascent transcripts associated with a given transcription unit increase in length toward the distal end. However, it is necessary to take into account the effects of the spreading procedure required for visualization. The detergent treatments involved in these preparations probably result in the removal of some proteins, and the surface tension to which the sample is exposed converts what might be imagined a solid ribonucleoprotein core of increasing diameter into the open, two-dimensional form illustrated. A useful calculation of Macgregor (1980) shows that average dimensions of the solid ribonucleoprotein matrices from which the typical transcription complexes visualized in the electron microscope would have derived are consistent with those actually observed in the phase microscope. The relation between the number of loops and of transcription units is not straight-forward, though many loops probably contain-single transcription units. However, it is not uncommon to observe two transcription units separated by a short nontranscribed region. This is particularly striking when the transcription matrices are oriented in opposite directions, indicating initiation sites on opposite strands. A given region may also contain multiple transcription units oriented similarly, and occasionally a series exmatrices of identical size and orientation are observed that probably indicate the transcription of a tandemly repeated gene family. Careful light microscope observations of discontinuities in the matrices of giant loops have also provided indications that some loops bear multiple transcription units. It follows that the number of transcription units is by some unknown factor greater than the number of loops.

The number of loops, or of loops and associated chromomeres was classically assumed to provide an approximation of the number of genetic loci, or at least of active loci. Given that the physical manifestation of an active locus is a transcription complex, this venerable correlation is clearly incorrect. Furthermore, while at electron microscope resolution the active transcription units of lampbrush chromosomes are easily defined, the distinction between the loop and chromomere domains inferred from light microscopy in most regions is not all obvious. Closely apposed transcription units no doubt derive from the same loop, while at the other extreme the long inactive axial fibrils that can also be observed could represent the chromomeric DNA of the classical model. After spreading by the Miller procedure all of the inactive regions retain a similar beaded structure, indicative of nucleosomal conformation. Nucleosomes can be perceived even within active transcription units, where the nascent fibrils and polymerase molecules are not so densely packed. An implication of the latter observation is that nucleosomal structures must be able to reform within less than a minute following the passage of a polymerase and the associated nascent transcript.

A reasonable interpretation of the structure of lampbrush chromosomes as classically observed in the light microscope is that chromomeres occur where there are long regions of inactive chromatin that *in situ* are condensed into higher order structures. These are disaggregated during the preparation of transcription spreads by the Miller procedure. Shorter inactive regions may bunch together, giving rise to complex aggregate that would also be recognized as a chromomere in the light microscope, from which protrudes an array of multiple loops, all containing active transcription units. Hill (1979) pointed out that the alternative interpretation would require that much of the length of an average loop be occupied by inactive intermatrix spacer sequences, contrary to indications from light microscope autoradiography. A scanning electron microscope study of Angelier et. al. (1984), in which the same lampbrush chromosomes were also visualized in the phase microscope, supports the view that chromomeres may consists of associated regions of condensed nucleoprotein from which many loops extend. It follows that the chromomere is not an invariant structure equivalent to a single genetic locus.

It is rather to be considered a compacted aggregate of transcriptionally inactive DNA, the manifestation of which depends on the spacing of the surrounding active transcription units. In different organisms among which the mode of genomic sequence organization varies, the prominence and distribution of chromomeres might also vary. Thus the differences in the number of chromomeres and the size of loops observed in comparing the lampbrush chromosomes of various species probably reflect primarily differences in the distribution and lengths of the nontranscribed DNA sequences that define the termini of the loops. This interpretation clears away the paradox that develops from the classical view that such differences would indicate fundamental variation in the number of the functional genetic loci among related species.

Phylogenetic Occurrence of Lampbrush Chromosomes

For no other organisms do there exist ultrastructural or molecular observations on lampbrush chromosomes comparable to those available for amphibians. Yet these structures occur in the oocytes of many other vertebrates and in a great many invertebrates as well, and at least as perceived in the phase microscope they appear completely homologous in their cytological organization to the amphibian examples. Here are displayed lampbrush chromosomes from an orthopteran insect. *Decticus albifrons*; a squid, *Sepia officinalis* : a snail. *Bithynia tentaculata* and a starfish, *Echinaster sepositus*. The general structural similarity of lampbrush chromosomes is illustrated in the detailed study of DeLobel (1971) on *Echinaster* lampbrush chromosomes. The map constructed of the chromosomes of this organisms displays the same kinds of special structures, such as giant loops of unusual conformation, that serve as landmarks in amphibian lampbrush chromosomes, and the average loop dimensions are similar to those of *Xenopus* lampbrush chromosomes.

The collected data regarding the overall distribution of lampbrush chromosomes, and where possible the duration of the lampbrush stage. It is clear that lampbrush chromosomes occur in many major groups, both deuterostome and protostome. Thus, like the process of oogenesis itself, lampbrush chromosomes are probably of very ancient evolutionary origin. Lampbrush chromosomes evidently perform some fundamental function in oogenesis, since they have been retained throughout most of metazoan evolution. The list of organisms in which lampbrush chromosomes have been reported is of course limited by the choices made by investigators and the difficulty of observing them in some material. Nonetheless, it is now established that lampbrush chromosomes are not ubiquitous.

A generalization suggested by their known distribution is that lampbrush chromosome occur in large oocytes that contain relatively huge pools of heterogeneous maternal RNA, while they are absent from small oocytes. The mouse egg, for example, contains less than 1/3000 the amount of maternal poly(A) RNA that is present in the egg of *Xenopus*, and the sea urchin egg less than 1/500 this amount. The *Xenopus* egg, in turn, is small compared to some other anuran eggs and to urodele eggs. The same relation pertains among the echinoderms. Thus the egg of the starfish *Echinaster sepositus*, which contains lampbrush chromosomes, is several hundred times the volume of the egg of the sea urchin *Strongylocentrotus purpuratus*, which does not. The correlation suggests a simple interpretation of lampbrush chromosome function. This is that these structures exist where the logistic demands of supplying the growing oocyte with maternal transcripts in the allowed time require that almost all transcription units function at maximum rate. As we have seen, widespread and intense transcription is the definitive property of oocyte lampbrush chromosomes. A logistic function for lampbrush chromosomes is also suggested by the exclusive relationship between meroistic oogenesis and the presence of lampbrush chromosomes in the oocyte nucleus. That is, lampbrush chromosomes probably perform the same function that the polytene nurse cell chromosomes do in meroistic oocytes, viz. the provision of maternal RNAs. This function appears to be exercised continuously over the relatively long period required for oocyte growth. Thus, the lampbrush stage lasts for weeks, months, or even longer in species for which estimates are available. In some amphibian species a mechanism has evolved that almost approaches the meroistic mode of oogenesis. Here each oocyte contains multiple nuclei Macgregor and Kezer (1970) showed that in the tailed frog *Ascaphus*, for example, the growing oocyte has sight germinal vesicles, formed initially by oogonial nuclear divisions that are not accompanied by cytokinesis. Each nucleus is endowed with a complete set of lampbrush chromosomes, and thus there are 32 copies of every active locus functioning per oocyte.

Even more extreme examples have been described by del Pino and Humphries (1978) in studies of oogenesis in marsupial frogs. In these forms development proceeds directly from egg to juvenile, with a reduced or nonexistent larval tadpole stage. In several genera of marsupial frogs the previtellogenic oocyte can be seen to contain hundreds of nuclei. Each oocyte is the product of many oogonia, formed by the disappearance of cell membranes within an oogonial cyst. During the growth phase each of the multiple nuclei is endowed with lampbrush chromosomes and all are active in RNA synthesis. Later in oogenesis all but one nucleus disappears. It may be relevant that the eggs of species carrying out direct development are relatively enormous, ranging up to 9 mm in diameter. The egg of *Xenopus*, for comparison, is about 1.2 mm in diameter. In summary, the comparative biology of occurrence, when combined with ultrastructural evidence for intense transcriptional activity in all or most of the loops, suggests that the basic function of lampbrush chromosomes to provide the growing oocyte with maximum possible flow of newly synthesized transcripts.

Complexity, Average Structural Characteristics, and Synthesis Rates for Lampbrush Chromosome RNAs

Size of Primary Transcripts and Association with Specific Proteins

Newly synthesized germinal vesicle RNAs have been extracted from the oocytes of several amphibian species, and the distribution of their molecular lengths determined under denaturing conditions. In one study carried out on vitellogenic oocytes of the newt *Pleurodeles poireti* transcripts as large as 10-30 kb were observed, though most RNAs recovered were smaller, RNAs up to at least 60 kb in length have been extracted from *Triturus* oocytes. Scheer and Sommerville (1982) also reported measurements on extracted germinal vesicle RNAs carried out by electron microscopy. Length distributions obtained in this study extended upto about 20 kb for both *Triturus* and *Necturus* oocyte in nRNAs, and to about 10 kb for *Xenopus* oocyte nRNA. However, the number average size of the molecules recovered from the isolated germinal vesicles of all three species is around that of mature mRNA, about 2 kb. The RNAs examined in these studies could have suffered strand scissions during their preparation, despite all possible precautions, and whether for this reason, or because endonucleolytic cleavages occur in the course of processing while the transcripts are still nascent, the *extracted* molecules clearly fail to match the enormous dimensions of the larger transcription units observed in the electron microscope.

As expected for a high complexity population of nuclear transcripts, lampbrush chromosome RNA has a DNA-like base composition, except for a still unexplained bias towards unusually high uridylic acid content reported for both *Xenopus* and *Triturus* oocyte nRNAs. The nascent transcripts of the loop matrix are associated with proteins, which results in a significant compaction. Hill (1979) noted that even after preparation for electron microscopy the apparent nascent RNA length is less than half the length of the DNA from which it is transcribed, and in native RNA the degree of compaction is undoubtedly much higher. In high voltage electron microscopic images of thick sections though the loop matrices RNA fibrils can be seen connecting particles of about 20 nm diameter. The RNP assemblages have been isolated from *Triturus* oocytes by differential centrifugation and their structure analyzed *in vitro*. Linear chains of the 20 nm ribonucleoprotein particles in different states of aggregation, and their disaggregation into monomers after mild RNase treatment. The particles consist largely of proteins, and more than 20 distinct polypeptides have been isolated from such preparations. Antibodies prepared against these polypeptides generally react with the matrices of an loops, even though different loop often display individual morphologies, due to the various conformations into which the 20 nm particles are arranged, perhaps a function of the primary RNA sequences.

However, there are certain proteins that are present only on a small number of specific loop pairs out of the whole germinal vesicle complement. The same proteins are included in the ribonucleoprotein complexes released into the nuclear sap from these loops when transcription has been completed. In their physical properties and constitution the ribonucleo-protein particles that contain newly synthesized chromosomal RNAs of the amphibian oocyte directly resemble the heterogeneous nuclear RNP complexes of somatic cells. This suggests that the functions mediated by these assemblages, whether transcript packaging, transport or processing, may also be similar.

Rates of RNA Synthesis in Lampbrush Chromosomes

The lampbrush chromosome stage begins in *Xenopus* in previtellogenic, early diplotene oocytes only 50-60 μm in diameter. According to the staging criteria of Dumont (1972), these are stage 1 oocytes just beginning their growth phase. Hill and Macgregor (1980) showed that the chromosomes of even these very young oocytes contain heavily loaded transcription matrices typical of those present in midvitellogeneis lampbrush chromosomes. Transcription is initiated even earlier, in oocytes that are only 25-40 μm in diameter. Though the nascent transcripts are at first 5-10 fold more widely separated than in the later stage 1 oocytes the average length of the earliest transcription units is already the same as at the maximum lampbrush stage. This study yields the important inference that a high rate of chromosomal RNA synthesis is instituted very soon after the stage 1 oocyte begins to grow, rather than only at the midlampbrush stage, as earlier assumed. Martin et. al., (1980) showed, furthermore, that mature lampbrush chromosome transcription units persist even in fully grown oocytes. Lampbrush chromosomes are thus present almost from the beginning to the end of oogenesis, although in the light microscope the early and late oocyte lampbrush chromosomes are less easily resolved. The main difference in morphology between the transcription matrices of very early and very late oocytes, on the one hand, and of the Dumont stage 3-5 vitellogenic oocytes classically described as "maximum lampbrush stage," on the other, reflects the degree to which the nascent transcripts are compacted in their RNP assemblages. The lateral loops of Dumont stage 6 oocytes appear shorter, and in Dumont stage 1 and 2 oocytes the nascent transcripts are about 2-fold less extended in preparations spread for electron microscopy than in stage 3 oocytes.

A valuable series of direct synthesis rate measurements carried out on stage 3 and stage 6 *Xenopus* oocytes. Both stage 3 and stage 6 oocytes synthesize high molecular weight, unstable, nucleus-confined RNAs. The accumulation kinetics of these transcripts are illustrated. Transcripts of the same size range, which turn over with a half-life of about 20-45 min are of course found in somatic nuclei as well. In addition both stage 3 and stage 6 oocytes synthesize a somewhat greater amount of unstable, nucleus-confined RNA of smaller size which accounts nicely for the relatively low number average size of nRNA isolated by Scheer and Sommerville (1982) noted above. In vitellogenic stage 3 oocytes the unstable 4-40S RNA fraction turns over rapidly, at a rate similar to the >40S RNA, while in stage 6 oocytes the rate of turnover of the 4-40S nRNA class is somewhat lower. This results in a large increase in the steady state nRNA content in the stage 6 oocyte nucleus. Anderson and Smith (1977) suggested as a possible explanation a decrease in processing efficiency late in oogenesis, a proposition discussed further in a following section.

Only 11% of the total heterogeneous RNA synthesized in the stage 3 *Xenopus* germinal vesicle is destined for the cytoplasm, and in stage 6 oocytes this value has fallen to 5%. The RNA transported to the cytoplasm is detected kinetically as the stable 4-40S component of total newly synthesized heterogeneous RNA. Thus Anderson and Smith (1978) prepared enucleate cytoplasms by manual dissection and showed that the rate of entry of the 4-40S RNA into the cytoplasm equals its rate of synthesis. The accumulation kinetics for the stable 4-40S RNA component, compared to those for rRNA. In absoluted terms the rates of heterogeneous RNA synthesis are enormous. For comparison the rate at which total heterogeneous nuclear RNA is synthesized in postgastrula *Xenopus* embryo cells as about 0.015 pg min^{-1} per (2C) nucleus. It is clear that the lampbrush stage oocyte nucleus is a thousand times more active. The difference lies mainly in the number of polymerases transcribing each functional region in lampbrush chromosomes. As calculated earlier the density of nascent heterogeneous nuclear RNA molecules is typically less than one per 10^4 na in somatic cells, compared to one per 10^2 nt in the active lampbrush transcription units. Transcription units may also be longer in the lampbrush chromosome, as further discussed below, and the 4C oocyte genome contributes an additional factor of two to the comparison. However, the fractions of newly synthesized heterogeneous RNA exported to the cytoplasm from the lampbrush chromosome stage germinal vesicle, here 5-12%, are not very different from those found in embryonic or other somatic cells. It follows that during the lampbrush stage the rate of flow of stable transcripts into the cytoplasm is also as much as 1000 times greater than in other cells. The significance of the conservation that heterogeneous RNAs are being exported at high rates to the cytoplasm throughout the midlampbrush chromosome stage is that it directly relates lampbrush chromosome transcription to the function of supplying such RNAs to the cytoplasm. Dolecki and Smith (1979) showed, furthermore, that much of this heterogeneous RNA flowing into the cytoplasm is polyadenylated. The 4-40S RNAs are the least moderately stable, since no turnover can be detected in the form of accumulation curve during the labeling period.

Complexity of Germinal Vesicle RNA

There have been no direct experimental estimates of the complexity of the nuclear RNA of *Xenopus* oocytes. Though the complexity and sequence content of the cytoplasmic RNAs stored in the mature egg are relatively well known. However, the measurements of transcription rate can be utilized to calculate the approximate total length of the DNA included in the transcription matrices of the stage 3 oocyte chromosome, and from this the single copy sequence complexity may be estimated. The result is that around 30% of the *Xenopus* genome appears to be transcribed into newly synthesized in nRNA. This fraction is not significantly different from that observed for many somatic cell nuclear RNAs, also indicates that the stably accumulated cytoplasmic RNA of the *Xenopus* oocyte includes only about 4-6% of the calculated complexity of the nuclear RNA. This ratio falls at the lower end of the range commonly observed for somatic cells, about 5% to 20%. The conclusion that the RNA transcribed in lampbrush chromosomes is not more complex than the nuclear RNA of somatic cells is also consistent with several early hybridization studies which showed that only a minor fraction of the diverse genomic repetitive sequences is represented in the newly synthesized RNA of stage 3 *Xenopus* oocytes. Speculations regarding lampbrush chromosome function that require the whole genome to be transcribed in lampbrush chromosomes are thus not supported. Furthermore RNAs coding for hemoglobin and vitellogenin, both expressed specifically in terminally differentiated adult tissues, have been shown to be absent from *Xenopus* egg RNA as are mRNAs for certain though not all of the heat shock proteins that can be expressed in somatic cells. The possibility is not excluded, however, that the rapidly decaying nuclear transcripts synthesized in lampbrush chromosomes include these particular sequences. What is clear is that a *specific, though large, set of sequences is transcribed and that these in turn give rise to a specific array of stable cytoplasmic RNAs.*

Also presents estimates for the length of sequence represented in *Triturus* lampbrush chromosome RNA. The basis in measurement is here almost complementary to that available for *Xenopus*. While there are no synthesis rate measurements for *Triturus* oocytes, direct attempts have been made to estimate the nRNA complexity by the single copy saturation hybridization method. Furthermore, *Xenopus* lampbrush chromosomes are small, thus rendering quantitative cytological examination at the light microscope level difficult, while extensive cytological measurements have been carried out on *Triturus* lampbrush chromosome.

A visual comparison between *Xenopus Triturus* lampbrush chromosomes, photographed at the same magnification. There are about 20,000 loops in the whole chromosome set of *Triturus*, or about 5000 loops per haploid set. From direct length measurements these are estimated to contain about 5% of the genomic DNA. In the length of DNA transcribed in *Triturus* oocytes on this basis is also listed, i.e., assuming that on the average each loop consists only of transcription matrix. The value obtained, i.e. 1.5×10^9 nt, is not significantly different from the measured single copy complexity, i.e. > 9.6×10^8 nt, taking into account that about 40% of the *Triturus* genome is repetitive.

The estimated lengths of genomic sequence represented in the nRNA of *Xenopus* and *Triturus* oocytes probably differ only by a factor of about 1.7, though the genome size of *Triturus* is 10 times large than that of *Xenopus*. On the other hand as much as a six times greater fraction of the *Xenopus* genome is apparently being transcribed (30%) than of the *Triturus* genome (5%). This conclusion is directly contrary to the earlier assumption that because the lampbrush chromosomes of organisms of greater genome size have larger loops, an equivalent *fraction* of the genomic DNA is always being expressed. From that assumption has risen the famous mystery known as the "C-value paradox." The question posed in statements of the "C-value paradox" is why animals of similar biological organization should display manyfold differences in the length of genomic information transcribed in homologous cells, or in more extreme form, why the homologous expressed "genes" of one organism should on the average include manyfold longer DNA sequences than in a related organism. With respect to amphibian oogenesis this paradox in fact does not exist, though the enormous fraction of silent DNA accumulated during evolution in urodele genomes indeed remains a mysterious phenomenon. Thus the sixfold greater *relative* activity of the *Xenopus* genome reduces the ratio of expressed genome sizes to 16.5, which might suggest a slightly greater length of sequence utilized in *Triturus* lampbrush chromosomes. This last factor indicates that *Triturus* transcription units on the average include about twice as much noncoding sequence that is confined to the nucleus as do *Xenopus* transcripts. However, Sommerville and Scheer (1982) found no striking difference in the fraction of oocyte nRNAs from these organisms that consists of repetitive sequences.

The significance of the difference in loop lengths among species requires reexamination as well. For one thing, contrary to the premise of the "C-value paradox". Macgregor (1980) showed that average loop lengths in fact vary far less than proportionately with genome size. It is indeed the case that in *Xenopus* the loops average only 5-10 μm and large loops are about 10-15 μm and the largest range up to 20μm. In urodeles with even larger genomes that *Triturus cristatus*, e.g., *Necturus maculosus* the average loop length is slightly greater. However, since average loop length is overall a complex function of sequence organization, i.e., the distance between the relatively long regions of silent DNA that are compacted in chromomeric structures, together with the spacing of genes active in *oogenesis*, the most reasonable conclusion is simply that transcription units are arranged differently in anuran and urodele genomes. In the latter a particularly large fraction of the genome would appear to consist of very long blocks of silent sequences, but the active regions bunched together in the loops include an amount of single copy genomic sequence is within a factor of two the same as is expressed in the genome of *Xenopus*.

Transcription of specific Sequences in Lampbrush Chromosomes

Transcription of specific sequences has been visualized by *in situ* hybridization of labeled DNA probes with the nascent matrix RNA of lampbrush chromosome loops. Such hybridization is abolished by prior treatment with RNase, and the signal obtained depends directly on the dense packing of the nascent transcripts. The first application of this procedure to lampbrush chromosomes was attempt to identify the locus of the active 5S rRNA genes in *Triturus* lampbrush chromosomes. The probe consisted of labeled 5S DNA that had been purified from genomic DNA by isopycnic centrifugation. As it may have contained some other sequences, the results of this particular investigation remain equivocal. However, the *in situ* hybridization method when utilized with cloned probes, has provided useful information in regard to both the locus of particular genes, and greater importance, the general properties of transcription in lampbrush chromosomes.

Transcription of Repetitive Sequences

A great variety of repetitive sequence is transcribed in amphibian oocyte nuclei, and is also included in the interspersed poly(A) RNAs that by mass form the dominant fraction of the heterogeneous maternal transcripts stored in amphibian egg cytoplasm. Heterogeneous nuclear RNAs that have an interspersed sequence organization have also been extracted directly from the germinal vesicles of several amphibian species. When renatured and spread for electron microscopy these RNAs form partially duplexed multi-molecular structures similar to those observed in studies of renatured oocyte cytoplasmic poly(A) RNA. In addition many low molecular weight RNAs are transcribed from repeated genes. Aside from 5S rRNA these include U1 and U2 snRNAs which are synthesized in the *Xenopus* oocyte and the 181 nt cytoplasmic species known as OAX. There are other, yet unidentified tandemly repeated sequences transcribed in lampbrush chromosomes as well, among them the cluster of short, densely packed transcription units visualized in the electron microscope by Scheer (1981).

Molecular measurements carried out on genomic DNA show that the member of most short repetitive sequence families in amphibian genomes are widely interspersed, rather than tandemly repeated. Since many different interspersed repeat families are actively transcribed during oogenesis *in situ* hybridizations to loop matrix RNA carried out with probes representing such sequences as a class would be expected to reveal reactions at multiple loci. Just this result was obtained by Macgregor and Andrews (1977), whose probe consisted of a low C_0t DNA fraction labeled *in vitro*, and also by Sommerville and Scheer (1982) who utilized an RNA probe prepared from renatured RNase digested nRNA, labeled *in vitro* with ^{125}I. An unexpected observation in the study of Macgregor and Andrews (1977) was that only certain regions of some loops reacted with the repeat sequence probe. These loops are interpreted to contain more than one transcription units, and some of these lack sufficient repetitive sequence to react detectably. Exactly the same pattern of labeling is present in the homologous loops of chromosomes from different oocytes, from both the same and different animals. This particular observation refutes an early theory of lampbrush chromosome function, according to which the genomic DNA is constantly being spun out into the loops from the chromosomes, so that at any specific time a different set of sequences is being transcribed is given loops.

As expected a widely distributed pattern of hybridization is also observed with cloned probes representing single interspersed repeat sequence families. From a study of Jamrich et al. (1983) carried out on *Xenopus* lampbrush chromosomes. The probe represents a sequence present in about 10^3 copies per haploid genome. These appear to be so extensively interspersed that they are present in a significant fraction of the lampbrush loops visible. A similar result was reported by Kav et al. (1984), who found that about 100 pairs of lampbrush chromosome loops react with a different cloned repetitive sequence probe. In another experiment a probe representing satellite 2 of the *Notophthalmus viridescens* genome was hybridized to the lampbrush chromosomes. Tandem sequence blocks of this satellite are scattered throughout the genome. However, very little hybridization was observed with the loop matrix RNAs. Satellite 2 may thus be an example of the large, widely distributed silent DNA regions inferred above to be characteristic of urodele genomes.

Much of the repetitive sequence hybridization observed by Macgregor and Andhres (1977) in *Triturus cristatus carnifex* lampbrush chromosome 1. These arms are invariably distinct in structure in the maternally and paternally derived genomes of any one animal and they bear several recognizable landmark loops. According to Giemsa staining and additional cytological criteria they appear to consist mainly to satellite DNA and other repetitive sequences. A surprising result-obtained by Varley et al. (1980a, b) is that satellite DNA is actively transcribed in certain loops located on the heteromorphic arms. Experiments were carried out with both satellite DNA probes and probes prepared by direct isolation of satellite sequences from genomic DNA. Satellite transcription is unexpected on the basis of the lack of expression of this type of DNA sequence in somatic cells its usual presence in inactive heterochromatic regions of the chromosomes ; and the extremely low complexity and tandem repetition of these sequences. Thus it would appear that in lampbrush chromosomes sequences may be transcribed that are not expressed in other cell types, though of course this result in itself provides only a suggestion to this effect, as measurement on transcription of the same satellites in somatic *Triturus* cells have not been carried out. Further exploration of satellite transcription in lampbrush chromosomes has provided a new insight into the functional characteristics of these unique structures.

Transcription of Satellite Sequences at the Histone Gene Loci of Newt Lampbrush Chromosomes

A series of investigations on histone gene transcription in oocytes of *Notophthalmus viridescens* has provided direct evidence that initiation from histone gene promoter can result in the transcription of downstream satellite DNA sequences. There are about 600-800 copies of each of the histone genes per haploid genome, organized primarily in 9 kb clusters that contain all five of the genes plus intragenic "spacer" DNA in the order H1, H3, H2b, H2a, H4. The 9 kb gene clusters are separated by long stretches of satellite 1 DNA, some of which extend for more than 50 kb. This satellite consists of tandem repeats of a 222 nt sequence. Gall et al. (1981) showed by *in situ* hybridization to the DNA of both somatic mitotic chromosomes and oocytes lampbrush chromosomes that the histone genes are located predominantly in two major clusters, on chromosomes 2 and 6. Additional minor sites may exist as well. The number of histone genes at each major site is approximately equal, i.e.., each includes 300-400 of the 9 kb clusters. *In situ* hybridization at the chromosome 2 locus. From the size and number of the labeled loops it may be surmised that in this oocyte a significant fraction of the histone genes present on chromosome 2 are being transcribed. This may not be the case at the other histone gene locus or in different animals, since the DNA/DNA hybridizations revealed some chromomeric labeling. A significant correlation is observed between the chromosomal locations of the histone gene cluster and the presence of landmark bodies, known 60 "as spheres," which are attached to the chromosomal axis. The spheres are composed of an acid protein, conceivably accumulated for the special purpose of transporting or processing histone transcripts. The sphere loci are the sites of histone probe hybridization in *Triturus cristatus* and *Triturus alpestris* as well as in *Notophthalmus*, though in each case they are found on different chromosomes. Sphere loci also exist in *Xenopus* lampbrush chromosomes but in this organism the location of the histone genes is not yet known.

Lampbrush chromosome loops at the *Notophthalmus* sphere loci react with cloned satellite 1 probes as well as with histone. This is shown. The most important observations have been obtained with asymmetric probes transcribed from M13 templates. These experiments demonstrate that within the certain loops there are multiple transcription units, each many kilo-bases in length, since only sharply defined regions of these loops are labeled by the probe. This transcription units of opposite polarity can be seen to about directly within the same loops. Both strands of the satellite sequence are represented in the loop RNAs in different transcription units. Transcription initiates at a histone promoter within the gene cluster, and continues in the downstream direction across the remaining histone genes within the cluster and on into the flanking satellite DNA. The large size and the morphology of the transcription matrices implies that transcription fails to stop until another gene cluster is reached, or until a transcription until initiated in the opposite direction from an adjoining cluster is encountered. Thus, in *Notophthalmus* Lampbrush chromosomes, termination fails to occur at the end of the histone gene sequences. Unfortunately, nothing is known in regard to histone gene transcription in somatic cells of this newt, and thus it is not yet demonstrated that inefficient termination is a peculiarity of the transcription of these genes in lampbrush chromosomes. Mature histone messages accumulate in amphibian oocyte cytoplasm during oogenesis and must be derived from such readthrough transcripts. Processing of these transcripts thus must occur, probably by a strand scission at the 3' terminus of the initial histone mRNA in each transcript, since the satellite 1 sequences are not found outside the nucleus. The satellite transcripts together with the readthrough histone gene transcripts are presumably degraded within the germinal vesicle.

It remains to be determined whether read through transcription is a special feature of the histone genes or a general explanation for the enormous size of some urodele transcription units. Since there occur in lampbrush chromosome loops well defined transcription units separated by short regions of silent DNA, probably exist some means of termination within at least some loops other than encounter with an adjacent transcription unit. Nonetheless, it is possible that readthrough is a common feature of transcription in lampbrush chromosomes, perhaps an indirect consequence of the generally high rate of initiation, i.e., relative to the rate at which termination complexes could form. Readthrough transcription initiated at a normal gene promoter could account for the satellite sequences detected in the heteromorphic arms of the *Triturus c. carnifex* lampbrush chromosomes by Varley et al. (1980a, b). This region of the genome must contain some functional genes, since homozygosity for the heteromorphic arms is lethal in this newt species. Ribosomal RNA sequences are also transcribed in chromosome I, though this is not the site of the true nucleolus organizers, where the vast majority of ribosomal genes are located, and which appear silent in lampbrush chromosomes. The transcription of the ectopic rRNA sequences in lampbrush chromosomes is mediated by polymerase II rather than polymerase I, and therefore it is likely that it too is initiated at a functionally unrelated upstream promoter. It is reasonable to suppose in the absence of additional knowledge that transcription of any sequence that is found by *in situ* hybridization to be represented in lampbrush chromosome RNA could have been initiated at the normally regulated promoter of another nearby gene. Note that there is no evidence in the best studied example, the histone genes, that transcriptional initiation occurs at abnormal locations in lampbrush chromosomes : that is, the promoters at which initiation occurs are those of the various histone genes.

Processing of mRNA Precursors in the *Xenopus* Oocyte Nucleus

Several of the observations thus far reviewed reflect on the capacity of the amphibian oocyte nucleus to process and selectively transport mRNA precursors. For example there is the presence in the mature egg of a large quantity of nontranslatable poly(A) RNA that differs from most mature mRNA in its content of interspersed repetitive sequences. Further evidence reviewed below shows that in *Xenopus* oocytes RNA of this nature is actively exported to the cytoplasm during the lampbrush stage. It is clear on the other hand, that a stringent selection of those transcripts destined for the cytoplasm does take place, since as shown in Table the complexity of germinal vesicle RNA is 15-25 fold greater than that of cytoplasmic RNA. These observations completely exclude the possibility that the oocyte germinal vesicle randomly "leaks" heterogeneous nuclear RNA into the cytoplasm. The measurements summarized in Table indicate in addition that the flow of newly synthesized heterogeneous RNA into the cytoplasm includes only 5% to 12% of the nucleotides initially polymerized into heterogeneous nuclear RNA.

Processing has been demonstrated in the *Xenopus* oocyte nucleus in several studies on the molecular fate of specific precursors, either synthesized in the germinal vesicle from injected plasmids, or injected directly. One such experiment that may be relevant to the possible disposition of the readthrough histone transcripts was carried out by Krieg and Melton (1984). A histone precursor RNA was synthesized *in vitro* from a plasmid containing a chicken H2b histone gene, and introduced into the oocyte nucleus. Generation of the native histone message from this precursor requires a correct 3' endonucleolytic processing reaction, and this was observed. Furthermore, the appearance of a correctly terminated histone mRNA was not affected by the presence of several hundred nucleotides of additional vector sequence distal to the proper termination site. Birnstiel and associates also have demonstrated the production of correctly terminated histone mRNAs after injection of cloned sea urchin histone genes into *Xenopus* oocytes nuclei. Both a high conserved 3' terminal sequence element in the histone gene, and a small ribonucleoprotein appear necessary for the formation of normally terminated messages. These experiments indicate the existence of the mechanism that would be required for production of mature histone messages from the 5' terminal gene sequences of the readthrough histone transcripts synthesized in the lampbrush chromosomes.

Direct evidence that *Xenopus* oocytes also possess the capacity to carry out RNA splicing reactions has been obtained by injecting cloned genes that include introns. In several such experiments synthesis of the protein coded by the exogenous sequences is reported to occur. Examples include the ovalbumin gene, in which correct processing involves the precise removal of no less than seven introns, and the SV40 T antigen, requiring precise excision of one into Rungger and Turler, (1978). Accurate splicing of a yeast tRNA^{try} precursor in the *Xenopus* oocytes nucleus has also been demonstrated. Processing of the yeast tRNA^{tyr} precursor involves removal of 5' leader sequence and also of extra nucleotides at the 3' end and a sequence and also of extra nucleotides at the 3' end and a sequential series of base modifications, as well as the splicing reactions. All of the enzymatic machinery required for these activities is confined to the nucleus.

The observations of Anderson and Smith (1977, 1978) that in late oocyte there seems to be an increased retention of nuclear RNAs and a lower intranuclear turnover rate suggest that the efficiency of the processing reactions might be relatively low. By "efficiency" meant the fraction of precursor molecules that converted into mature messages. A kinetic study of the processing of SV40 transcripts in the stage 6 *Xenopus* oocyte nucleus suggests that the efficiency with which processing is carried out is indeed poor. Miller et. al. (1982) injected SV40 DNA into the oocyte nuclei, and found that 90% of the primary transcript is degraded within the nucleus. The half-life of these SV40 primary transcripts is only 20-40 min, just as for the endogenous nRNA. However, a small fraction of the transcription products appears to be converted into correctly processed 19S poly(A) mRNA that accumulates in the cytoplasm. An even smaller fraction of the precursor is converted into 16S mRNA. It seems unlikely that these results are artifactual consequences of overloading the processing capacity of the oocyte with SV 40 transcripts, since the amount of SV 40 transcript synthesized in the study of Miller et al. (1982) was less than the amount of endogenous nRNA transcript, and the total amount exported to the cytoplasm was only about 10-20% of the endogenous newly synthesized poly(A) RNA flow into the cytoplasm.

A further investigation of the processing of late SV 40 transcripts by Wickens and Gurdon (1983) demonstrates that the requirements for transport to the cytoplasm are endonucleolytic cleavage at the site of the 3' terminus of the mature message, and polyadenylation. Both spliced and unspliced derivatives with these terminal features are found in the cytoplasm, while nonpolyadenylated transcripts that include sequences distal to the mature 3' terminus are retained in the nucleus. These results suggest that the amount of processing may be a function of the *relative rates* at which transcription, degradation, preparation of the 3' terminus, and the splicing reactions occur. Where the rates of the latter are relatively insufficient, unprocessed transcripts will either be degraded, or if they have acquired mature 3' termini, exported to the cytoplasm where further processing reactions cannot take place.

An investigation of Green et al. (1983) in which human β-globin precursor RNA was injected into oocyte nuclei shows that molecules lacking the 5' cap structure are rapidly degraded. The β-globin precursor molecules were synthesized enzymatically *in vitro*, and only if capped prior to injection were they found to be stable. Accurate splicing out of both β globin precursor introns occurs in the oocyte germinal vesicle, and for this a terminal poly(A) sequence is not required. However, these splicing reactions are also inefficient in that the majority of the globin precursor molecules are never processed. A sequence-specific form of selective nRNA processing has been demonstrated in the oocyte nucleus as well. Thus Buzzoni et al. (1984) showed that two of the nine introns included in primary transcripts produced from an injected ribosomal protein gene are not spliced out, while all nine introns from a different ribosomal gene transcript are removed. Most of the incompletely protein processed precursor molecules are confined to the germinal vesicle, but again some are recovered four the cytoplasmic compartment.

A role for snRNPs in at least some splicing reactions carried out in *Xenopus* oocytes has been demonstrated by Fradin et al. (1984). These experiments showed that injection of various antibodies against U1 RNP sharply inhibit splicing of the SV 0 late transcript, resulting in an accumulation of unspliced precursor. Many of these antibodies had no effect on the splicing of the early transcript, however, suggesting that a variety of other snRNP cofactors could be required for the processing of different precursor molecules. It may be relevant that the *Xenopus* oocyte germinal vesicle lacks several snRNA species that are synthesized in embryos after the blastula stage. Among these are two particular U1 snRNA species found in somatic cells, though other U1 snRNA are synthesized during oogenesis. The mature *Xenopus* egg contains about 8×10^8 molecules of the oocyte types of U1 snRNA, sufficient to provision the several thousand nuclei that have been formed by the time transcription resumes at the blastula stage. It remains to be determined whether deficiencies in specific snRNRs or other characteristics of the germinal vesicle snRNA pool are in any respect responsible for the inefficiency of nRNA processing in the oocyte.

In summary, it is clear that the oocyte nucleus can carry out all known RNA processing reactions ; that it does so inefficiently, however ; and that if they have been polyadenylated, transcripts that are not mature messages may be exported to the cytoplasm. These conclusions provide a reasonable interpretation for the presence of the interspersed cytoplasmic poly(A) RNAs. However, this proposition must remain inferential, since the results have all been obtained with exogenously introduced sequences, many of which are foreign to the *Xenopus* oocyte. The posttranscriptional disposition of an endogenous primary transcript of the lampbrush chromosomes, readthrough or otherwise, has yet to be described in similar detail.

Messenger RNAs and Interspersed Heterogeneous RNAs of the Oocyte cytoplasm : the Logistic Role of Lampbrush Chromosomes

Accumulation and Turnover of Cytoplasmic Poly (A) RNA

We have so far regarded the newly synthesized oocyte transcripts mainly from an intranuclear perspective. However, the key issue, in considering the ultimate function of lampbrush chromosomes, is the relation between the chromosomal transcription products and the cytoplasmic RNAs accumulated in the course of oogenesis. This relation is by no means an obvious one. In *Xenopus*, the only organism for which extensive measurements exist the quantity of total poly (A) RNA present at the end of oogenesis is about 80-90 ng and the same amount is to be found in midvitellogenesis stage 3 oocytes. Relative measurements carried out at a series of different stages of oogenesis by Rosbash and Ford (1974) and Golden et al. (1980) show that in fact the final quantity of oocyte poly (A) RNA has already accumulated before the end of Dumont stage 2. Golden et al. (1980) also reported that the patterns of accumulation of several cloned maternal sequences follow that of the total poly (A) RNA, all achieving, their final levels, as judged from RNA gel blots, by the beginning of vitellogenesis. The only exceptions observed were mitochondrial poly(A) RNAs, which continue to accumulate throughout oogenesis. The mature *Xenopus* oocyte contains about, 10⁻ mitochondria, and about 4 ng of mitochondrial DNA or~12 mitochondrial genomes per mitochondrion. However, mitochondrial transcripts account for only about 15% of the newly synthesized cytoplasmic poly(A) RNA of the stage 6 oocyte. The early appearance in oogenesis of the final content of cytoplasmic poly(A) RNA seemed particularly paradoxical before the demonstration by Hill and Macgregor (1980) that even in growing stage 1 oocytes only 60 μm in diameter the chromosomes contain densely packed transcription units similar to those found in Dumont stage 3 oocytes.

A certain amount of confusion has been caused by the fact that continuous transcriptional activity takes place in the lampbrush chromosome in stages 26 oocytes, while the net mass of the cytoplasmic poly(A) RNA remains constant. Several discussions appear in the literature in which it is concluded that because there is no further accumulation, of cytoplasmic poly(A) RNA, newly synthesized poly(A) RNA must not be exported to the cytoplasm during most of the lampbrush chromosome stage. This is obviously a faulty deduction, however, in that it is directly contradicted by the measurements of Anderson and Smith (1977, 1978) and Dolecki and Smith (1979). As reviewed in Table these demonstrate a continuous flow of newly synthesized heterogeneous RNA into the cytoplasm in Dumont stages 3 and 6 oocytes, most of which is polyadenylated. The poly(A) RNA accumulation function shown in means simply that the poly(A) RNAs emerging from the germinal vesicle accumulate until a steady state content is reached and thereafter this is maintained by the balance between new synthesis and turnover. The lampbrush chromosomes provide synthetic activity that throughout oogenesis drives the kinetic process. The following calculations show that the maximum rate of transcription in virtually all the transcription units represented in maternal RNA is required in order for this function to carried out.

There are on the average about 4×10^6 copies of each maternal mRNA sequence in the *Xenopus* egg, representing 1-2 $\times 10^4$ different transcription units. The polymerase II translocation rate in *Xenopus* oocyte nuclei is 10-20 nt sec^{-1}. Given that the transcribing polymerases are packed only 100 nt apart, to synthesize 4×10^6 molecules of an average transcript on four single copy genes per nucleus would require about 80 days. The average duration of the growth phase of stage 1 and stage 2 when net poly(A) RNA is accumulating has been estimated to be as long as 8-9 months. However, 2-3 months is a reasonable minimal estimate for young frogs. Thus, vitellogenic oocyte appear four months after metamorphosis is complete, while at three months after metamorphosis no oocytes have yet advanced into stage 2. A complementary calculation based on the complexity of the maternal RNA was presented by Anderson et al. (1982). In stage 3 and stage 6 *Xenopus* oocytes about 1.7-2.2 pg min^{-1} of stable newly synthesized 4-40S RNA, most of which if polyadenylated, enters the cytoplasm. Of this perhaps 10% consists of repetitive sequence transcript and < 15% of mitochondrial transcript. Thus a minimum value of 1 pg mig^{-1} can be taken as the flow into the cytoplasm of newly synthesized transcripts representing single copy genomic sequence. The complexity of this RNA is about 4×10^7 ntp, and we assume for the limit calculation that all of the maternal species are continuously being synthesized. Thus the number of copies of each sequence exiting from the oocyte nucleus is about 45 per minute or 11 transcripts per minute attributable to each of the four active genes per nucleus. However, this rate requires one initiation per 5.5 sec. which is almost the same as calculated independently from the measured polymerase translocation rate, if one assumes only 100 nt between transcribing polymerases. Therefore, virtually all of the transcription units active in synthesizing cytoplasmic poly(A) RNA must be maximally packed with transcribing polymerases. Since the mean size of *Xenopus* oocyte poly(A) RNA is about 2.2 kb, the number of such transcription units must be close to the total number included in the lampbrush chromosome loops i.e., almost 2×10^4. It follows that to account for the measured flow of stable heterogeneous RNA into the oocyte cytoplasm the large majority of the transcription units active in the oocyte nucleus must be densely packed with polymerase. We may concluded that the transcriptional structure of lampbrush chromosomes is in fact required to explain both the rate at which the poly(A) RNA of the oocyte initially accumulates (stages 1-2) and the rate at which it is synthesized and exported at steady state (stages 2-6).

Given a steady state pool as large as 80 ng of poly(A) per oocyte, and a flow rate of 1 pg min^{-1}, the half-life of the newly synthesized cytoplasmic poly(A) RNA would be about 35 days. Of course a fraction of the poly(A) RNA may be turning over at a higher rate. Since no detectable turnover was observed in the kinetic experiments of Anderson and Smith (1978) and Dolecki and Smith (1979). The hair-life of such a fraction would still have to be greater than several days. In any case it is clear that many if not all of the newly synthesized cytoplasmic transcripts ultimately turn over at a rate which these calculations predict would be too low to permit detection in labeling experiments, and yet would still be sufficient to provide a kinetic steady state.

Transport of Interspersed RNAs to the Cytoplasm at the Midlampbrush Stage

The interspersed cytoplasmic RNAs were those that are stored in mature oocytes, where they constitute about 70% of the poly(A) RNA mass. This class of RNA can be experimentally defined by its ability to renature, forming multimolecular networks held together by repetitive sequence, duplexes. Anderson et al. (1982) showed by renaturing poly(A) RNA extracted from stage 3 oocytes that interspersed RNAs have been already been deposited in the cytoplasm by the midlampbrush stage. Electron micrographs of the partially duplexed multimolecular structures. These structures included about 500 of the total poly(A) RNA mass examined. The oocytes were labeled by microinjection of precursor, and after 2 days the newly synthesized poly(A) RNA was extracted from the cytoplasm. About 90% of the labeled molecules fractionated as interspersed RNA. Thus in quantitative terms, the major product of lampbrush chromosome transcription that is exported to the cytoplasm at least at this late stage of oogenesis is interspersed poly(A) RNA. These experiments confirm the view that newly synthesized lampbrush chromosome transcripts continue to enter the cytoplasm even in late oogenesis. The function and the ultimate disposition of this large class of heterogeneous maternal RNAs are still to be determined.

The Quantity of Functional mRNA in Xenopus Oocytes

During oogenesis the rate of protein synthesis per oocyte increases over 100-fold from about 0.18 ng hr^{-1} to about 22 ng hr^{-1}, as illustrated. The quantity of ribosomal RNA per oocyte increases to about the same extent. Thus the *fraction* of ribosomes engaged in protein synthesis remains approximately constant. L.D. Smith et al. (1984) report this as 1.6% at stage 1 and 2.4% and 2.3% at stages 3 and 6, respectively, in good agreement with the earlier estimation of about 2% made by Woodland (1974). Since the mass of total poly(A) RNA remains constant after stage 2, an increasing fraction of the poly(A) RNA pool must be involved in the translational apparatus as oogenesis proceeds. In the amount of mRNA engaged on the oocyte polysome at stages 3 and 6 is compared to the total amount of translatable message estimated to be present in the oocyte, calculated from the total poly(A) RNA content, after deducting mitochondrial poly(A) RNA and interspersed poly(A) RNA. The latter is known to be largely nontranslatable. Thus the mass of true maternal message in the stage 6 oocyte is estimated to be about 21 ng. To account for the quantity of polysomes present in the embryo up to the blastula stage, when mRNA begins to be synthesized a new, would require about the same quantity of *bona fide* message as by these calculations is apparently present in the stage 6 oocyte, i.e., assuming that the mRNA is stable after fertilization. Thus during maturation the polysome content increases about 2-fold, and a further gradual increase in the fraction of polysomes occurs as additional maternal mRNAs are recruited in early embryogenesis. By blastula stage about 15% of the ribosomes are engaged in polysomes. The mass of polysomes at blastula stage is therefore about six times that in the stage 6 oocyte while the mass of putative mRNA in the oocyte is about 5-fold in excess of that being utilized. The conclusion is that the unused *bona fide* mRNA in the stage 6 oocyte just constitutes a sufficient supply of maternal mRNA for the embryo, assuming no significant contribution from the interspersed poly(A) RNA class.

Demonstrates an impressive difference between stage 3 and stage 6 oocytes. There is at least as much putative mRNA in stage 3 as in stage 6 oocytes, and there could be almost twice as much. However only about 2% of the putative stage 3 mRNA i.e., about 20 pg, included in polysomes compared to almost 20% of the stage 6 mRNA. Thus throughout the early stages of oogenesis, all but a few percent of the putative mRNA present is sequestered and stored in a translationally inactive form. Richter and Smith (1983) and Darnbrough and Ford (1981) isolated poly(A) RNA binding proteins from early oocytes and showed that the quantity of these mRNAs per oocyte declines throughout oogenesis. Messages coding for one such poly(A) RNA binding protein also decrease during oogenesis. There are five major mRNA binding proteins that are regulated downward during oogenesis, 94, 68, 56, 55 and 50 kd in mass. Richter and Smith (1984) showed that if the mRNA binding proteins of the early oocyte are reconstituted with message *in vitro*, the resulting particles cannot be translated on injection into *Xenopus* oocytes, unlike particles similarly reconstituted with histones, or with a variety of message-associated proteins extracted from other cells. As the amount of the transitional repressor declines through oogenesis, and the quantity of message engaged on polysomes increases, mRNA molecules stored early in oogenesis are probably released for translation at later stages. Thus the sequestration proteins could serve the function of controlling the rate of biosynthesis during the growth phase of oogenesis.

The mechanism responsible for the failure of the remaining 80% of the putative stage 6 oocyte mRNA to be translated during oogenesis remains unclear. The bulk of this fraction of the poly(A) RNA clearly is capable of protein synthesis when extracted and tested *in vitro*, or injected into *Xenopus* oocytes. Perhaps within the later oocyte it remains translationally repressed by the same mRNA binding proteins are present earlier. However, this is an unnecessary hypothesis. A sufficient alternative explanation has derived from an exhaustive series of experiments that have demonstrated a limitation in some component(s) of the stage 6 oocyte translation apparatus, other than the supply of accessible mRNA. The key observation is that in stage 6 oocytes injected mRNA can compete for translational components with endogenous mRNA, or with other injected messages. One limiting factor required for translation of membrane-bound mRNA is associated with rough endoplasmic reticulum, since coinjection of a rat liver cytoplasmic membrane fraction enhances the ability of the oocyte to translate injected of this class. The translational component(s) that are limiting for cytosol messages remain unknown. Limitation of translational capacity is a specific feature of late oocytes, however. Thus injection of globin mRNA into stage 3 and stage-oocytes causes an *increase in total protein synthesis*, and no competitive depression of endogenous protein synthetic activity is observed, while the same exogenous mRNA injected into stage 6 oocytes competes with the endogenous messages. The highest rate of protein synthesis that can be induced by injection of mRNA in stage 4 oocytes approaches that of stage 6 oocytes, i.e., ~21 ng hr^{-1}. This suggests that by stage 4 the maximum translational capacity of the oocyte has been set but that this capacity is not fully utilized until late in oogenesis, perhaps because of the prevalence of repressive mRNA binding proteins in younger oocytes.

mRNA for Specific Proteins

Many diverse proteins have been identified in late *Xenopus* and *Rana* oocytes, stored in relatively enormous quantities for use after fertilization. Examples include histones, of which a sufficient supply exists to assemble the chromatin of over 20,000 cells, nucleoplasmin, which is apparently involved in the mobilization of histones for chromatin assembly involved in the mobilization of histones for chromatin assembly and which constitutes 10% of the total nuclear protein RNA polymerases, equal in amount to that contained in about 10^5 larval amphibia cells ; DNA ligases ; tubulins, also resident in a huge pool amounting to about 1% of all soluble oocyte protein ; some even more prevalent actin species that constitute >8% of the total soluble protein in stage 6 oocytes;

the ribosomal proteins included in the 10^{12} ribosomes of the mature oocyte, proteins binding specifically to snRNAs ; and a variety of other nucleo-proteins. Several of latter are considered in the following section, in connection with 5S RNA synthesis during oogenesis, and others, e.g., those associated with cytoplasmic mRNAs have already been mentioned. DNA polymerase provides a particularly striking example, since there is no nuclear DNA synthesis during the growth phase of oogenesis. The number of molecules of $\alpha_1 + \alpha_2$ DNA polymerase increase continuously from about 10^8 in the stage 1 oocyte, to about 10^9 in the stage 3 oocytes and 2×10^{10} in the unfertilized egg. For comparison, in the hatched tadpole, a level of about 7.5×10^4 molecules of $\alpha_1 +$ α_2 DNA polymerase is maintained per cell by new synthesis. The egg thus stores sufficient a-polymerases to provision over 2×10^5 embryonic cells.

Though sequences represented in the maternal RNA continue to be transcribed throughout the lampbrush stage there are only a few examples that provide evidence on the origins of specific maternal mRNAs. Anderson et. al: (1982) showed that transcripts labeled to high specific activity by in vitro incubation of isolated stage 6 oocyte nuclei reacted with six different cloned cDNAs derived from the maternal poly(A) RNA. However there is yet no direct demonstration in these or any other experiments, of the entrance of particular newly synthesized mRNA into the cytoplasm in late oocytes. The problem is technically difficult because specific activities adequate for the detection of individual mRNA species cannot be obtained in whole oocytes, even after injected of high specific activity precursor nucleotides, and thus only the overall flow rates have so far been measured. Quantitative studies of protein synthesis during oogenesis have shown by two dimensional gel electrophoresis that most protein species continue to be synthesized at the same relative rates throughout, which would be consistent with later utilization of a pool of mRNAs stored from stage 1-2. Though a considerable number of stage-specific changes have been reported in these studies, they are apparently the result of alteration in the fraction of mRNA of given species that are located on the oocyte polysomes. Thus King and Barklis (1985) reported that the *in vitro* translation products of poly(A) RNA extracted from stages 2, 3 and 6 oocytes are essentially identical.

We turn now to two sets of identified proteins for which cloned probes have been generated, the histones, and the ribosomal protein (rp). As summarized the *Xenopus* oocyte contains about 4×10^8 molecules of each nucleosomal core histone mRNAs, and on the average $8 \times 10^{-}$ molecules of each of the rP mRNAs investigated, though the mRNA quantities for individual rP species differ up to fourfold with respect to one another. The greater content of histone mRNAs is more than accounted for by the difference in the respective gene copy numbers. Thus there are about 40 genes for each histone species per haploid genome in *Xenopus*. The number of RNA polymerase mRNA molecules, estimated indirectly from the subunit synthesis rate measurement of Hollinger and Smith (1976), severalfold lower. A similar number of molecules of the mRNA for the 70 kd heat shock protein is also reported, i.e., about 3×10^6. For comparison, the prevalence of the average poly(A) RNA species in the egg of stage 6 oocyte is about 4×10^6 molecules.

The rP mRNAs and the histone mRNA the maximum level of accumulation is achieved by the beginning of vitellogenesis at Dumont stage 2, just as for the bulk of the poly(A) RNA. Pierandrei-Amldi et al (1982) and Baum and Wormington (1985) noted that the rP mRNAs have declined in quantity by stage 6, and Dixon and Ford (1982) also reported that synthesis of rP is much decreased in late oogenesis. On the other hand, the quantity of histone mRNA remains about the same throughout the later phases of oogenesis. This difference in behaviour is directly interpretable in terms of the temporal requirements for the respective proteins. The assembly of ribosome so occurs most rapidly in midoogenesis, and is largely completed by the end of oogenesis, mRNAs coding for ribosomal proteins achieve their peak concentrations just prior to institution of the maximum rate of rRNA synthesis, which occurs at stage 3. Thus, by stage 2.50% of the rP mRNA is loaded on polysomes, compared to only a few present of total oocyte mRNA and the high level of translational utilization of the rP messages persists throughout oogenesis. Much of the rP mRNA is degraded after fertilization, and some species of ribosomal proteins are not synthesized again until the neurula stage, though the mRNA begins to be produced several hours earlier at gastrulation. In contrast, both the large pool of histone protein synthesized during oogenesis, and the mass of maternal histone mRNA are destined for use in early development. The histones begin to be synthesized at a high rate immediately upon mutation, and maternal histone mRNAs thereafter support all new histone synthesis until the blastula stage.

Sufficient measurements on accumulation of both histone mRNA and histone proteins during oogenesis in *Xenopus* exist so that quantitative image of the kinetics of this process merges. We have seen that accumulation of the maternal histone mRNA pool occurs very early in oogenesis. Assuming that the polymerases synthesizing histone transcripts are packed only 100 nt apart ; that the translocation rate is 15 nt sec^{-1} Anderson and Smith (1978) ; and that there are 40 genes for each core histone in each of the four haploid genomes of the 4C oocyte nucleus ; a period of 4–5 months would be required to synthesize the 3.4×10^8 histone mRNA molecules that according to Table constitute for each core species the maternal histone mRNA pool. As discussed earlier, this is similar to the estimated time required for accumulation of the steady state content of an average poly(A) RNA coded by a single copy gene, about 3 months. This calculation requires the histone transcription units to be in the densely packed lampbrush chromosome configuration during early oogenesis. In addition, it requires that the histone mRNA be stable during its initial accumulation phase, and thereafter be sufficiently long-lived so that it decays no faster than it can be replaced by new synthesis. Woodland and Wilt (1980b) showed that sea urchin histone mRNA infected into oocytes turns over with a half-life of only a few hours. There are at least two known factors that could contribute to the stability of endogenous histone mRNA in the oocyte. Unlike these injected messages the endogenous histone mRNA might be complexes with protective proteins, such as the mRNPs analyzed by Darnbrough and Ford (1981) and Richter and Smith (1983). In addition, 50-75% of the histone mRNA in the *Xenopus* oocyte is polyadenylated. It has been demonstrated directly by Huez et al. (1978) that prior polyadenylation increases the stability of HeLa cell histone mRNA injected into *Xenopus* oocytes.

The four nucleosomal core histones are synthesized at a rate of about 50 pg hr^{-1} during oogenesis. This is only a very small fraction of the rate that could be supported by the mass of stored histone message, and indeed during maturation the rate increases 50-fold to about 2500 pg hr^{-1}. The final histone content is about 135 ng. This includes about 3-4 pmoles of each of the core histones, and 0.5 pmoles of H1. Van Dongen et al. (1983b) found that accumulation of H1a, the major H1 species, is completed during stage 3. Though some H1 synthesis continues after this, it must be balanced by turnover. At the rate of 50 pg hr^{-1} for the core histones, it would take about 100 days to accumulate the amounts of stored core histone proteins in the egg. Again, this is not inconsistent with the time usually required for Xenopus oocytes to progress from stage 2 onwards as discussed earlier.

The stored histones are mainly accumulated within the germinal vesicle of the oocyte, where they are stored in stoichiometric complexes of several kinds with the nucleophilic proteins N_1, N_2 and nucleoplasmin. N_1 and N_2 are synthesized exclusively in germ line cells, including lampbrush stage oocytes, but not after fertilization in the embryo, or in somatic cells. The embryonic role of these maternal proteins may be to be bind, neutralize, and transport into the blastomere nuclei the newly synthesized histones that are translated from maternal mRNA during maturation and in early cleavage. Nucleoplasmin and N_1 and N_2 are not the only important proteins destined for use after fertilization that are stored in the germinal vesicle. For example most of the a DNA polymerase, the RNA polymerase, the actins and the snRNPs are similarly localized in the enormous nucleus of the stage 6 oocyte. Germinal vesicle breakdown at maturation releases all of these components into the cytoplasm, where they remain until after fertilization, when they are redistributed to the newly formed nuclear compartments of the embryo.

The Functional Role of Lampbrush Chromosomes

A coherent interpretation of the data so far reviewed is as follows. Lampbrush Chromosomes are the cytological manifestation of a transcriptional apparatus in which the distinguishing–and entirely unique–feature is that almost all the active transcription units are operating at the maximum possible initiation rate. The basic role of lampbrush chromosomes is to provide the transcription products loaded into the oocyte cytoplasm. This function can be considered in two phases. Early in the lampbrush state net accumulation of maternal poly(A) RNAs occurs until the levels that will be maintained thereafter are achieved. This has been illustrated in many different ways, e.g., by measurements of the total poly(A) RNA content, analyses of the spectrum of proteins synthesized at different stages, and observations on several specific transcripts.

The second phase begins with vitellogenesis. Throughout this process the oocyte is synthesizing the enormous supplies of proteins that will be required during early development, of which the best understood example is provided by the histones. The role of the lampbrush chromosomes during this phase is evidently to maintain in kinetic steady state the required level of the maternal transcripts. This is demonstrated directly by measurements of the flow of newly synthesized poly(A) RNA into the cytoplasm in stage 3 and stage 6 oocytes. We have seen that when the observed flow rates are considered in conjuction with the complexity of the RNA, the maximum possible transcription rates would be required merely to preserve the levels of transcripts present, even were the half-lives of these in excess of a month. Thus it may be concluded that lampbrush chromosomes provide the transcriptional solution to the logistic problem of building and maintaining very large oocytes. The same problem is solved in a completely different manner in meroistic oogenesis. As the following sections of this chapter illustrate there are still other solutions utilized in the same oocytes in answer to the special logistic demands posed by the requirements for 5S rRNA and 18 and 28S rRNA.

There could exist adaptive advantages for a system of oogenesis in which the maternal constituents are continuously renewed, even through this entails the maintenance of an extremely high rate of synthesis right to the end of oogenesis. Many amphibians, for example the tropical temporary water breeder *Engystomops pustulosus*, store mature oocytes within the ovary until and appropriate environment for egg laying becomes available, and this probably also is characteristic of *Xenopus*. Continuous molecular turnover and renewal could provide a means of escaping a rigid requirement to complete oogenesis and shed eggs on a certain schedule, thereby improving adaptive flexibility in response to uncertain environmental conditions. In any case it should not seem peculiar to assume that the transcripts of the oocyte are not absolutely stable. In all living cells that have been examined, except for special examples of dormancy such as cryobiotic systems, seeds, and spores, mRNAs are found to decay, usually stochastically, and almost always with kinetic far faster than implied by the minimum half-lives required in the foregoing considerations of amphibian oocyte transcripts.

The subject of lampbrush chromosome function has a curious intellectual history. Lampbrush chromosomes have provoked a remarkable series of colourful, but wrong interpretations, among which are the once widely accepted idea that the loops are the *products* of the chromomeric genes the "master-slave" theory ; the proposition that new DNA sequences are progressively unwound into the loops from the chromosomes ; the "C-value paradox" ; and most recently, the assertion that the intense RNA synthesis in the loops is all nonproductive, in the sense that none of the newly synthesized RNA is exported to the cytoplasm. The value of these proposals has lain mainly in the efforts they have stimulated to disprove them, and none seem at this point likely avenues to further understanding. But it is scarcely the case that we have reached the end of this curious progression. Though we may now understand in logistical terms what lampbrush chromosomes do, we clearly do not understand why they are doing it. Thus the majority of the heterogeneous transcripts exported from the lampbrush stage oocyte nucleus are nontranslatable, interspersed RNAs, of unknown function. Perhaps these are useless products of readthrough transcription, exported to the cytoplasm as a side effect of some yet unknown characteristic of the post-transcriptional processing apparatus of the oocyte. In this case there remains to be discovered an unexpected feature of the mechanism by which transcripts are handled in the oocyte germinal vesicle, a mechanism so valuable that it is retained even at the expense of loading the cytoplasm with 70% nonfunctional poly(A) RNA. Or these transcripts may serve as unprocessed nRNAs destined for modifications and translational use after fertilization, or perhaps they constitute some totally different class of maternal RNA required in developing systems about which we yet understand nothing. A further solution to the problem of lampbrush chromosome function, and perhaps a basic new understanding of oogenesis, could lie somewhere among the answers to these newly arisen mysteries.

The Accumulation of Ribosomal RNAs

The mechanisms by which 5S rRNA and the 18S and 28S rRNAs are accumulated are for Xenopus relatively well described. About 10^{12} ribosomes must be assembled per oocyte, and this logistical requirement is met by three completely different strategies. The synthesis of ribosomal proteins has already been discussed, and in this section we compare the means by which the heavy ribosomal RNAs and 5S RNA are synthesized and accumulated. Though they can only be summarized briefly here, the molecular details of the processes by which these two classes of ribosomal genes are controlled provide interesting new insights into the mechanisms by which the genes transcribed by polymerase I and III are developmentally regulated.

5S rRNA Synthesis in Xenopus Oocytes

Low molecular weight RNAs, mainly tRNAs and 5S rRNA, are the major transcription products of Dumont stage 1 and 2 oocytes. Thomas showed that after 24 hr of labeling about 24% of the radioactive RNA in the cytoplasm of these previtellogenic oocytes is tRNA and 39% is 5S RNA, and data of Rosbash and Ford indicate that 5S RNA and tRNA account for about 70% of the mass of total RNA then present. In previtellogenic oocytes the molar ratio of newly synthesized tRNA to heavy ribosomal RNA is as high as 25, while for 5S RNA it is over 100. Both tRNA and 5S RNA continue to be synthesized, but as the rate of synthesis of 18 and 28S rRNA accelerates after stage 2, these ratios shift dramatically. Previtellogenic oocytes are thus very unusual in the large fraction of their total accumulated RNA accounted for by stable low molecular weight species.

About 5% of the 5S RNA synthesized in previtellogenic oocytes is stored in a 7S particle, where it is complexed with a single 38.5 kd protein. The remaining 50% is found in a multimeric 42S particle, which contains in addition to the same 38.5 kd protein and 5S RNA, tRNA and two other protein molecules of ~50 kd mass. The latter proteins are believed to be bound to the tRNA, which is present in the aminoacyl form and in the 42S particles, of previtellogenic oocytes turn over with half-lives of 4-36 hr, depending on the tRNA species. Both forms of storage particle are prominent only in previtellogenic oocytes. The proteins of the 42S particle are reported to be involved in transport of 5S RNA to the nucleolar site of ribosome assembly. This occurs as newly synthesized 18S and 28 rRNAs become available. In the ribosome the 5S RNA is complexed with a ribosomal protein that differ in molecular weight and amino acid composition from the storage particle proteins. The proteins of the 7S and 42S particles are synthesized at the highest rate in stage 1 oocytes. The same general mechanism, viz., early synthesis of 5S RNA and storage in 7S and 42S RNP particles, followed during vitellogenesis by incorporation of the stored 5S RNA into ribosomes, may occur in many large vertebrate oocytes that also bear lampbrush chromosomes. Thus homologous 42S and 7S particles have been reported in privitellogenic oocytes of anurans, urodeles, and teleost fish.

There are two classes of 5S RNA genes in Xenopus, those active in oocytes and those active in both somatic cells and oocytes, termed the somatic 5S RNA genes. Oocyte-type 5S RNA differs from somatic 5S RNA at 6 out of 120 nt. Only the oocyte-type of 5S RNA is stored in the 7S and 42S RNP particles. The products of the somatic 5S genes are apparently unstable in the oocyte, since though they are these transcripts fail to accumulate. After fertilization and up until the mid-blastula stage, both classes of 5S RNA genes are silent, as are all other nuclear genes. When transcription resumes some oocytes 5S RNA synthesis occurs, as does somatic 5S RNA synthesis. After the gastrula stage, however, synthesis of oocyte-type 5S RNA dwindles to an insignificant level. Thus the oocyte-type gene are regulated, in that they are mainly transcribed only during oogenesis, while the somatic 5S RNA genes are expressed constitutively. The very high rat of oocyte 5S RNA synthesis in early oogenesis is accounted for by the large number of oocyte-type genes. Thus there are about 20,000 oocyte 5S RNA genes and about 400 somatic 5S RNA genes per haploid genome. During oogenesis each oocyte-type 5S RNA gene is responsible for the synthesis of about 107 5S RNA molecules.

Chromatin containing the 5S RNA genes can be transcribed with fidelity *in vitro*, by added polymerase III. Exploitation of this advantage has led to a relatively advanced understanding of the chromatin regulatory factors involved in developmental control of these genes. A DNA sequence within the gene, extending from nucleotide +45 to +97 and known as the "internal control region," is required for the formation of a transcription complex and for initiation of transcription by polymerase III. The active transcription complex contains three protein species, aside from the polymerase itself. One of the most important properties of transcription complex is that once formed it remains stable and functions for many rounds of transcription. Though little is so far known of the transcription factors B and C, it is established that a single molecule of TEIIIA a transcription complex cannot be assembled and the genes instead may from a stably repressed complex with histone. This is apparently the fate of the oocyte-type 5S RNA genes in the nuclei of somatic cells such as adult erythrocytes or kidney tissue culture cells, since these genes can be reactivated and assembled into active transcription complexes following treatments that remove histone H1. Furthermore, oocyte type 5S DNA chromatin, i.e., the DNA in nucleosomal from, can be reconstituted in vitro together with histone H1 to reproduce the repressed structure. In tissue culture cells the somatic 5S RNA genes are instead included in stable transcription complexes, and they require only the addition of polymerase III for transcription. The oocyte 5S RNA genes are found in such complex only during oogenesis.

A simple binary choice between TFIIIA binding or histone H1 binding may constitute a sufficient explanation for the regulation of the 5S RNA gens. *In vitro* experiments with TFIIIA demonstrate a 4 : 1 preference for somatic over oocyte-specific genes, apparently the result of the three nucleotide difference in the oocyte and somatic gene sequences within an important area of the internal control region. However, when a mixture of oocyte and somatic type 5S gene derivative is injected into fertilized *Xenopus* eggs a preference for the somatic 5S gene sequences of up to 200-fold is observed. In vivo the discrimination between the endogenous somatic and oocyte-type genes for a limiting amount of regulatory factors may indeed be the cause of the 100-fold greater activity observed on the average) for each of the somatic-type 5S RNA genes in somatic cells, relative to each of the oocyte-type genes. This type of explanation depends on the decrease in TFIIIA concentration that occurs during embryogenesis. In the fertilized egg there are about 3×10^9 molecules of TFIIIA : in the gastrula stage embryo about 9×10^4 molecules per cell ; and in tadpole stage embryos and adult kidney about 1.2×10^4 per cell. The relative *kinetics* with which there forms either an active transcription complex with TFIIA or an inactive complex with histone may be determinant, since the somatic cell content of TFIIIA is still sufficient to bind about 1/4th of all the cellular 5S RNA genes, i.e., including the oocyte-type genes. Thus Gargiulo et.al. (1984) showed that an coinjection of excess oocyte and somatic 5S RNA genes into the *Xenopus* oocyte nucleus, the greater affinity of the somatic genes for a limiting factor, presumably TFIIIA, results in the irreversible formation of inactive histone complexes with the oocyte type genes. This result mimics the events that occur in newly formed somatic cells, and demonstrates that in the presence of histones the binding preference of TFIIIA for the somatic 5S RNA gene control sequences is apparently sufficient to account for the activation of somatic-type 5S RNA genes and the stable repression of oocyte-type 5S RNA genes. Furthermore, Brown and Schlissel (1985) demonstrated that injection of TFIIIA itself into syncytial Xenopus embryos that have been prevented from undergoing cytokinesis results in a sharp increase in the fraction of endogenous 5S rRNA transcript deriving from the oocyte-type genes, i.e., after transcriptional reactivation at the blastula stage. This experiment provides a direct indication that a determining parameter controlling the selective expression of the somatic type genes is the *limited concentration of TFIIIA* in late embryonic and somatic cells. The observation noted above that in normal late blastula stage embryos when there are still about ten molecules of TFIIIA per 5S RNA gene there is a transient reactivation of the oocyte-type 5S RNA genes is also consistent with this interpretation.

The changing prevalence of the positive regulatory factor TFIIIA also helps to explain directly the course of 5S RNA synthesis during oogenesis. A key observation made by Pelham and Brown (1980) and Honda and Roeder (1980) is that TFIIIA is the 38.5 kd protein that is complexed with the 5S RNA in the 7S particles of previtellogenic oocyte. In other words, this protein binds to both the anticoding strand of the 5S RNA gene and to the identical sequence of the RNA product. In the presence of the large excess of TFIIIA that exists in early privitellogenic stages, when there are about 10^{12} molecules of this protein per oocyte, the 5S RNA genes are all actively transcribed. However, as the available TFIIIA is progressively sequestered together with the newly synthesized 5S RNA in the 7S and 42S storage particles, a decrease of 5S RNA transcription occurs. When these particles disappear later in oogenesis the total amount of TFIIIA in the oocyte has decreased by a factor of about two. The TFIIIA gene has been cloned, and estimates of the level of TFIIIA mRNA provide by Ginsberg et. al. (1984). There are about 5×10^6 molecules of TFIIIA message in the early oocyte, classifying this transcript as a member of the low abundance sequence class. This falls to about 1×10^6 molecules by stage 4-6 of oogenesis, and to only about 1 per cell in the swimming tadpole.

In summary, the 5S RNA of the oocyte ribosomes is provided by the active transcription of about 80,000 genes in the 4C oocyte nucleus, occurring mainly during the previtellogenic period of oogenesis. The transcription of the these genes is regulated by the formation of a specific transcription complex including three protein factors plus polymerase III. As the concentration of 5S RNA builds, one of these factors. TFIIIA, becomes bound instead to the 5S RNA product. Competitive interactions between a limiting amount of TFIIIA and histones are probably involved in the regulatory events that result in the extremely accurate discrimination between the oocyte and somatic types of 5S RNA genes later in development.

Ribosomal RNA Synthesis during Oogenesis

Though the same enormous number of the large rRNAs are required to be synthesized during oogenesis as of 5S rRNA, there are typically no more than a few hundred genes for these rRNAs per haploid genome. In organisms whose oocytes contain lampbrush chromosomes the ribosomal genes are amplified early in oogenesis, and are present as extra-chromosomal circular DNA molecules that are packaged into nucleolar structures. Transcription of ribosomal DNA during oogenesis occurs in the amplified rDNA. Ribosomal DNA amplification has been observed in a number of urodele and anuran-amphibians, in teleosts, and in orthopteran insects all of which, as shown in Table utilize lampbrush chromosomes in their oocytes. Ribosomal gene amplification is also reported in coleopteran and in a few dipteran insects, which carry out meroistic oogenesis, and to a low level in an echiuroid worm and in lamellibranch molluces. As reviewed in the following section ribosomal gene amplification probably does not occur in the mouse, though it does in primates, and it is absent in those sea urchin oocytes that have been examined, as well as in some other invertebrates, e.g., the nematode *Panagrellus*.

Replication of the rDNA occurs mainly at the pachytene stage in amphibian oocytes though the process begins in the premeiotic oogonia. In orthopteran oocytes amplification also takes place during the pachytene stage. The original copies used for rDNA replication are of chromosomal origin and usually but probably not always the amplified rDNA of each original circle derives from a single genomic rRNA gene. This inference is supported by the observations that spacer sequence heterogeneity, which is observed frequently in adjacent genomic rDNA repeat units in only occasionally seen within the tandem repeats of a given *amplified* rDNA molecule. Most of the amplified rDNA is synthesized by some form of rolling circle replication of preexistent extrachromosomal rDNA circles.

In *Xenopus* oocytes the extrachromosomal rDNA formed by the amplification process is packaged in a large number of extrachromosomal nucleoli, which can be observed in the germinal vesicle sap throughout oogenesis. On germinal vesicle breakdown the extrachromosomal rDNA is released to the cytoplasm, where it remains during early embryogenesis. By the gastrula stage it has disappeared. Thiebaud (1979) showed that the number of extrachromosomal nucleoli per oocyte varies from 500 to 2500 in oocytes of the same female. The maximum number is attained at stage 4, and after this some nucleoli apparently fuse, resulting in a lower value in later oocytes. The number of rRNA genes per nucleolus also varies, ranging from 500-11,000. On the other hand, the *total amount of amplified rDNA per nucleus is always the same*, about 30-35 pg or approximately 2.5×10^5 rRNA genes. Since there are about 450 rRNA genes per haploid genome this represents for the oocyte about a 1400-fold increase in gene number. However, even after amplification a long period is required for the accumulation of the large rRNAs, and rRNA synthesis is likely among the rate limiting process in the growth of the amphibian oocyte, just as in the growth of the *Drosophila* oocyte.

In the earliest previtellogenic oocytes there is little rRNA synthesis, though extrachromosomal replication of the rDNA has been completed. Scheer et al. (1976b) showed that in the newt *Triton alpestris* previtellogenic oocytes synthesize rRNA at only about 0.01-0.5% of the rate measured in vitellogenic oocytes. This relatively low synthetic rate is correlated with sparse packing of transcripts in the extrachromosomal nucleolar genes, and with the presence of many totally inactive gene regions. In contrast, 90-95% of the nucleolar ribosomal genes are being transcribed in midvitellogenic oocyte, and the transcripts visible on these are tightly packed, with as many as 130 transcripts per gene region. From the known length of the amphibian 40S rRNA precursor, the spacing between polymerases in the extrachromosomal nucleolar genes is calculated to be only about 100 ntp.

The rate apparently varies greatly depending on the state of gonadotropic hormonal stimulation of the female. As ribosomes are a major constituent of the cytoplasm, hormonal determination of their rate of accretion, and also of the rate of yolk uptake appear to provide direct physiological controls over the kinetics of oocyte growth. The difference in synthesis rate between stimulated and nonstimulated females is due to the frequency of initiation in the active extrachromosomal nucleolar genes, and in the fraction of these genes that are transcribed at all, since hormone treatment does not affect the fundamental polymerase translocation rate. Under optimal stimulation, the rate of rRNA production in midvitellogenic oocytes is close to that expected were all of the 2.5×10^6 extrachromosomal nucleolar genes functioning at the maximum rate. Thus over 3×10^5 rRNA molecules are being synthesized per second, compared to 10-100 sec^{-1} in somatic cells. The instantaneous rate of rRNA synthesis exceeds the rate of total heterogeneous RNA synthesis in the lampbrush chromosomes, and is many times the rate of stable heterogenous RNA flow into the cytoplasm. The newly formed rRNA molecules exit into the cytoplasm through "pore complexes," which occupy 25% of the area of the nuclear membrane. From the number of these structures Scheer (1973) calculated that 1-2 molecules of rRNA pass through each pore per minute.

Net rRNA accumulation continues at least under some conditions in stage 6 *Xenopus* oocyte. Under conditions of hormonal stimulations synthesis of 40S ribosomal precursor occurs in these oocytes at an average rate about 50% of that in stage 3 oocytes. However, this rate varies greatly from female to female, controlled by unknown physiological factors. The 40S ribosomal precursor synthesized in stage 6 oocytes seems not to be processed with the normal efficiency. On germinal vesicle breakdown the precursors are released into the cytoplasm where they persist into early cleavage. In species other than *Xenopus* more sharply decreased levels of rRNA synthesis are observed towards the end of oogenesis. For example, in postvitellogenic mature oocytes of *Titon alpestris* the nucleolar genes contain only about 15% as many transcripts per ribosomal gene region as do growing oocytes, and the rate of synthesis is about 13% of that observed in midoogenesis. In mature oocytes of hibernating *Rana pipiens* at least 70% of the extrachromosomal ribosomal gene units appear completely devoid of nascent transcripts when examined in the electron microscope. In summary, the rate the rRNA synthesis on the extrachromosomal nucleolar genes increases from a low initial level to maximum at midvitellogenesis and then falls again at maturity, to an extent that varies according to species. These regulatory changes reflect large differences in the frequency of chain initiation and in the fraction of the rRNA genes utilized. In has been noticed that in either very young or mature oocytes when not all the extrachromosomal genes are functioning, adjacent transcription units may display large differences in activity. The rate of initiation of *each gene* in the tandem nucleolar arrays thus appears to be regulated independently during oogenesis.

Transcription of the ribosomal genes is controlled by two upstream regions, viz, a proximal promoter, and a set of regulatory elements located distally in the intergenic spacer. The promoter located immediately at the 5' end of the gene has been mapped by observing the effects on transcription of various 5' deletions. These experiments were carried out either by injecting synthetic deletion mutants into *Xenopus* oocyte nucleus, or by transcribing them in a cell-free polymerase I system that consists essentially of an oocyte nuclear homogenate. Transcripts of exogeneous origin were distinguished by utilization of mutant *Xenopus laevis* gene constructs in the *Xenopus borealis* transcription systems. The proximal promoter is found to extend from nt –142 upstream of the start of the transcript to nt –6. The 13 nt sequence from –7 to –6 is the major requirement for accurate transcription in injected oocyte nuclei, and this sequence element is conserved among the rRNA genes of several Xenopus species. This region is evidently directly utilized for polymerase 1 initiation. At –27 is a T_6 sequence that occurs in an analogous position to the "TATA box" of genes transcribed by polymerase II. The remainder of the promoter region is apparently required at least under conditions of limited initiation in order to obtain maximal transcription rates. There is some evidence that portions of the promoter region extending out to –142 serve as the binding site for a transcription factor. An interesting indirect light on the mechanism by which these genes are activated derives from an analysis of the effects of topological constraint on rDNA transcription. A supercoiled configuration was found to be required for continued expression of injected plasmids, but not for transcription of the endogenous rRNA genes in the oocyte nucleus. Thus on injection of restriction endonucleases, both classes of gene are digested, but while this abolishes transcription from the injected plasmids it does not effect endogenous rRNA synthesis. The torsional strain to which the supercoiled plasmids are subjected may thus mimic the effects of the assembled transcription complex on the endogenous rRNA.

A novel aspect of regulation in the tandem rRNA gene sets is the role of sequences, located between about 200 and 2500 nt upstream in the non-transcribed spacer. The importance of these spacer sequences for transcription of the rRNA genes has been shown in several ways. Moss (1983) found that in oocyte nuclear injection experiments deletion of a portion of the spacer region reduces the amount of transcription 20-fold compared to a coinjected wild-type gene, and a similar conclusion was drawn by Busby and Reeder (1983) from experiments in which deletion mutants lacking far upstream sequences were injected into Xenopus eggs, so as to observe the effect on developmental regulation in the embryo. After the midblastula transition the injected genes are activated, and deletion of the spacer sequence reduces their transcription in the embryo 5-10 fold.

Additional experiments show that when different rDNA plasmids are coinjected into the oocyte nucleus in sufficient quantity to induce competition, those plasmids containing the more extensive spacer sequences always produce the most transcript. The key spacer elements responsible for these effects are the repetitive sequences identified as the 60/81 base pair repeat. Both forms of this repeat include a 42 ntp sequence also found in the proximal promoter at position –73 to –114. The 42 ntp elements have the properties of enhancer sequences. Thus they endow a plasmid carrying them with the ability to compete successfully with a second plasmid that lacks them, and their effect occurs exclusively on genes to which they are in cis relation.

Furthermore, they operate equally well in either orientation, or several kb away from the gene. This interpretation also explains an old mystery of *Xenopus* species hybridization experiments, When X. *laevis* and X. *borealis* are crossed, the X. *laevis* rRNA genes are always utilized predominantly in the hybrid embryos, irrespective of the direction of the cross X *laevis* rRNA spacers turn out to contain more than 20 copies of the 42 ntp enhancer elements while the X. *borealis* genes contain only four. Thus "nuclear dominance" can be recreated experimentally by injecting genes from these two *Xenopus* species into oocyte nuclei of either, under conditions where competition will occur. The intergenic spacer regions also contain several nearly perfect duplications of the complete proximal promoter sequence. These upstream promoters are capable of initiating transcription themselves under certain circumstances, both experimentally and naturally occurring and they are thus to be regarded as potentially functional. However, they are followed by a "fairsafe" terminator at position –243 that prevents readthrough from them into the gene proper. The role played by the duplicated upstream promoters is not yet apparent, though it is known that the requirements for their function are subtly different than for the proximal promoter. They could serve in some way as ancillary elements necessary for the function of the contiguous enhancer sequences.

These detailed researches into the regulation of the genes that synthesized the rRNAs may provide a preview of the immense molecular complexity of the transcriptional apparatus that functions during oogenesis. Though major in amount, the 5S and 40S rRNAs are but two of the 1-2 $\times 10^4$ diverse RNA species that are being transcribed in the lampbrush chromosome stage nucleus. A lesson that we have encountered several times in this chapter, as well as earlier, is that the detailed mode of regulation is particular to each gene, or functionally related set of genes. It is likely that we can anticipate the discovery of many new regulatory structures of diverse properties when equivalent studies are carried out on individual single copy transcription units of the lampbrush chromosomes.

Nonmeroistic Oogenesis Without Lampbrush Chromosomes

In the oogenesis of many species neither nurse cells nor lampbrush chromosomes are utilized. Nurse cells and lampbrush chromosomes are best interpreted as alternative devices for the supply of a large variety of transcripts at elevated rates, as we have seen. The requirement for high synthesis rates is not an invariant feature of oogenesis, however. Such requirements do not exist in animals in which the final quantities of maternal RNAs are relatively low, and the allowed time for completion of oogenesis is sufficient. In such animals the structure of the oocyte transcriptional apparatus as well as the kinetics of RNA synthesis are very different from those we considered earlier. Yet in qualitative terms the end result, viz, the transcript pool of the mature oocyte, has the same distinctive characteristics as in the species discussed earlier. Thus where investigated, oocytes that develop without either nurse cells of lampbrush chromosomes also contain sufficient ribosomes and stored maternal mRNAs to support protein synthesis after fertilization, and at least in sea urchins, interspersed maternal poly(A) RNA as well. The characteristics of transcription in oocytes that do not utilize lampbrush chromosomes provide interesting comparisons, which illuminate those aspects of oocyte nuclear function that are intrinsic to the process of creating an egg, irrespective of any particular logistic strategy.

RNA Synthesis in Urechis and Sea Urchin Oocytes

The echiuroids represents an unsegmented lower protostome grade of organization, while the echinoderms constitute a major branch of deuterostome evolution. Yet the strategies utilized in these most unrelated creatures for the preparation of the maternal transcript pools are strikingly homologus. Although the evidence is still incomplete it appears that the oocytes of neither form posses lampbrush chromosomes. In the light microscope the diplotene meiotic prophase chromosomes of growing *Urechis* oocytes appear to assume a diffuse configuration that is difficult to resolve. However, it does not resemble a true lampbrush chromosomes structure. Electron microscope observations on chromatin spread from *Urechis* oocytes have not been reported. Lampbrush chromosomes were thought to be present in sea urchin oocytes on the basis of light microscopy. However, despite determined attempts to visualize them, typical lampbrush chromosome transcription matrices could not be demonstrated in electron microscope preparations of *Strongylocentrotus purpuratus* oocyte chromatin. Except for densely packed ribosomal gene matrices, only sparsely distributed nascent transcripts were observed. Since the prevalence of maximally packed transcription matrices provides a definitive ultrastructural characteristic for the true oocyte lampbrush chromosomes, it can be concluded that such structures are absent from the growing oocytes of at least this sea urchin species.

Both the *Urechis* egg and the sea urchin egg are relatively small and they contain amounts of rRNA that are proportional to their respective volumes. The amount of poly(A) RNA or heterogenous RNA stored in the *Urechis* egg is not reported, though from measurements both of hybridization kinetics and of the content of poly(A) tracts it would appear to be less than 1% of the total RNA. However, Rosenthal and Wilt (1986) found that most maternal mRNAs in fully grown *Urechis* oocytes lack poly(A) tracts, though after fertilization they are rapidly adenlyated. In the egg of the sea urchin *Strongylocentrotus purpuratus* there is about 30 pg of mRNA, about half of which is polyadenylated, plus approximately twice this amount of interspersed poly(A) RNA that is not translatable. The *average* prevalence of each RNA species of the heterogeneous sequence class is only 1-2 × 10^3 molecules per egg. The date is suffice to illustrate the quantitative contrast between the *Strongylocentrotus* and *Urechis* eggs on the one hand, and on the other, eggs such as that of *Xenopus*. The volume of the *Xenopus* egg is about 3000 and 1000 times greater than the volumes of the *Strongylocentrotus* and *Urechis* eggs, respectively ; and the prevalence per egg of an average complex class maternal poly(A) RNA species in Xenopus is again about 1000 times the corresponding value for the sea urchin, *Urechis*, and *Xenopus* eggs, as well as the length of the time required to complete the growth phase of oogenesis, are all very similar. A difference of about three orders of magnitude therefore exists in the quantitative demands exerted on the transcriptional apparatus in *Xenopus* compared to se urchin or Urechis oocytes.

In echiuroid worms and several other groups of marine organisms, oogenesis is of the solitary type, in which follicle and nurse cells are absent, and the oocytes develops autonomously in the coelomic cavity. In the case of labeling, accessibility of oocytes of diverse stage, and the possibility of removing and then reimplanting growing oocytes in the coelom, these forms offer some unique advantages for the study of gene activity during oogenesis, though as yet they have been relatively little exploited. For example, immature oocytes of the polychaete annelid *Platynereis dumerilli* can be transferred between the coeloms of animals of different genotype and mature eggs recovered that are able to undergo embryological development.

Ficher (1977) utilize this method to demonstrate maternal inheritance of larval eye color, which is evidently the consequence of the autonomous expression during oogenesis of genes affecting the pigmentation pathways. Measurements carried out by Lee and Whiteley (1984) on growing oocytes of another polychaete annelid. *Schizobranchia insignis*, show that the rate of RNA synthesis during oogenesis decreases sharply as the oocyte grows, and that the egg RNA appears to include high complexity components similar to those observed in other forms. In addition, a variety of molecular measurements that are directly relevant to the transcriptional processes carried out by relatively small solitary oocytes have been reported for *Urechis*.

Patterns of Transcription during Oogenesis in Urechis

The course of RNA synthesis in *Urechis* oocytes differs in several ways from that in amphibian and other oocytes that bear lampbrush chromosomes. As in the amphibians the ribosomal RNA genes of *Urechis* oocytes are amplified, but only to the moderate extent of about six times the number of rRNA genes included in the 4C chromosomal complement. The excess rDNA is located in a single very large nucleolus present in the germinal vesicle throughout oogenesis. In *Urechis* oocytes rRNA is the major transcript species synthesized all stages of oogenesis. The rate of rRNA synthesis actually increases 4-5 fold in late vitellogenic oocytes and rRNA continues to be produced, though at a lower rate, even in mature unfertilized eggs. There is no evidence for preferential accumulation of 4S and 5S RNAs in previtellogenic oocytes, as in amphibian and teleost oogenesis. These low molecular weight species are instead synthesized continuously in *Urechis* oocytes at all stages.

In *Urechis* about 8% of the growing ribosome pool is utilized for protein synthesis throughout the growth phase of oogenesis. Thus as the oocyte develops it engages an increasing mass of mRNA. Heterodisperse RNA that is detectable after 2 hr of labeling also continues to be synthesized at all stages of oogenesis, and poly(A) tracts, presumably associated with the heterogeneous RNA molecules, accumulate. Some specific examples have been reported by Rosenthal and Wilt (1986), in a study of the representation is total RNA of a series of transcripts identified by cDNA clones. Certain of these transcripts accumulate continuously throughout the period of oocyte growth, though the quantity of others, presumably those required only for translation during oogenesis, has decreased sharply by the final stage.

At least for many of those species that remain polyadenlyated throughout, this provides another contrast with *Xenopus*, where as we have seen, the final net quantity of poly(A) RNA is attained very early in oogenesis. In quantitative terms, the rate of accumulation of poly(A) RNA in the cytoplasm of Urechis oocytes is probably less than 1% of the steady state flow into the cytoplasm of newly synthesized poly(A) RNA in midlampbrush chromosome stage *Xenopus* oocytes.

Transcription in Sea Urchin Oocytes

Most sea urchins have an annual reproductive cycle and from observations on oocyte size and number during this cycle it is possible to estimate the approximate lengths off the various phases of oogenesis. Oogonial multiplication occurs a new each year. Intertidal populations of *Strongylocentrous purpuratus* spawn intermittently from late November to May or June, and in these animals oogonial multiplication takes place primarily between April and November, when the latest oogonial DNA synthesis is observed in each cycle. The number of previtellogenic oocyte increases during the winter months to a maximum value that is attained in March, and remains constant until the following September, when the first group of oocytes inters the vitellogenic growth phase. Between March and September the newly formed oocytes grow slowly in volume, from about 10-15 µm diameter to about 30 µm. Their rate of growth then increases about 2-fold, and mature oocytes 80 µm in diameter are found about 2-3 months later. Oocytes in all stages of the growth process can be observed in the ovary at any one time during the winter months. It may be concluded that the previtellogenic growth phase requires at least six months (March to September), and the more rapid vitellogenic growth phase about 2-3 months (September to November-December). Oocytes that begin vitellogenesis later in the spawning period must spend an even longer time in previtellogenic stages, though the length of the previtellogenic *growth* phase may depend largely on environmental conditions such as nutrient supply. The previtellogenic oocytes are located in the walls of the ovarian tubules, where they exist in close contact with other cells, though no specific follicular structures are evident. As the oocytes increases in size they lose their associations with the epithelial wall, and the mature and late vitellogenic oocytes accumulate in the lumen of the ovarian tubules. Both meiotic reduction divisions are completed in this location, i.e. prior to release from the ovary.

RNA synthesis has not been studied in any detail in *previtellogenic* oocytes, and it is known only that they synthesize transcripts of all classes, including low molecular weight RNAs, 18 and 28 rRNAs, and heterogeneous RNAs. Hough-Evans et al. (1979) showed that the nuclear RNA of previtellogenic oocytes has a complexity of at least 1.60×10^8 nt. i.e., close to 30% of the single copy genome. Note that this is about the same as the

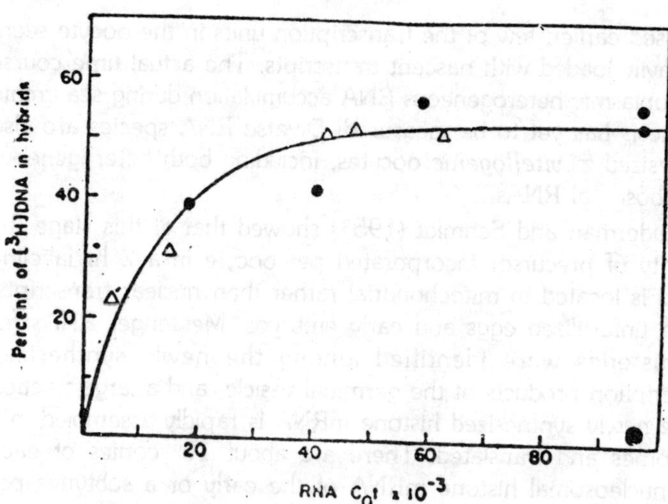

Fig. 6.1. Hybridization with sea urchin gastrula and previtellogenic oocyte nuclear RNAs of single copy ^{3}H-DNA recovered from duplexes with previtellogenic oocyte nuclear RNA. A pseudo first order function was used to fit data from the reaction of this ^{3}H-DNA with previtellogenic oocyte nuclear RNA (•), and with gastrula nuclear RNA (Δ), with the assumptions that there is a single kinetic component and that the ordinate intercept is zero.

complexity of embryo and adult see urchin nuclear RNAs. Furthermore, the particular single copy sequence set represented in the previtellogenic oocyte nuclear RNA is within the limits of experimental detection wholly represented in gastrula stage embryo nuclear RNAs. The significance of this experiment is that it shows that *most if not all the transcription units operative in midembryogenesis have already been activated at the previtellogenic stage of oogenesis.*` In the sea urchin stable heterogeneous maternal transcripts accumulate *sequentially* during oogenesis. The cytoplasmic fraction of previtellogenic *S. purpuratus* oocytes thus contains only about 45% of the ultimate maternal RNA sequence set.

The prevalence of typical maternal sequences is about 1-2 × 10^3 molecules per egg and even for sequences present at ten times this concentration less than an hear would be required for their transcription from single copy genes and accumulation, were these genes being initiated at the maximum possible rate. This calculations assumes the polymerase translocation rate measured for sea urchin embryo nuclei, about 9 nt sec^{-1}. However, as discussed earlier, few of the transcription units in the oocyte seem to heavily loaded with nascent transcripts. The actual time course of cytoplasmic heterogeneous RNA accumulation during sea urchin oogenesis has yet to be measured. Diverse RNA species are also synthesized in *vitellogenic* oocytes, including both heterogeneous and ribosomal RNAs.

Ruderman and Schmidt (1951) showed that at this stage the majority of precursor incorporated per oocyte in a 2 hr labeling period is located in mitochondrial rather than nuclear transcripts, just in unfertilized eggs and early embryos. Messenger RNAs for the histones were identified among the newly synthesized transcription products of the germinal vesicle, and a large fraction of the newly synthesized histone mRNA is rapidly assembled into polysomes and translated. There are about 10^6 copies of each core nucleosomal histone mRNA of the early or a subtypes per egg. However, Angerer et al., (1984) showed by *in situ* hybridization with early histone gene probes that a-histone mRNAs do not accumulate in vitellogenic oocytes, but appear only after maturation, when they are synthesized and stored within the egg pronucleus. Were all 400 histone genes coding for each early histone species in the haploid pronucleus as active as are these genes in cleavage stage embryos, the accumulation of $\sim 10^6$ mRNA molecules of each histone species would require less than two days. The histone mRNAs synthesized in vitellogenic oocytes are thus likely to be of the cleavage stage (CS) Type. CS histones are stored in the unfertilized egg in sufficient quantity to accommodate the DNA of a number of embryonic division cycles.

In contrast to the α-histone messages, the total poly(A) RNA, and several other specific mRNAs, are distributed throughout the cytoplasm of the mature egg, and total poly(A) RNA is accumulated both in the germinal, vesicle and the cytoplasm during the vitellogenic phase of oogenesis. It is likely, by analogy with the case in *Xenopus*, that the α-histone mRNAs are sequestered in the pronuclear compartment by some specific protein to which they are bound. Messenger RNAs for actins and tubulins are also prevalent in growing oocytes, and these proteins are actively synthesized during oogenesis. However, at maturation these proteins cease to be translated at a high rate as we have seen only a small amount of maternal mRNA coding for these proteins is stored in the unfertilized egg.

A different picture emerges in considering rRNA accumulation during sea urchin oogenesis. Measurements of absolute rRNA synthesis rate were carried out in *Tripneustes gratilla* oocytes by Griffith et al. (1981). The average rate of synthesis per oocyte is about 1.1 × 10^5 molecules or rRNA precursor per hour. The mature egg contains about 4 × 10^8 ribosomes, and thus 4 to 5 months would be required for accumulation. Since all stages of oocyte were included in these experiments, it is possible that previtellogenic oocytes synthesize rRNA less actively, and that an even longer time is necessary, e.g., the whole of the oocyte growth period. In echinoids rDNA amplification does not occur. The 4C *T. gratilla* oocyte genome contains about 200 rRNA genes and in order to account for the measured rate of synthesis initiation must occur on each rRNA gene about every 6.5 sec. These rRNA synthesis measurements lead to conclusion that essentially all of the rRNA genes are operating at close to the maximum possible rate throughout oogenesis. Thus if the polymerase were packed only 100 n apart, a translocation rate of about 15 nt sec^{-1} would be required. Aronson and Chen (1977) determined rates of 13 nt sec^{-1} for both polymerase I and II in feeding *S. purpuratus* larvae, and 6-9 nt sec^{-1} for embryos of this species, which are cultured at a temperature several degrees lower than are *T. gratilla* embryos. It follows that for sea urchins, as for other diverse creatures considered in this chapter, e.g., *Drosophila* and *Xenopus*, the rate of accumulation of rRNA may be among the parameters that determine the overall pace of oogenesis.

Seventy percent of the poly (A) RNA stored in sea urchin eggs consists of nontranslatable, interspersed transcripts, just as in the amphibian eggs. The structural characteristics of the interspersed RNAs that can be recovered from mature sea urchin eggs, and their disposition in the embryo. From their complexity it is clear that the interspersed RNAs are the stored products of a great many discrete transcription units operative during oogenesis Posakony et al. (1983) estimated that the prevalances of a set of individual interspersed poly(A) RNAs fall in the rang 10^3 to 10^4 molecules per egg, and accurate molecular titration data for two cloned examples indicate a few hundred to about a thousand copies per egg. Thus the abundance of these sequences is about 2-3 orders of magnitude lower than for the interspersed poly(A) RNAs of *Xenopus* eggs, just as are the respective abundances in these two systems of histone mRNAs. rRNAs total poly(A) RNA, or average complex class transcripts. Yet both the structure and the amount of interspersed transcripts relative to total egg poly(A) RNA is the same in the eggs of these two species. This surprising correspondence clearly does not favour the specific explanation that interspersed poly(A) RNAs are incidentally released to the cytoplasm because processing in the germinal vesicle cannot keep pace with the exceptionally high rate of transcript production, e.g., as in lampbrush chromosome. That is, the rate of heterogeneous RNA transcription in sea urchin oocytes is probably quite low. Perhaps the flow of interspersed poly(A) RNAs into the cytoplasm is a tangential consequence of a basis inefficiency of processing in the oocyte nucleus that is yet unexplained, though very general in occurrence. Or perhaps their storage in the cytoplasm in the course of two such different forms of oogenesis implies that like other maternal transcript species, these also perform a functional role after fertilization.

Transcription and the Accumulation of Maternal RNAs during Oogenesis in the Mouse

Overall Rate of Heterogeneous RNA synthesis and the Absence of Lampbrush Chromosomes

Oogonial multiplication, premeiotic DNA synthesis, and the initial stages of the meiotic prophase occur in eutherian mammals during fetal life. In the neonatal mouse the primary oocytes are arrested at early diplotene in a nongrowing stage, termed the dictyate phase. Beginning a few days after birth and throughout the reproductive life of the female small groups of oocytes periodically withdraw from the dictyate pool, and enter the growth phase of oogenesis. Dictyate mouse oocytes may thus persist in a quiescent though viable state for anywhere from a few weeks to 18 months or more, and in the human where a similar course of events obtains, for over 40 years. Once activated the process of oocyte growth in the mouse requires only two weeks, during which the diameter of the oocyte increases from about 12 μm to 85 μm. a greater than 300-fold change in volume. The maternal RNAs found in the mature egg are synthesized and accumulated during this period, and many cytological and biochemical changes that lie outside the scope of this discussion occur as well. Within several days after growth is completed the follicular structure surrounding the oocyte undergoes an enormous expansion, and the meiotic prophase terminates at ovulation with completion of the first reduction division.

Midgrowth phase oocytes denuded of follicle cells can be cultured in the presence of a feeder layer of somatic follicular cells or in excised follicles, under conditions in which growth continues at about 70% of the normal rate. The synthesis and turnover rates of RNAs synthesized in cultured oocytes that had been labeled for various periods of time were measured by Bachvarova (1981). As in *Xenopus* lampbrush chromosome oocytes, by far the major fraction of the newly synthesized heterogeneous RNA is unstable, displaying a typical nuclear RNA half-life estimated at about 20 min. The data shows that less than 2% of the heterogeneous RNA synthesized is destined to become stable transcription products. This value is similar to those estimated from the similar comparisons for amphibian oocytes. The low ratio of stable RNA transcript production to total heterogeneous RNA transcription is thus clearly not dependent on the overall *rate* of chromosomal RNA synthesis, but is rather an intrinsic aspect of transcription and processing in the oocyte nucleus. The major result is for our purpose the *low absolute rate* of heterogeneous RNA synthesis. The comparable rate was estimated as 18.3 pg min^{-1} in the lampbrush chromosome stage Xenopus oocyte, which is over forty times the rate measured for the mouse oocyte. Yet these organisms have about the same haploid genome size, approximately 3 pg.

A light and electron microscope investigation of the cytological state of mouse and Rhesus monkey oocyte chromosome during the growth phase was carried out by Bachvarova et l. (1982). At light microscope magnification the oocyte chromosomes appear fuzzy and diffuse, with central proteinaceous cores surrounded by loose bundles of fibrils. In basic organization these structures bear no homology with the lampbrush chromosomes of amphibian oocytes. The key observation is that when examined in the electron microscope the transcription units of growing mouse oocyte nuclei are generally populated only sparsely with nascent RNA molecules. As discussed earlier, this provides direct evidence that true lampbrush chromosomes are absent where the rate of heterogeneous RNA transcription is relatively low.

Lampbrush chromosomes are to be considered the cytological manifestation of a condition in which the maximal transcriptional initiation rate is established in essentially all active regions of the oocyte genome. The complexity of the stored maternal RNA has not been measured for any mammalian egg.

If it is assumed to approximately that of sea urchin and *Xenopus* maternal poly(A) RNAs, and thus that something over 104 diverse transcription units are represented the rate of synthesis of 0.005 pg min^{-1} per nucleus for stable poly(A) RNA implies than the average no more than one or a few nascent transcripts would be found on the active genes of the growing oocyte. Thus 0.005 pg min^{-1} would represent a production of about 5000 molecules min^{-1} would represent a production of about 5000 molecules min^{-1} per nucleus, assuming a mean length of 2000 nt for the poly(A) RNA distributed among at least 40,000 transcription units in the 4C oocyte nucleus. It follows that initiation would occur on each productive transcription unit, on the average, only once every eight minutes, and thus for a 37°C polymerase translocation rate on the order of 20-50 nt sec^{-1}, the nascent transcripts would be spaced 1-2.5 $\times 10^4$ nt apart. This result predicts exactly the low transcript frequency, though of course were processing very inefficient, or if there are transcription units that produce RNAs wholly confined to the nucleus, the transcript density in some regions could be markedly higher.

Accumulation of Stable Maternal RNAs

The general pattern of RNA synthesis appears not to change during the growth phase of oogenesis. A number of investigations have shown that all stages mouse oocytes synthesize 18 and 28S rRNAs. 5S rRNA, tRRNAs and poly(A) RNAs, and that throughout, the ratio of thee various components in the newly synthesized RNA remains constant. An early growth stage characterized primarily by synthesis of tRNA and 5S RNA thus does not exist the mouse, which in this respect resembles the sea urchin and *Urechis* examples, and differs from those vertebrates that utilize lampbrush chromosomes. By day 14-15 the oocyte has accumulated its final maternal complement of rRNA, about 300 pg, as well as of stable poly(A) RNA, though its volume is only 5--60% of that of a mature 21 day oocyte. Though it continues to grow and to accumulate protein for the following 7 days, 14-15 oocytes are already functionally mature, in the sense that they have attained competence to resume the meiotic division process RNA synthesis, but no accumulation, also continues in late oocytes, and becomes undetectable only at germinal vesicle breakdown and meiotic maturation.

For the first 9 days of the growth phase the rate of rRNA synthesis is about 2/3 of that shown in Table or about 0.01 pg min^{-1}, while the rate of 0.015 pg min^{-1} obtains between days 9 and 14-15 Ribosomal gene amplification has not been reported in the mouse, though a fourfold rDNA amplification has been noted both in diplotene baboon oocytes and in human oocytes. However, ribosomal gene amplification is not required to account for the measured rate of rRNA accumulation in mouse oocytes. Thus Kapian et al. (1982) calculated that the rate of 0.15 pg mig^{-1} represents the synthesis of about one rRNA molecule per 15 sec for each of the 1200 genomic rRNA genes present in the 4C oocyte nucleus. This rate is below that measured in somatic mouse tissue culture cells.

The mouse egg is unusual in the large quantity of maternal poly (A) RNA that it contains, relative to total RNA. Bachvarova and De Leon (1980) labeled oocytes at all stages of the growth period by injection of radioactive precursor into the bursal sac, and on recovery of matured eggs from 4 to 19 days later found that about 8.3% of the mass of labaeled RNA retained is poly(A) RNA. The length of the interval between injection of precursor and collection of ova did not affect this result. Calculations based on the mass of poly(A) per egg indicate the quantity of stored poly(A) RNA is about 25 pg. or 5.5% of the total RNA. This represents about 2.4 × 10^{7} poly(A) RNA molecules of average length. Since the bulk of the RNA accumulating during growth is rRNA, which is completely stable, the close agreement between the calculated mass of poly(A) RNA and the fraction of total egg RNA labeled during oogenesis that is poly(A) RNA shows that the poly(A) RNA synthesized in the oocyte is largely stable as well. This has also been demonstrated directly in pulse-chase experiments carried out on cultured follicles by Brower et al. (1981). As it accumulates the poly(A) RNA is deposited in the cytoplasm of the growing oocytes, according to *in situ* hybridizations carried out ^{3}H-poly(U).

About half the poly(A) RNA accumulated in the course of oogenesis disappears during the maturation period, though it is not clear whether in general it is degraded or is only deadenylated. The oocyte acting messages, however, have been shown to be deadenylated rather than degraded during maturation though total stable RNA content does decline during this period by about 20%. In any case, the fraction of *newly synthesized* oocyte RNA that is poly (A) RNA is about twice the fraction of mature, unfertilized egg RNA that is poly(A) RNA. Thus about 18-20% of precursor incorporated in the relatively stable transcripts of *growing oocytes* is found in poly(A) RNA. The newly synthesized poly(A) RNA on the oocyte distributes into two functionally distinct compartments. About 23% is assembled onto polysomes and utilized directly for oocyte protein synthesis, and the half-life of this fraction is about 6 days, according to measurements carried out on cultured oocytes in contact with follicle cells or in whole follicles. The remainder is stored, and it displays no detectable turnover. The pool of poly (A) RNA present in the oocyte at termination of growth is thus the sum of the steady state quantity of polysomal poly (A) RNA, and the quantity of nontranslated poly (A) RNA that has been synthesized and accumulated during oogenesis. Given that the rate of synthesis of total poly (A) RNA is 0.005 pg min^{-1}, the steady state amount of polysomal poly (A) RNA about 78 pg for a total of 92 pg a fifth of all the RNA in the oocyte. This value may be compared with the equivalent fraction in the mature *Xenopus* oocyte, which is about 2.2%.

Protein Synthesis during Oogenesis in the Mouse

The absolute rate of protein synthesis increases proportionately with oocyte diameter from shortly after the beginning of the growth phase to its termination. Protein synthesis rate measurements were obtained on cultured oocytes, and are based on determination of the methionine pool specific activity. The maximum rate of synthesis is about 42 pg hr^{-1} in full grown oocytes of about 80 μm diameter, which is close to forty times the rate observed in 10-15 μm oocytes. Of the total newly synthesized protein only 1-2% represents proteins encoded by the 10^5 mitochondrial genomes present per egg. De Leon et al. (1983) showed that throughout the period of oocyte growth about 35% of the ribosomes are engaged in polysomal structures. The remainder may be aggregated in large assemblages utilized for storage of ribosomes in an inactive form, as they can be sedimented from homogenates at very low centrifugal forces. Assuming a ribosomal translocation rate of 3 codons sec^{-1} per ribosome (value for 37°C vertebrate systems range from about 2-8 codons sec^{-1} per ribosome ; the rate of protein synthesis measured in fully-grown oocytes is close to that expected, providing that 35% of the ribosomes are included in the translocational apparatus. However, this rate of translation is 4-8 fold *lower* than expected for the steady state quantity of *polysomal* poly (A) RNA, i.e., 14 pg. The major cause would seen to be a low rate of translational initiation, since the ration of the mass of mRNA to the mass of polysomal rRNA is about four times greater than the 4% value obtained for fully-loaded polysomes (i.e., 14 pg mRNA 0.35 × 300 pg rRNA=0.13. The rate of ribosomal translocation may also be depressed about 2-fold, caveat pointed out by Kaplan et al. (1982) is that the protein synthesis rate measurements of Schultz et al. (1979b) were carried out under culture conditions inadequate to support net growth, and this could conceivably have adversely affected the translational activity of the oocytes. In any case, assuming the rates measured, it can be calculated that if all the protein synthesized during oogenesis were stably accumulated, this would account for only about 60% of the 25 ng of protein in the fully-grown oocyte. This is in fact an overestimate, since only about 60% of the newly synthesized protein is stable. The oocyte this probably absorbs the balance of its protein from the blood, by endocytosis, just as yolk is taken up in lower vertebrates.

Qualitative analysis by two-dimensional gel electrophoresis has indicated that most of the ~400 species resolved are synthesized throughout oogenesis, and that the newly synthesized species are the same as the species detected by sensitive staining procedures. However, there are a few specific proteins whose rate of synthesis increases sharply in fully-grown oocytes. At meiotic maturation, the overall pattern of protein synthesis changes noticeably, and these changes are programmed at a posttranscriptional level since there is little new mRNA synthesis at this time. The stored maternal mRNA accumulated during oogenesis does not appear to be utilized until maturation, when a qualitative pattern of protein synthesis is established that persists until well after fertilization.

Synthesis of a number of known protein species has been studied in mouse oocytes. In Table are presented synthesis rate data for several examples. Though the estimates for the number of functioning mRNA molecules are of limited accuracy they suffice to indicate the minute accumulations of mRNA in the mouse oocyte compared, for example to the specific message contents in *Xenopus* oocytes. These estimates do not include *stored* maternal mRNAs coding for the same protein species. However, in all the cases shown, the rate of synthesis *falls* rather than increases during maturation, when the maternal mRNA begins to be utilized. The histones and ribosomal proteins, which were also considered in respect to protein synthesis in *Xenopus* oocytes, again provide interesting contrasts. In the mouse, the 60 pg of core histones accumulated during oogenesis would suffice for the chromatin of only about 10 cells. As will be recalled, mature *Xenopus* oocytes contain histone protein sufficient for over 10^4 cells. The leisurely pace of cell division in the mouse imposes no urgency on new histone synthesis, and whereas in *Xenopus* there is a 50-fold increase in the rate actually declines about 40% during maturation. Thus it is unlikely that there is a very large store of maternal histone mRNA, such as is found, e.g. in both sea urchin and *Xenopus* eggs.

An interesting aspect of the data included in Table for ribosomal proteins is that many of these are encoded by mRNAs that are distinctly of the rare message class. For example. LaMarca and Wassarman (1979) showed that synthesis of several individual ribosomal proteins represents less than 10^{-4} of total protein synthesis, and on the basis of the calculation in Table there would be no more than about a thousand molecules of the messages for such individual species functioning in the oocyte. The quantity of each functional ribosomal protein mRNA species in the mouse oocyte is thus equivalent to <10 molecules per typical somatic cell (i.e. normalizing 1.3×10^7 polysomal poly(A) RNA in a somatic cell, $\sim 10^5$ molecules).

The molar rates of synthesis of the individual ribosomal proteins differ by as much as a factor of four. However, this imbalance is compensated by a differential accumulation of newly synthesized molecules of the various species in the germinal vesicle, so that within this compartment an equimolar ratio is established. The total amount of newly synthesized protein that is distributed to the germinal vesicle is about 0.9 pg hr^{-1} in the full-grown mouse oocyte, and of this histone accounts for about 10% and the ribosomal proteins about 25%. After maturation ribosomal proteins continue to be produced in the absence of any new rRNA synthesis. It is not clear whether the maternal messages utilized for this synthesis are the same as were already functioning in the polysomes prior to maturation, or alternatively, are recruited from the previously inactive stored poly(A) RNA pool.

Three secreted proteins, the major constituents of the zona pellucida, account for at least 10% of the newly synthesized proteins in growing oocytes. The zona pellucida is the thick glycoprotein coat that surrounds the oocyte, and its constituents are synthesized exclusively during the growth phase of oogenesis. Denuded oocytes actively incorporate amino acids or labeled fucose into these proteins. An O-linked oligosaccharide component of these glycoprotein species, designated ZP3 (~83 kd) apparently serves as the specific sperm receptor molecule. A second protein, ZP2 (~120 kd), which constitutes over 50% of the total zona pellucida protein, is involved in the hardening of the egg coat on fertilization. This process is mediated by proteolytic enzymes released from cortical granules, just as in the sea urchin. The messages for these proteins must be relatively very prevalent, and the genes for ZPI-3 are likely to be transcribed much more intensely than the genes coding for typical maternal poly(A) RNAs required after fertilization.

A complementary example has been described by Cascio and Wassarman (1982). This is a set of proteins designated the fertilization proteins. FP1-4 are synthesized at very low levels, each accounting for about 0.05% of total protein synthesis in growing oocytes, and FP5 and FP6 are not detectable at all. However, after fertilization. FP1-6 become major species, together representing 3-5% of the total protein synthesis in one and tow cell embryos. The dramatic change in their rate of synthesis which occurs as a result of maturation and fertilization. The fertilization proteins are among those coded primarily by the stored, *inactive* poly(A) RNA of the oocyte. This is demonstrated in the cell-free translation experiment. The partition between polysomal and stored poly(A) RNA in the oocyte thus encompasses some sharp *qualitative* delineations, illustrated respectively by the zona pellucida proteins and the fertilization proteins.

7

Ascidian Embryology

All the solitary ascidians are hermaphrodites. The solitary ascidians commonly used for study of developmental biology, such as *Ciona*, *Ascidia*, *Phallusia*, *Styela*, *Boltenia*, and *Halocynthia*, are oviparous, but viviparous forms are found in *Corella* and *Molgula* species. Fertilized eggs of the viviparous species develop in the cloacal cavity. Many ascidians have distinct breeding seasons, though there is no special season in the genera *Ciona* and *Molgula*; they become sexually mature and breed particularly all year around.

Primordian Germ Cells

Little is known about the origin of germ cells in solitary ascidians, although in the colonial ascidian *Botryllus* it has been demonstrated that certain kinds of coelomic cells or *hemoblasts* give rise to primordial germ cells. In *Ciona* presumptive primordial germ cells first become distinguishable in the coelom of juveniles at 2-3 days after the initiation of metamorphosis. The germinal epithelium in 6-7-day-old *Ciona juveniles* contains small cells with electron opaque cytoplasm, called *clear cells* by Sugion, Tominga, and Takashima (1987), and cells with electron-dense cytoplasm (dark cells). Recently, an antibody has been obtained that specifically recognizes both young oocytes and spermatocytes of *Ciona* juveniles. Immunological studies with the electron microscope to elucidate the origin of germ cells, with the aid of the antibody, have shown that the antigenicity is detected in the dark cells, suggesting that the dark cells are primordial germ cells.

Oogenesis

Growth and Maturation of Oocytes

Many solitary ascidians, *Halocynthia roretzi* for example, show distinct, seasonal changes in germ-cell *proliferation* and *maturation*. During the inactive period soon after the end of the spawning season, the ovary is occupied by oogonia and small oocytes. Toward the onset of the next spawning season and/or the growth of juveniles, the oocytes begin growth. The oocytes become covered with two layers (outer and inner) of follicle cells, and yolk granules appear in the oocyte cytoplasm. Then test cells appear, as though they are embedded in the oocyte cytoplasm, and finally the vitelline coat or chorion appears, to separate the follicle layers and test cells. Several solitary ascidians that the commonly used for studies of developmental biology. Depending on the species, the sizes of eggs vary from 100 to 500 mm in diameter usually around 170 mm. Detailed descriptions of the size and colours of the eggs of various ascidian species are available in a review by Reverberi (1971).

The eggs is enclosed in a tough, noncellular vitelline coat. On the outer surface of the vitelline coat are attached many follicle cells, within the perivitelline space between the egg and the vitelline coat there are many test cells; the existence of these test cells within the perivitelline space is a unique phenomenon in the animal kingdom. In this species, oocytes grow surrounded by inner and outer follicle cells with the germinal epithelium. At the time when oocytes are delivered from the germinal epithelium into ovarian cavity, the outer follicles rupture to release the oocytes, which are surrounded by, from proximal to distal, test cells, vitelline coat, and (inner) follicle cells. The outer follicle cells are left behind at ovulation. Detailed description of ultrastructural changes in developing ascidian oocytes, with careful consideration of the functions of organelles, are provided in a review by Kessel (1983) and the references cited there. According to him, a characteristic feature of young *Ciona* and *Molgula* oocytes is an active nuclear-cytoplasmic exchange through the nuclear pores. Granular masses of high electron density migrate from the nucleus to the cytoplasm the extruded cytoplasmic masses of granules frequently are associated with mitochondria.

These are similar to the so-called *nudge* that has been described in oocytes of a great many animals. The quantity of these granular masses is greatest in young oocytes, and the masses subsequently undergo fragmentation or dissolution, so that the material generally is not longer evident by the time vitellogenesis is initiated. In young oocytes, rough endoplasmic reticulum is actively formed from the outer layer of the nuclear envelope. Annulate lamellae that are thought to be cytoplasmic stocks of porous cytomembranes are common in the ascidian oocyte at the stage of porous cytomembranes are common in the ascidian oocyte at the stage of *previtellogenesis* and at *vitellogenesis*. Intranuclear annulate lamellae have also been reported in ascidian oocytes. The mitochondria are widely distributed within the oocyte cytoplasm. A quantitative relationship is noted between mitochondria and lipid droplets through much of the period of oocyte growth.

Vitellogenesis is mainly attributed to the function of *Golgi complexes*. The Golgi complexes consist of variable number of cisternae that often are arranged in stacks. The Golgi components increase in number as the oocytes grow and become widely distributed throughout the ooplasm. *Pinocytotic* vesicles are found to be added to the growing yolk granules later in vitellogenesis, because *micropinocytotic* pits and vesicles associated with the oolemma and cortical ooplasm are often observed in vitellogenic oocytes. Further, micropinocytotic vesicles of variable sizes, shapes, and internal densities often can be observed in abundance surrounding immature yolk platelets. Therefore, in addition to autosynthetic vitellogenesis, heterosynthetic mechanisms may function as other elements are incorporated into the oocyte from sources perhaps external to the ovary, and these components are mixed with those synthesized in the oocyte itself.

The phelobobranch ascidians, such as *Ciona*, have long genital ducts, and the ripe gametes can be seen within the separate male and female ducts. The *Ciona* oocyte begins its maturation divisions within the ovary, and then immediately moves into the oviduct, where it remains for 24 h or longer. In the oviduct, meiosis proceeds to metaphase of the first maturation division. The first polar body, however, is not ejected until after fertilization. Therefore, in the case of phelobobranch ascidians, the ripe gametes usually are obtained by cutting the gonoducts. On the other hand, the stolidobranch ascidians and *Corella* usually do not store ripe gametes in the genital ducts. In such species, naturally spawned eggs usually are used for investigations. Illumination after the critical time period of dark adaptation is convenient method to obtain ripe gametes. The fully grown ovarian oocytes of *Styela plicata*, *Styela clava*, *Halocynthia roretzi*, and *Ciona intestinalis* will mature in seawater; if the ovary is cut and the fully grown oocytes are removed into seawater, germinal-vesicle breakdown will occur within a few hours.

Origin and Function of Accessory Cells

As mentioned earlier, an ascidian egg is covered by two completely different types of *accessory cells* (follicle cells and test cells) and by the acellular vitelline coat that exists between them. Recent studies of material functions for *Drosophila* early embryogenesis have disclosed a significant role for follicle cells and nurse cells in the establishment of the egg axis that is associated with embryonic pattern formation. This suggest the need for careful reinvestigation of the function of ascidian accessory cells with respect to the pattern formation of the embryo.

There has long been controversy regarding the origins and functions of accessory cells. There are three hypotheses for the origins of *germ cells* and *accessory cells*. First, according to Julin (1983), Tucker (1942), and others, the eggs, *follicle cells*, and *test cells* all arise from undifferentiated cells of the germinal epithelium. They believe that oogonia and primary follicle cells are first formed from undifferentiated cell. The primary follicle cells then differentiate further into inner and *outer follicle layers*, and the cells of the *inner follicle layer* finally differentiate into inner *follicle cells* and *test cells*. The vitelline coat is formed by oocytes to separate the follicle cells and test cells. The second hypothesis, proposed by Knaben (1936) in *Corella parallelogramma*, suggests that the egg originates from the germinal epithelium, whereas follicle cells and test cells arise from *amoeboid cells* or *coelomic cells* of mesenchyme origin. The amoeboid cells penetrate into the ovary, surround the oocytes, and then differentiate into the follicle cells of *C. intestinalis* are derived from the germinal epithelium, whereas test cells are formed from amoeboid cells. This disagreement may be due to species differences. However, recent studies using autoradiography and *transmission electron microscopy* (TEM) have demonstrated that in the *Ciona* ovary, cells of the inner follicle layer differentiate into the inner layer of follicle cells and test cells.

Follicle cells

The morphology of *follicle cells* varies considerably from species to species, but follicle cells usually are vacuolated. Although several functions have been suggested for follicle cell, it seems difficult, at present, to attribute general functions to the cell. Follicle cells of *C. intestinalis* produce chemotactic substances of sperm. Follicle cells of *Corella inflata* are as large as 65 mm in diameter and contain a considerable amount (~350 mM) of ammonium ion. When *Corella* eggs are removed from follicle cells, the eggs that had been floating will sink to the bottom of a culture dish, suggesting a role for the follicle cells in egg flotation. In *Molgula pacifica,* follicle cells each contain a single, large adhesive vacuole. Shortly after spawning, these vacuoles rupture, causing the follicle cells to secrete a sticky mucus coat that causes the eggs to adhere to the substratum. In addition, recent studies have suggested a significant role for follicle cells in fertilization, which will be discussed later.

Test cells

The functioning of test cells is not completely understood either. Several functions that have been suggested are to nourish the oocyte to provide the oocyte with yolk precursors to produce pigment to transmit vanadium compounds to the oocyte to produce hatching enzyme and to function in the formation of the larval tunic. A series of electron microscopic studies of *Styela* oocyte formation has suggested that cells produce pigment granules, which are translocated to peripheral cytoplasm of oocyte. The test cells contain abundant protein and acid mucopoly saccharide; the secretory function of test cells has been reported by several authors.

In *Ascidia* synthesis is same a layer of test cells, attaches to the entire surface of the egg. After fertilization, they leave the egg surface and are scattered within the inner wall of the vitelline coat, On the other hand, fertilization in *A. anodori* does not affect the test-cell location. During Halocynthia embryogenesis, test cells show a unique behavior: Communicatively with the formation of tailbud embryos, test cells that had been scattered within the inner surface of the virelline coar move to attach the embryo extending and shortening pseudopodia, the test cells creep around the outer surface of the developing embryo, suggesting a role in the larval test formation. In many species the outer surface of the larval tunic formation, Cloney suggested that the test cells make the tunic hydrophilic. The presence of the accessory cells mentioned earlier mandates caution in biochemical and molecular biological studies.

First, it is know that radioactive tracer chemicals that are applied for analysis of protein or nucleic acid synthesis frequently become tapped by the follicle cells However, follicle cells can easily be removed from oocytes by pipetting them in Ca^{2+} -free seawater or Ca^{2+} -free seawater containing ethylenediamineteraacetic acid (EDTA). Second, it is estimated that single oocytes of *Ascidia nigra* and *C. intestinalis* contain approximately 200–300 and 1,000–1,200 tet cells within the perivitelline space, respectively. Therefore, smash-and-spin experiments using eggs and embryos with intact accessory cells do not always reflect what happens with the egg proper Removal of test can be achieved only after dechorination. *Dechorination* can be performed manually with sharp steel needles or chemically with the tadpole larva's hatching enzyme with a 0.1–3.0% solution of proteinase in seawater or with seawater containing 1% proteinase and 0.055% sodium thioglycolate, at high pH.

Spermatogenesis

Development of the testis is first indicated by the formation of *seminiferous tubes*, *spermatogenesis* proceeds in the tubule wall. The primary spermatocyte has periodic-acid-Schiff-positive (PAS-positive) cytoplasm and conspicuous nucleoli in the nucleus. Concomitantly with the growth of juveniles, spermatogenesis proceeds to form the primary and secondary *spermatocytes*, and then *spermatids*. The seminiferous tubes enlarge, and mature sperm appear to occupy the central region of the tube. The tubule lumen leads to the sperm duct, which runs alongside the oviduct and opens into the atrial cavity. Since the pioneer work of Retizus (1904, 1905), ascidian sperm have been investigated at the levels of the light microscope (e.g., Franzen, 1956) and the electron microscope. These studies of the structure and function of ascidian sperm have been summarized in recent reviews.

The sperm of solitary ascidians are similar in structure (e.g., Franzen 1956). Each is usually 50–60 mm long and consists of a short head (3–9 mm) and a long filamentous tail (48–53 mm). The tail contains a single axoneme exhibiting the usual 9+2 pattern of microtubules. The *axoneme* of the tail issues from the single basal body that adheres to the posterior tip of the nucleus. There is no paired proximal *centriole*; instead of it, dense material is located on one side of the basal body of the nucleus. As first described by Retizus (1904, 1905), ascidian sperm have no middle piece, the piece that connects the sperm head to the tail in other animals.

The head region is more or less elongated and contains the prominent nucleus and a single mitochondrion. The nucleus extends the entire length of the head. Nuclear pores are present apically, although in its lower two-thirds the *nuclear envelope* is devoid of pores. The mitochondrion is a broad, flattened structure that generally wraps about half way around the nucleus. The plasmalemma in the head of the sperm is covered by a prominent glycocalyx. In many species there is a structure called the "apical wedge" at the apical region of the head; as the nucleus and head taper toward the apical end, they form a wedge shape. In this region, the membranes of the nuclear envelope seem to fuse, forming a thick pentalaminar structure called the dense plate. It has been suggested that these plates function as sites for chromatin organization and play a role in the organization of apical vesicles that originate from the Golgi apparatus.

Nearly all marine invertebrates have an acrosome, a Golgi-derived, membrane-bound vesicle in the anterior sperm head that fuses with the sperm plasmalemma during fertilization, releasing its contents. There has long been disagreement regarding the existence and function of acrosomes in ascidian sperm; see De Santis and Pinto (198), Lamber and Koch (1988), and Fukumoto (1990a) for recent reviews. Many researchers have questioned the presence of an acrosome in fully differentiated ascidian spermatozoa, but others have insisted on its presence. It was in 1980 that Cloney and Abbott clearly demonstrated a membrane-bound vesicle in the anterior tip of the head of *A. callosa* sperm; they referred to it as the putative acrosome. Thereafter, it became the consensus that structurally the ascidian spermatozoon has an *acrosome*, albeit a small one; the existence of *acrosomes* has been confirmed in all of the 26 species of ascidians examined in the orders *Pleurogona* and *Enterogona*. During spermatogenesis, small vesicles appear in the blister at the anterior region of the spermatid and fuse to form an acrosome during further differentiation.

Spermatogenesis of ascidians has been studied by several investigators. Woollacott (1977) has studied spermatocytes of *C.*

intestinalis by freeze fracture methods. In the cytoplasm, a Golgi complex is apically situated. A crescentic structure is formed by the Golgi membranes, and numerous vesicles are seen situated within the concave surface. A well developed *Golgi complex* has been reported in spermatids of *H. roretzi* the complex forms an electron-dense vesicle in spermatids, but in mature sperm this vesicle becomes undetectable from the anterior part of the nucleoid. In early spermatid stages of *Corella parallelogramma* the nucleus has scattered chromatin. During spermiogenesis the nucleus goes through a series of processes that result in elongation and condensation of the chromatin into fibers and lamellae. Concomitantly with this process, a number of microtubules lie near the nuclear envelope; the occurrence of microtubules seems a rather common phenomenon in ascidian spermiogenesis. The mitochondria begin to form a single large mitochondrion at an early stage of spermatogenesis in this species. The ultrastructural features of sperm of colonial ascidians were reported by Tuzet et al., (1972, 1974).

Fertilization

Sperm approaching the egg are trapped in or on the egg envelope as a result of their randomly oriented directions of swimming. In ascidians it has been suggested that there may be a factor that acts as a sperm attractant; follicle cells produce attractive substances that induce sperm movement toward the egg. Species specificity of sperm chemotaxis exists in many ascidians species, but low levels of cross-specific are more general. A role for egg jelly coat substances in sperm activation, as demonstrated in sea urchin eggs, has not yet been shown in ascidian eggs.

In order for the sperm to reach the egg surface, it must penetrate several barriers of the egg vestments; usually this strip is quite species-specific. As explained previously, ascidian eggs are enclosed in a vestment complex that includes a layer of follicle cells surrounding the fibrous vitelline coat, a *perivitelline space*, and a layer of test cells on the surface of the oocyte. In addition, because ascidians are hermaphroditic and usually spawn eggs and sperm simultaneously, the gametes of self-sterile species must recognize self and non-self. The sperm first encounters a layer of follicle cells. The bases of follicle cells are associated with one another in polygonal fashion, and the boundaries between the cells are not continuous, but are marked by narrow clefts. The sperm passes through clefts at the bases of the follicle cells; then it binds to glycosides of the *vitelline-coat* surface exposed between the follicle cells. The sperm then penetrates the vitelline coat, leaving the single mitochondrion bound to the outside of the vitelline coat. Mitochondrial translocation is thought to be responsible for movement of the sperm from the *vitelline-coat* surface to the oocytes surface; see Lambert and Koch (1988) for a review. After its passage through the chorion, the sperm enters inside the perivitelline space, which contains the test cells, to reach the egg surface.

The time required for these events was first reported by Conklin (1905a) to be several minutes, in contrast to the few seconds needed in the case of the sea urchin. Using the sodium dodecylsufate (SDS) sperm-inactivation method, Lambert and Epel (1979) and Lambert (1986) showed that fertilization occurs in more than half of Ascidia eggs by 1 min. The first indication of sperm attachment upon the egg plasma membrane is the appearance of a fertilization potential and associated inward fertilization current. The current is due to the activation of relatively nonspecific fertilization channels around the point of sperm entry and reaches a peak after 30. Glycosidase activity is also released a few seconds after fertilization. A waste of elevated calcium concentration becomes detectable about 20 after the start of the fertilization current then a cortical contraction wave spreads across the egg surface.

During the period when the sperm moves from its point of entrance toward the posterior side of the egg, changes occur in the sperm head to form the sperm *pronucleus*. After extrusion of the second polar body, the egg pronucleus moves downward to fuse with the sperm *pronucleus*. The two pronuclei come into contact in the vegetal hemisphere. The synkaryon thus formed then shifts to the center of the egg. Before syngamy, a well-defined aster is formed at one side of the sperm pronucleus. Before the synkaryon divides, this aster cleaves, the ant the two resulting asters become located at the two poles of the first cleavage spindle.

Sperm Binding to the Vitelline Coat

After passing through the barrier of the follicle cells, the sperm binds to the surface of the vitelline coat. Sperm binding is distinguishable from lose attachment by *centrifugation* or *pipetting*, which removes any sperm attached to the egg covering.

It has been suggested that the sugar moiety on the vitelline coat play s a role in *sperm binding*. In sea urchin eggs, sperm binds to fucose sulfate residues on the vitelline coat via a protein called *binding*, which is exposed during acrosome reaction at the tip of the sperm. In mammals, a *sperm-surface* glycosyl transferase and zona-pellucida glycosides play important roles in sperm binding to the zona (e.g.., Wassermann, 1987). Involvement of such a form of enzyme-substrate complex has also been shown in sperm binding to the vitelline coat of ascidian eggs.

The *vitelline coat* of *Ciona* eggs contains fucose residues and sperm binding is prevented by *L-fucose*, but not by other sugars such as *D-fucose*, mannone, N-acetygalactosamine, and N-acetyglucosamine. The process is also very efficiently inhibited by fucose glycosides, and α-L-fucosidase has been demonstrated in Ciona sperm. Hoshi (1984, 1985) suggested that α-L-fucosidase molecules at the sperm surface bind to terminal fucose residues of sperm receptors in the vitelline coat. The α-Lfucosidase has been purified from Ciona sperm; its size has been estimated to be 105 kDa by SDS-polyacrylamide-gel electrophoresis (SDS-PAGE), or 112 kDa by high-performance liquid chromatography (HPLC), and its pH to be 5.3, with the optimal pH being 3.1. The enzyme, having an acid pH optimum, would form a stable enzyme-substrate complex not hydrolyzed at the alkaline pH of seawater.

On the other hand, in *Phallusia mammillata*, *Ascidia nigra A. collasa*, and *A. ceratodes*, N-acetyglucosamine (NAG) is likely to be involved in sperm binding to the *vitelline coat*. Fluorescent wheat germ agglutin (WGA) binds to the vitelline coat of eggs in these species, demonstrating the presence of NAG or sialic acid on the surface of the vitelline coat. In addition, sperm binding is inhibited by WGA and NAG glycosides, but not by the free sugar. Sperm N-acetyglucosaminidase has been demonstrated in *A. ceratodes* and *A. paratropa*. Recently, Godknecht and Honegger (1991) have isolated and characterized *P. mammillata* sperm N-acetyglucosaminidase; this enzyme is a dimer of molecular size about 158 kDa. In addition, they have shown that the dimer of molecular size about 158 kDa. In addition, they have shown that the enzyme is present at the sperm tip, over the mitochondrion, and at the sperm head-tail junction and that it is indispensable for fertilization of intact eggs.

Mitochondrial Movement during Sperm Reaction

A characteristic behaviour of the ascidian sperm mitochondrion is known as the *sperm reaction*. During *sperm penetration* through the *vitelline coat*, the mitochondrion remains outside the *vitelline coat* as the sperm slide past it and into the *perivitelline* space; in face, only sperm deprived of this organelle can be found. Lambert and Epel (1979) showed that this process can be performed in vitro. That is, mitochondrial translocation is a sperm property and might be responsible for generating the force to drive the sperm through the vitelline coat into the perivitelline space.

Mitochondrial translocation is independent of sperm motility, because the translocation occurs in the presence of methyl cellulose, a complete inhibitor of sperm motility, whereas, N-ethlmaleimide or cyclic adenosine monophosphate (cAMP) inhibits mitochondrial translocation without affecting sperm motility. Calmodulin inhibitors block the mitochondrial translocation, indicating that the movement is triggered by calcium through interaction with calmodulin. In fact, calcium is taken up during activation by the sperm cells and is released from internal stores as well. The actual machinery for the driving force is supposed to be actin and eosin; localization of these proteins has been shown, by indirect immunofluorescence, on the surface of the mitochondrion. The argument for the presence and functions of actin and myosin in the mitochondrial translocation is also supported by high-voltage electron microscopy and electron microscopic immunological studies.

Acrosome Reaction

As explained earlier,, the existence of acrosomes in ascidian sperm is certain at the ultrastructural level, although their function is still a matter of controversy. Fukumoto (1988, 1990a,b) inferred that the acrosome reaction in *C. intestinalis* occurs on the surface of the *vitelline coat* or after passage through the vitelline coat into the perivitelline space, although the exact site of the *acrosomal reaction* remains to be determined. During the reaction the acrosomal outer membrane fuses with the plasmalemma enclosing the acrosome, resulting in exocytosis of the acrosomal substance. It has been proposed that the fuzzy extracellular material (surface ornamentation) at the tip of the sperm head in ascidians is the site where the lysins are found and that it plays an important role in sperm–vitelline-coat interactions at fertilization. According to

Fukumoto (1990a), in the perivitelline space, apical processes protrude from the apex of the sperm head. Gamete fusion occurs between some of these processes and egg plasmalemma resulting in incorporation of the sperm into the egg from the anterior tip of its head, in the same way that it occurs in other marine invertebrates. Therefore, fertilization in *C. intestinalis* has characteristics in common with fertilization processes in both mammals and marine invertebrates.

Sperm Penetration of the Vitelline Coat

The vitelline coat is a *noncellular* coat. Ultrastructurally, the vitelline coat of Ciona eggs appears as a single, 46–85-nm thick network of interwined fibrous material having tufts that project between the follicle cells as its outer surface. In *Phallusia* the vitelline coat is 150-200 nm thick and is made up of three layers; the central layer is the most dense and contains randomly spaced electron dense particles. Evidence has been accumulating to suggest the participation of trypsin- and chymotrypsin-like proteases and high-molecular-weight multicatalytic proteinase (proteasomes) in the processes of ascidian fertilization. *Lleupeptine* (an inhibitor of trypsin) and chymostain (an inhibitor of chymotrypsin) block fertilization of *Halocynthia* intact eggs, although neither inhibitor has any effect on fertilization of investment-free eggs. Using Peptidyl-e-methylcoumaryl-7-amides as substrates, the presence of a trypsin-like enzyme and a chymotrypsin-like enzyme has been demonstrated in halocynthia sperm; both enzymes have strong substrate specificity and chymotrypsin-like enzyme have been isolated from crude extracts of freeze-thawed *H. roretizi* sperm.

More refined purification procedures revealed two acrosin-like fractions, one being *homologous* to mammalian acrosin (hence called *acrosin*), and the other having novel characteristics, and being spermosin. For each enzyme a preferred artificial substrate was determined; each substrate was subsequently shown to interfere with fertilization of intact eggs. Chymotrypsin-like enzyme has also been purified from *Ciona* sperm; it is also involved in sperm penetration of the vitelline coat. Partially purified chymotrypsin-like enzyme is thought to be a kind of multicatalytic proteinase (*proteasome*), and in combination with acrosin it performs some lytic activity on the vitelline coat. A multicatalytic proteinase has also been isolated from Halocynthia eggs. Because chymostatin blocks cleavage, the egg *proteasome* may function in some process for the intimation of cleavage. Involvement of egg prolyl endopeptidase has also been suggested for this process. Fertilization of Halocynthia eggs is followed by an expansion of the vitelline coat. Trypsin-like enzyme and calmodulin have been shown to be involved in this expansion process.

After penetration through the vitelline coat, sperm swim in the perivitelline space prior to the final approach to the egg plasma membrane. The structure of the perivitelline space, however, has not received investigative attention commensurate with its importance in fertilization. In *Phallusia* the vitelline coat expands immediately after removal from the oviduct, resulting in an increase in the *perivitelline space*. Although this space appears to be fluid-filled, it is not structureless, as demonstrated by the central location of the egg in most species even after mild centrifugation. Furthermore, in most members of the genus *Ascidian*, where the *test cells* are adherent to the egg surface, a delicate enclosure can be seen in living eggs, but it has not yet been identified in electron micrographs.

Self-sterility and Self-fertility

As mentioned earlier, mature adults usually release eggs and sperm from the gonoducts nearly simultaneously. This may be advantageous for enlargement of the population, because simultaneous release of gametes of both sexes increases their opportunities to meet each other. However, this situation may raise a question regarding the mechanisms by which the eggs must distinguish their own sperm from those of other individuals in other to avoid self-fertilization. The problem of *self-sterility* and *self-fertility* seems quite complex. Many solitary ascidians, including *Styela plicata*, *Boltenia villosa*, and *H. roretzi*, are self-sterile. On the other hand, some are completely self-fertile; eggs of *Ascidia callosa* and *Phallusia mammillata* can fuse with their own sperm and undergo normal embryogenesis. Other species, such as *Ciona intestinalis* and *Styela partira*, are incompletely self-sterile.

It is well known that T. H. that T.H. Morgan carried out a series of experiments using mainly *C. intestinlis* to elucidate the mechanisms of *self-sterility* about 40% of individuals are not *self-sterile*, about 45% of them show low levels of self-fertility (1–10% of eggs), and the remaining 15% show high levels of self-fertility (90–100% of eggs). In Nepales, about 15% of the Ciona natural population can self-fertilize. In Kochi, Japan, about 20% of naturally spawned *Ciona* eggs are self-fertile, although the ratios fluctuate considerably with various spawnings. Morgans (1939) also reported that treatment of eggs with HCl-seawater for 2-5 min increases the rate of self-fertilization and that naked eggs (i.e., these deprived of all their investments) show nearly 100% self-fertilization.

Recently, the question of self-sterility in *Ciona* eggs has been investigated again by De Santis, Rosati, and their colleagues. Eggs deprived of follicle cells are fertilized by non-self sperm, but not by self sperm, whereas naked eggs without the vitelline coat accept not only non-self sperm but also self sperm and give rise to normal embryos. This result suggests that the vitelline coat has a key responsibility for self-recognition. Autologous sperm do not bind to isolated and glycerinated vitelline coat, but heterologous sperm do. A difference in ability to blind to the vitelline coat between *autologous* and *heterologous* sperm was confirmed by Kawamura et al. (1987). In addition, they showed that concanavalin A (Con-A) can modify these self-specific and non-self-specific sperm-egg recognitions; pretreatment of eggs with con-A facilitates attachment of sperm to the autologous vitelline coat.

Kawamura et al. (1991b) attempted partial purification of the components responsible for self–non-self recognition activity. *Ciona* eggs of self-sterile individuals were extracted with acid seawater, and the extract was found to contain self-non-self recognition activity. This activity was heat-stable and insentive to trypsin, but it was destroyed by V8 protease from Staphylocccus aureus and a-glucosidase. The activity is found in both the hydrophobic and hydrophilic components. On thin-layer chromatography, the hydrophobic components gave a major spot of glucose (Glc) and peptide spots containing mainly glutamic acid and/or glutamine (Glx). The glucosyl conjugate was purified by HPLC and was shown to block sperm-egg binding to various extends. Individual peptide subfractions had to inhibitory activity, but in combination they showed inhibitory activity. These findings suggest that the acid extract of *Ciona* eggs contains a Glc-enriched nonspecific inhibitor of sperm–egg binding (which could be the primary effector of self-incompatibility) and Glxen riched modulators (which serve as acceptor of allo-sperm). The cooperative interactions of these components may be responsible for the diversity of allo-recognition in *Ciona* gametes.

A significant role for follicle cells in fertilization has been suggested in other ascidians. For example, the egg of *Ascidiella aspersa* has large follicle cells and floats. Naked eggs of this species sink to the bottom of a culture dish and do not accept any sperm. Removal of follicle cells form *H. reretizi* and *V. villosa* eggs destroyed their fertilizability, irrespective of self and non-self. Fuke (1983) removed follicle cells from eggs of two Halocynthia individuals, A and B, and costructured chemiric A eggs with B follicle cells and B eggs with A follicle cells. She found that he former became fertilizable with B sperm but not A sperm, whereas the latter was fertilizable with A sperm but not B sperm, suggesting that the *vitelline coat* is responsible for self and non-self recognition and that *follicle cells* are indispensable for sperm *penetration*.

The helper function of follicle cells in sperm–egg interactions has also been shown in *Ciona*; removal of the cells reduces the number of sperm that bind and penetrate the *vitelline coat* (Kawamura et al., 1988a). In A.nigra, removal of the *follicle cells* depresses the rate of fertilization. A recent study by De Santis and Pinto (1991) has shown that ovarian oocytes accept self sperm for fertilization. Self-discrimination, which occurs on the egg vitelline coat, is likely to be established in late oogenesis and is contributed by or controlled by products of the overlying follicle cells.

Block to Polyspermy

Because sperm are usually released first, ascidian eggs are often released into dense masses of sperm, both from the same individual or other individuals in the population. Inspite of that, polyspermy is rare. In order to avoid polyspermy, ascidians have evolved effective mechanisms for ensuring *monospermic* fertilization. *A. nigra* both have effective blocks to *polyspermy* that form rapidly after fertilization. A major constituent of this block is rapid modification of the *vitelline-coat* surface so that supernumerary sperm fail to bind. This is the result of the egg releasing N-acetyglucosaminidase into the seawater in response to fertilization, which reduces the ability of the vitelline coat to bind sperm. *A nigra*, *A. certodes*, *A. mentula*, and *P. mammillata* eggs all release N-acetyglucosaminidase at fertilization, and this release is Na^+ -dependent. However, the block forms so rapidly the glycosidase release is not likely to be the entire picture. Perhaps there are other processes operational that are yet to be discovered.

Artificial Activation and Cross-fertilization between different Species

Lauric acid, urea, thymol, hypertonic seawater and other substances that induce *artificial activation* of sea urchin eggs are not able to activate ascidian eggs. *Artificial activation* of the ascidian can be achieved by treatment with the calcium ionophore A23187 or with the lectin con-A and WGA. Bevan et al. (1977) induced artificial activation, assessed by polar body formation, by treatment of naked Ascidia eggs with theophylline, as well as artificial seawater in which Na^+ was replaced by Ca^{2+}, suggesting that an increase in the intracellular concentration of Ca^{2+} triggers egg activation. Eggs artificially activated with these agents form polar bodies and perform the ooplasmic segregation, but usually do not divide.

Cross-fertilization has been extensively studied among *C. intestinalis*, *A. mentula*, *A. malaca*, *Ascidiella aspersa*, and *P. mammillata* by Italian ascidiologists. According to them, successful cross-fertilization usually required removal of the chorion, although they found that interspecific hybridization between *Corella inflata* and *Corella willmeriana* could be readily produced in the laboratory. Eggs that had been aged for more than 24 h after *dechorionation* showed successful cross fertilization more frequently than did those tested immediately after *dechorionation*. Cytological examinations, however, revealed that those hybrids usually were not real, but rather the result of diploid *gynogenetic* development. The capacity of *Ciona* sperm to develop Ascidia enucleated fragments or merogons was further examined; the cleavage timing of the andromerogon hybrids was of the maternal type. Some of the hybrids developed into swimming larvae, which in turn failed to complete metamorphosis.

One of the most intriguing *cross-fertilization* is that between species of indirect developer (*Molgula oculata*) and direct developer (*M. occulta*), which will be discussed later.

Polarity of Sperm Entry

Since Conklin's famous description of ascidian embryogenesis, it has generally been accepted that sperm tend to enter near the vegetal pole of the egg. However, subsequently it has become clear that animal fragments as well as vegetal fragments of unfertilized eggs can be fertilized. In addition, recent studies on the mechanisms of *ooplasmic* segregation, which will be discussed later, suggest that irrespective of the point of sperm entry, the myoplasm moves toward the vegetal pole during the first phase of ooplsmic segregation, eventually translocating the sperm nucleus to the same region of the egg. Speksnijder, jaffe and Sardet (1989b) have recently demonstrated in denuded eggs of *P. mammillata* that the sperm shows a string tendency to enter the animal *hemisphere* rather than the vegetal hemisphere. Within a few minutes after during the cortical contraction that accompanies the first ooplasmic segregation.

Activation Wave of Calcium and Egg Cytoplasmic Contraction

It is well established that fertilization results in a large transient increase in the concentration of free calcium ions in the egg initiate the physiological and structural changes associated with development; see Jaffe and Gould (1985) for a review. Free calcium has been shown to increase dramatically soon after insemination of *P. mammillata* and *C. intestinalis* eggs. This increase occurs from a resting level of about 90 nm in unfertilized eggs to a peak level of about 7 mm in *Phallusia* and 10 mm in *Ciona*, over a time period of 2-3 min. This is immediately followed by a series of 12-25 brief calcium transients, with peak levels of 1–4 mm. These calcium transients stop at about 25 min, as soon as the second polar body is formed. The bulk of the calcium seems to be released from internal stores. These calcium increases with a peak velocity of 8-9 mm/s. The first few pulse start in the animal hemisphere, whereas the later ones are mostly initiated near the *vegetal pole*. Some 30-40s after the calcium wave starts, a slow (1.4 mm/s) wave of cortical contraction begins near the animal pole. It carries the subcortical cytoplasm to a contraction pole near the vegetal pole.

8

ECTODERMAL DERIVATIVES

There are three germinal layers—the ectoderm, the mesoderm, and the endoderm. These layers are formed during the pregastrular process and give rise to all body organs of the related organism. We are taking the various derivatives of these germinal layers are by one. By the time the primitive body form develops, the neural ectoderm segregates from the epidermal ectoderm by a process called *neurulation*. It lies along the dorsal surface of the embryo as a primary organ rudiment, the neural tube, where it undergoes extensive differentiation. Epidermal ectoderm forms a continuous covering over the surface of the cylinder-shaped embryo. This ectodermal layer is only one cell thick in such animals, as the shark, chick, pig, opossum, or human, but is two cells thick in the fish and amphibian. The one-called epidermis soon divides into two layers so that the epidermal layer of all vertebrates eventually consists of two layers: an inner one, the *stratum germinativum*, that has the ability to proliferate, and an outer transitory, protective layer referred to as the *periderm*. The periderm first appears in the human embryo at $1^1/_2$ months of age. In general, the ectodermal layer gives rise to the skin.

The vertebrate skin, including all its accessory structures, is defined as the *integument* or covering of the organism. Its functions are obvious, since it separates the internal environment of the individual from its surrounding environment. It protects the body from physical injury, acts as a barrier to microorganisms, insulates the body and aids in the maintenance of its temperature. It may also act as an organ of excretion and respiration. It helps the individual to find food, or a mate, and to escape from predators by means of mechanoreceptors, radioreceptors, and chemoreceptors located either over the entire integument or in localized areas. As the members of the various chordate classes adapted to the diverse niches available to them, the adult integument also changed. It varies from a one-cell-thick covering in amphioxus to a stratified layer of several cells in those terrestrial animals in which desiccation presents a threat to survival. The multilayered skin of land animals is more impermeable than that of animals bathed by an aquatic medium.

In any case, the surface cells of the vertebrate skin are not exposed to the environment, but are protected by some kind of seal. A cuticle or a layer of mucus covers the body of aquatic animals; keratin, deposited in the cells of land animals, protects the surface of terrestrial forms. The accessory structures of the integument develop mainly from ectoderm, but often a papilla of mesoderm is present. Such accessory structures include unicellular and multicellular glands, scales, nails, beaks, horns, antlers, feathers, and hair and reflect the adaptation of the organism to its environment. This chapter is limited to a discussion of the skin and its accessory structures. Specialized structures such as the lens of the eye, the cornea, and certain cranial ganglia also form from epithelium. Thickenings, called *placodes*, develop into the organs of special senses: the olfactory organ, the ear, and the lateral-line organs. Epidermal ectoderm also gives rise to the lining of the orifices of the body and to some of the structures located within the cavities.

After the formation of primary organ rudiments in the embryo, the mouth and anus form. The ectoderm gives rise to a depression, the *stomodeum*, in the anterior region of the embryo near the cranial tip of the notochordal process. When the stomodeum contacts the endoderm, or oral plate forms and soon ruptures, giving rise to a mouth and an oral cavity. In the human embryo this occurs about the twenty-sixth day. After the rupture of the oral plate, it is difficult to determine the depth of stomodeal ectoderm penetration. Since it is almost impossible to distinguish the exact boundary line between ectoderm and endoderm, the specific embryonic origin of some of the structures located in the oral cavity cannot be described with certainty. In general, most embryologists agree that ectoderm contributes to the lips, cheeks, gums, teeth, tongue, palate, and salivary glands, although the endoderm may also contribute to these same structures. Since these organs are usually discussed in most anatomy books as part of the digestive system, for the sake of convenience and conformance with the generally accepted policy. In some holoblastic vertebrates a part of the blastopore becomes the anal opening.

In others, it forms in a similar fashion to that of the mouth.. The skin ectoderm in the posterior region forms a depression, referred to as the *proctodeum*. A proctodeal plate results when the cells forming the depression come in contact with the hindgut endoderm. The further development of this plate in the mammal is intimately related to the division of the cloaca into the dorsal rectum and ventral urogenital sinus. Eventually the plate breaks down, establishing continuity between those structures derived from the cloaca and the outside. It is as difficult to distinguish germ layer boundaries in the formation of the cloacal cavity as it was with the mouth. The ectoderm, however, is acknowledged to contribute to the lower part of the anal canal and to the terminal portions of the genital and urinary tracts. Since the development and evolution of the cloaca is usually considered a part of the urogenital system.

Skin

Development of Skin

The adult skin is composed of three layers: a superficial stratified epithelium, the *epidermis*; a dense fibrous *dermis*; and beneath these layers, a loose, fatty *subcutaneous layer*. The skin has a dual origin and any discussion of it must include not only its ectodermal but its mesodermal component as well. The stratum germinativum forms directly from ectoderm and gives rise to the epidermis, but the dermis differentiates from a layer of mesodermal tissue.

The origin of the dermis is not always clear but seems to vary, depending on the species involved. It may arise from head mesenchyme, from dermatome, possibly from neural crest cells, or from mesenchyme derived from the more lateral and ventral somatic mesodermal regions of the body. In general, the vascular dermis supports, cushions, and nourishes the avascular epidermis. Neural crest cells, considered ectodermal in origin, invade the epidermis and dermis and give rise to certain pigment-containing cells referred to as *chromatophores*. The formation of the integument is an excellent example of tissue interaction during development since the ectoderm depends on the mesoderm for the direction of its differentiation.

If the epidermal cells are separated from the underlying mesodermal cells and grown in tissue culture, the germinativum layer loses the ability to divide. If dermis is then added to the culture, cell proliferation begins again. When undifferentiated epidermis is combined experimentally with different types of dermal cells, it will respond according to the type of dermis present. For example, it is possible to induce beaks, feathers, scales, or mucus secretions to form from the epidermis by placing the appropriate dermis adjacent to the maturing epidermal cells. Thus, the dermis determines whether the epidermis clears and forms the cornea of the eye, becomes keratinized and gives rise to the skin, or whatever. However, the presence of mesoderm is probably not the only answer to epidermal differentiation. Further studies indicate that some kind of substratum also appears necessary for epidermal development.

On the other hand, although it needs the contact of a substratum on one of its surfaces, the outer surface of the ectoderm must be free of cellular contact for it to develop normally. The differentiation of vertebrate skin is a very complex subject and involves many factors, some of which we do not understand at the present time. As is true of most organs, at an early stage of development the ectoderm has greater developmental potentiality than at later stages. It acquires specialized characteristics about the time of the gastrulation and begins to show regional specialization.

The fates of the different areas become definitely determined at various times depending on the organ and the species involved. Usually this occurs soon after gastrulation, when new interrelationships established by the tissues play a vital role in their further organization. As is true for most inductive interactions, we do not understand the exact mechanisms by which the various dermis influence the epidermis. Presumably the dermis acts either by affecting the genes of the ectodermal cells directly or by influencing the cytoplasm of the epithelial cells. In any event, in vitro studies demonstrate that *isolated* epidermis eventually becomes necrotic (dies)

Evolution of Vertebrate Skin

The skin of amphioxus is an example of a chordate covering in its simplest form, probably representing the state of the early chordates living 400 million years ago in the Silurian period. It is similar to the skin of some invertebrates. It differs from the skin of all other chordates since the ectoderm in the adult animal develops into a single layer of ciliated columnar cells. These cells rest on a thin layer of mesodermal tissue, the basement membrane or cutis.

A non-cellular cuticle containing many pores and presumably secreted by the epidermal cells covers their outer surface. Amphioxus has only a very thin dermal layer and no epidermal derivatives. This thin barrier, no doubt, allows some exchange of materials between the blood and the aquatic medium. There is a large gap in our knowledge between the present day protochordates, of which amphioxus is an example, and the earliest known vertebrate fossils, the ostracoderms, since the ostracoderms were heavily armored with bone, a dermal (mesoderm) derivative. One theory postulates that the invertebrates of the period, the eurypterids, preyed upon the tiny ostracoderms, and the presence of a bony armor offered the ostracoderms definite survival value. A second theory attempts to associate the presence of a bony armor to a seal of the body.

The earliest vertebrates left the saltwater of their origins and invaded the fresh water streams and ponds where absorption of water became a problem unless kept out by some kind of a seal. The kidneys of these animals, evolved to handle saltwater, are believed to have been incapable of coping with the hypotonic pond water. The bony armor of the ostracoderms may have helped in keeping the pond water on the outside of the animal. The dermal portion of the integument gave rise to bony plates of the ostracoderms and placoderms and should not logically by discussed in a chapter devoted to ectoderm. However, it is impossible to gain an insight into the structure of the vertebrate skin without some knowledge of its evolutionary development. This, of course, includes a consideration of the mesodermal component, the dermal contribution to the integument, so important during the early days of vertebrate history. Many biologists consider the presence of a bone-filled dermis in fossil vertebrates to represent the vertebrate primitive state.

The mesodermal component of the dermis of these animals is apparently formed bone, a type of connective tissue, which was covered by a thin layer of epidermis. Later in evolutionary history, as the kidneys and an impervious skin evolved along with a biting mouth, the animal changed from a sedentary to a predacious animal. The lack of a cumbersome armor and an increase in motility were a definite advantage. Mesenchyme cells are capable of differentiating into a variety of tissues. Under normal circumstances they give rise, not only to bone, but also to different types of connective tissue. It is not a great change, therefore, for mesenchyme to stop forming bone and start forming dermis. For example, instead of producing a bony matrix in which salts are deposited, the mesenchyme probably differentiated into a tissue whose matrix lacked a deposition of salts.

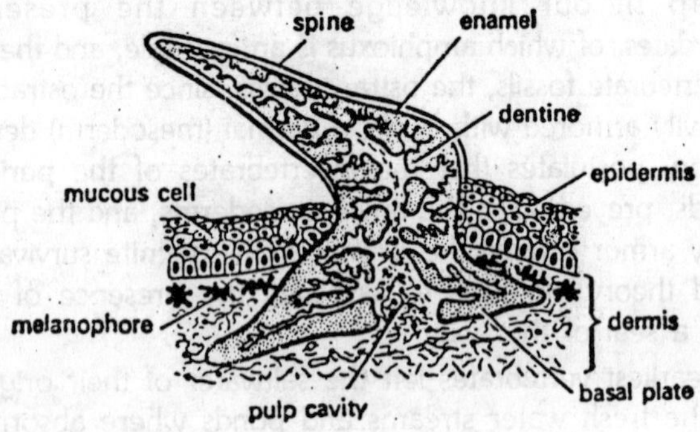

Fig. 8.1. V.S. of skin of Shark.

As a result, the dermis evolved consisting of fibres of different types intertwined in different degrees of denseness and producing the leathery types of integument associated with modern animals. Dermal degeneration (that is, the absence of bone in the integument) is almost complete in modern animals. The skin of most cyclostomes lack bone, and only dermal denticles, toothlike projections known as placoid scales, remain on the surface of sharklike fishes. These scales projecting through the epidermal layer. Bony fishes retain dermal scales, but they are, lost in most land animals, with the exception of the abdominal ribs (gastralia) of lizards, crocodiles, and sphenodon, the bony plates of the turtles, and the flat bones of the vertebrate skull. In summary, when one considers the integument from an evolutionary standpoint, a decrease in ossification and a decrease in thickness of the skin occurs as anamniotes evolved. On the other hand in land forms the epidermal layer and especially the dermal component increase in thickness. These changes prevent the evaporation of water and help to maintain body temperature.

Basic Integument

The basic integument of vertebrates consists of an epithelium of several layers (a stratified epithelium) resting on a dermal layer. It ranges from the simple, thin skin of the cyclostomes to the thick hides of mammals. In cyclostomes, all cells remain alive covered by a thin, nonliving cuticular layer. On the other hand, in mammals several layers of cells compose the skin, with the outer layer consisting of dead keratinized cells. In fish the basal layer (stratum germinativum) of the epidermis contains three different cell types. They differentiate into mucus-secreting cells (the major cell type), club cells, and granular cells containing some keratin filaments. However, the stratum germinativum of most vertebrates above the fish appears to be morphologically uniform and to consist of only one type of cell. It gives rise to cuboidal cells that are the only cells in the epidermis to undergo mitosis.

Cell division occurs in the stratum germinativum either continuously or periodically in all vertebrate classes. It may go on at a fast or slow rate; in either case, after they divide, the daughter cells follow one of three alternatives. Either they both migrate to the surface of the integument, both remain in the basal layer where they act as reserve cells giving rise to future skin cells, or one daughter cell remains and one moves toward the surface. During their migration to the surface of the integument, the cells usually become keratinized, a process that involves the stepwise deposition in the cell of a number of fibrous proteins all loosely referred to as keratin.

Keratinization occurs when the proteins are cross-linked through disulfide bridges: however, because of its insoluble properties, the characterization and identification of the specific protein composition of keratin is incomplete. The cells of fish lack morphologic keratinization, but keratin is present in amphibians and is most pronounced in land animals : the reptiles, birds and mammals. In the human embryo it occurs by the fourth month of pregnancy. As a result of keratinization, the cuboidal cells flatten, the nucleus degenerates, the cell loses its capacity to divide, and the fibrous keratin replaces the cytoplasm. X-ray diffraction patterns that reflect macro-molecular organization distinguish between two types of keratin, alpha or beta. For example, keratin that has the beta pattern composes avian feathers, but the cells of mammalian hair, stratum corneum, and nails are filled with keratin of the alpha type. The fully keratinized cell consists of not much more than a tough cell membrane filled with an abundance of filaments. It acts as a barrier, keeping such materials as water inside the body and preventing other substances from entering.

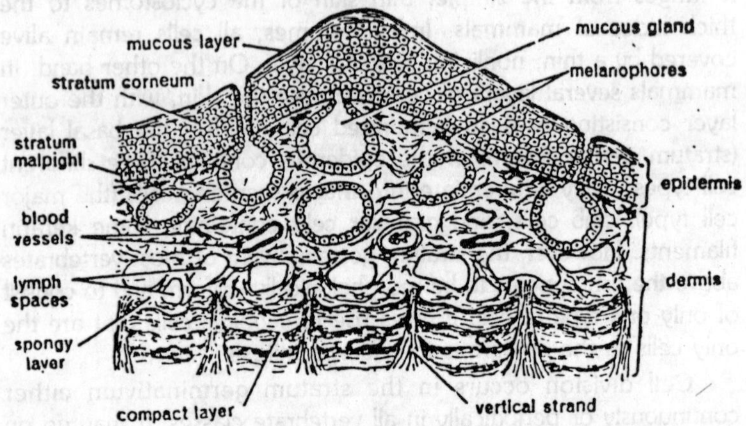

Fig. 8.2. V.S. of skin of frog.

It is resistant to abrasive substances. As a result of keratinization, a layer of dried, horny cells known as the *stratum corneum* covers the surface of the skin. These cells are constantly lost (desquamated). Replacement of the cells composing the epidermis, lost either through wear and tear or actual shedding, occurs continuously. Either the epidermis sloughs off in the form of small fragments, such as dandruff, or the loss may involve practically the entire outer covering of skin, scales, feathers, or hair. If the loss is extensive. It is usually referred to as a *molt*. The periodic shedding of integument allows the animal to change from a heavier to a lighter coat or from dull to bright plumage, or merely to repair those cells that have worn away. Sometimes the cells are shed periodically and are associated with seasonal and reproductive cycles.

Genetic and neuroendocrine factors interact in the animal's response to environmental stimuli. In some animals, such as the amphibian, the thyroid gland plays an important role in molting.

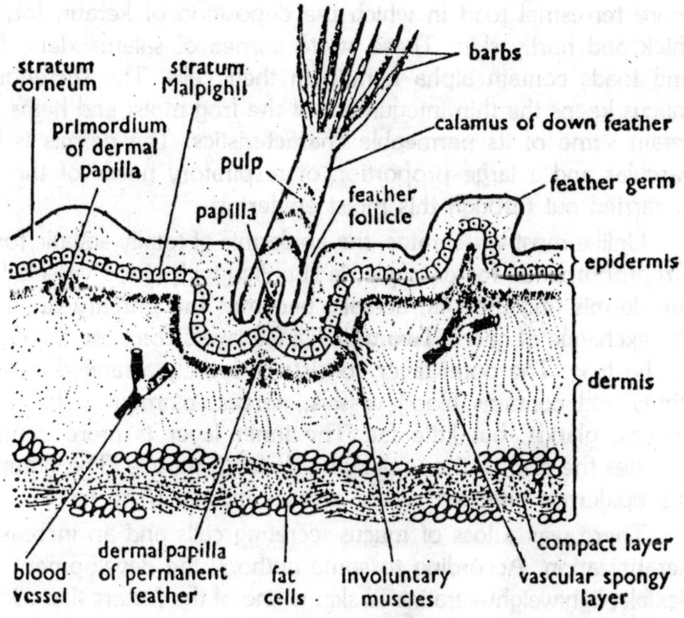

Fig. 8.3. V.S. skin of a bird.

Also, the gonadal hormones may act in a synergistic fashion with other hormones to influence growth of the integument or its derivations. Progesterone, for example, stimulates molting in birds. The integument of the vertebrate represents one of the few areas of the body in which growth and cellular differentiation continue into adulthood. The epidermis of the amphibian, unlike that of the fish, contains only one type of differentiated cell. It produces both mucus and keratin filaments. After cell division as the cells migrate from the basal layer to the surface of the skin, they secrete and release mucus into the intercellular spaces. Along the way to the surface of the animal, however, the mucus-secreting cells transform into horny cells filled with keratin. Although bony scales covered the body of the earliest amphibians, the labyrinthodonts, most living forms lack scales.

Only in the Apoda do remnants of these ancestral scales remain. Instead in modern amphibians, those that are primarily land dwellers, a dead outer *stratum corneum* appears, an adaptation to terrestrial life. The integument of the amphibians varies from the aquatic *Necturus*, in which a cuticle covers the skin, to the frog in which a few keratinized cells appear, to the more terrestrial toad in which the deposition of keratin forms a thick and horny skin. These strata cornea of salamanders, frog, and toads contain alpha keratin in their cells. The secretion of mucus keeps the thin integument of the frog moist and helps it to retain some of its permeable characteristics. The dermis is very vascular and a large proportion of respiratory needs of the frog is carried out through this moist epidermis.

Unlike most vertebrates, the epidermis of totally aquatic forms, *Cryptobranchus* for example, is supplied with blood vessels from the dermis and the skin actually becomes the primary organ for the exchange of gases. Two well-defined layers compose the dermis of the frog. The outer layer consists of loosely arranged collagen fibres and contains blood vessels, nerves, pigment cells, lymph spaces, glands, but little fat. The inner layer is more compact and ties the skin to the underlying skeleton muscle. The nature of the epidermis changed as the reptiles moved on to land.

There was a loss of mucus-secreting cells and an increase in keratinization. According to some authors, the development of a flexible, lightweight keratinized skin is one of the factors that allowed aquatic animals to evolve into terrestrial forms and attain the large size of many species. The keratinized scales of the reptilian skin provides a flexible covering that protects the animal against abrasion and desiccation and aids in locomotion. The skin allows for growth, since it is periodically shed and replaced. The main disadvantage of keratinized scales, however, is that they do not control the loss or gain of heat. This problem was not handled until the evolution of feathers and hair, both epidermal structures.

The epidermis of reptiles and birds, in general, consist of a living germinative layer, an intermediate zone of cells and an outer covering of flat, dead cells, the *stratum corneum*. The dermis layer, upon which the epidermis rests, varies in thickness, depending on the species. The integument of reptiles and at least one bird, the ostrich, is thick enough that it can be prepared for leather. The epidermis of mammals may contain as many as four different layers depending on the location in the animal. In man, the skin of the soles of the feet and palms of the hands is much thicker and exhibits all four layers, as compared to the thinner skin covering the eye lid. A vertical section through the thick part of the epidermis of man shows the first (or deepest) layer to consist of the cuboidal to columnar cells of the *stratum germinativum*.

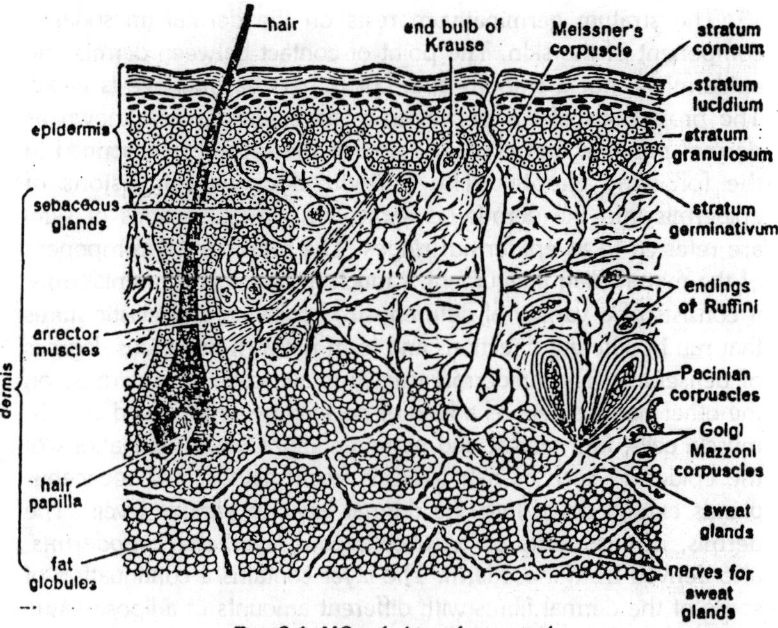

Fig. 8.4. V.S. of skin of mammal.

These are associated with another layer of cells that have short processes or spines that meet and to end and are attached by desmosomes, plate-like thickenings on opposing cell surfaces.

As mitosis continues in the stratum germinativum, the cells move outward and form the next general layer, the *stratum granulosum*. This layer is three to five cells thick and may contain granules of keratohyalin, a substance whose function has not been agreed upon by authorities in the field. As the cells continue to push outward, the *stratum lucidum* appears. It consists of several layers of flat cells that have lost their nuclei and contain droplets of material known as eleidin, believed by some biologists to be modified keratohyalin granules. This layer is present in the thicker regions of the skin but may be entirely lacking in the thinner portions. Several layers of flat cells that lack nuclei compose the surface layer or *stratum corneum*. The loss and replacement of these keratin-filled cells occur constantly. The outer surface of the epidermis is not smooth but contains grooves and ridges that create patterns; the most well-known are responsible for the fingerprints. These patterns are so variable that they can be used to identify the individual.

The stratum germinativum rests on the dermal (mesoderm) component of the skin. The point of contact between dermis and epidermis is not a smooth one, but rather the interface is away. The projections of the dermis into the epidermis are known as *dermal papillae* and contain many of the receptors described in the following chapter. The periodic downward invasions of epidermis into the dermis (as shown in vertical sections of skin) are referred to as *epidermal ridges*. The mesodermal component of the mammalian integument is much thicker than the epidermis, it consists of a variety of different-sized collagen and elastic fibres that run in different directions and form network of various degrees of denseness. The epidermis usually is avascular; the dermis, on the other hand, serves as a kind of packing tissue for blood vessels, muscle cells, fat, nerves, and nerve ending. Certain derivations of the epidermis such as hair follicles, sweat glands, and sebaceous glands extend down into the dermal layer to various levels. The dermis, in turn, rests on a subcutaneous layer, the *hypodermis*, also derived from mesoderm. The layer contains a continuation of some of the dermal fibres with different amounts of adipose tissue (fat) and muscle. The hypodermis anchors the integument in a loose fashion to the muscles, bones, and tendons of the body.

Epidermal Derivatives

Any discussion of the integument must include not only a consideration of the skin but of its derivatives as well. These structures can be roughly divided into three groups. Some form solely from epidermis although the mesoderm usually plays an inductive role in that some sort of papilla or core is present. Some of the more major epidermal structures are hair, feathers, horns, reptilian scales, scutes, beaks, claws, nails, chromatophores, and glands. A second group of structures involves the participation of both epidermal and dermal components. Examples of this type of co-operation may be seen in dermal denticles, vertebrate teeth, and some scales. The third group contains derivatives formed solely from the dermis (mesoderm) with no epidermal contribution. Such structures as antlers and the dermal bones of vertebrates may be classified in this manner. The first group, those derived primarily from ectoderm, are considered in this chapter, and those structures derived from a combination of mesoderm and ectoderm or of mesoderm only are discussed later in the appropriate chapters.

Keratinized Epidermal Structures

The epidermis of the reptile folds upon itself about small dermal projections, the papillae, to form *honey scales*. The epidermis on both sides of a papilla becomes keratinized. The living mesodermal component then mostly withdraws leaving the two upper and lower epidermal layers in close apposition. These scales vary from the small projections of the chameleons to the overlapping, thin, keratinized scales of snakes. Horny scales are a product of the stratum corneum layer of the epidermis and should not be confused with the bony scales of the fish that arise from the dermal layer. In the turtle the scales form flat scutes or plates and cover the bony plates of the carapace and plastron. Muscles attach to the transversely arranged bands of ventral scales in the snake and elevate and depress them. Reptilian scales, therefore, can be used in locomotion. The remnants of the reptilian scales remain in the two higher groups of vertebrates. They persist on the legs of birds and the tails of certain rodents, insectivores, and marsupials and offer evidence of the reptilian ancestry of these forms.

The stratum corneum also gives rise to *claws* in the reptile and some mammals, which become modified to form *nails* and *hoofs* is most mammals. All three structures cover the digits and resemble epidermal scales as illustrated. They possess a long hard nail plate (*unguis*) and a softer less cornified pad of tissue, the *subunguis*. They differ from one another, depending on the varieties of unguis and subunguis. Hoofs, for example, differ from nails in that they possess more than one shortened nail plate and the subunguis forms a pad within the curve of the hooftip. The horny plate that makes up the nails covers the *nail bed* and is surrounded laterally and proximally by *nail wall*.

A *nail groove* separates the bed from the wall. The nailfold usually covers a whitish crescentic portion, the lunula (or "moon") located near the roof of the nail. This fold has all the layers of the skin. Its underlying region, usually referred to as the *nail matrix*, forms new nail substances from the dead residues of the cornified epithelial cells. Thus the matrix forms the nail. Only the stratum germinativum covers the surface of the nail beds. Most authors agree that it does not contribute to the new nail substances but merely provides a surface for the nail to glide over. The skin of the jaw in some amniotes becomes modified. The stratum corneum that covers the upper and lower jaw gives rise to a *bill* or *beak* and takes the place of teeth in turtles, tortoises and birds. The bill of the mammalian duckbill platypus is soft and pliable, unlike that of the birds. Horns are found only among mammals. These are non-living structures composed of keratin and have no nerves or blood supply. If horns are removed, they do not regenerate. The horns of sheep, goats, cattle, and antelopes have a core of bone, the *os cornu*, sheathed in keratin. They are not branched like the solid bony antlers of the deer and are never shed. Rhinoceroses have the simplest and most primitive horns composed of hardened, solid keratin. The pronghorn antelope (*Antilocapra americana*), a native of North America has a horny covering over a bony protection. In this type of horn, the covering is periodically shed.

Feathers

Although the reptilian scales persist on the legs and feet of birds and become specialized to form beaks and claws, they are modified as feathers over the rest of the body. Feathers are composed of cornified epithelial cells. They represent the modification of the integument for flight and are the bird's most notable distinguishing feature. They occur in no other class of vertebrates. They are lightweight, have great strength, provide a water-repellent surface and effectively insulate the body. Muscles at their bases control the feather's position, an important aid in a bird maintaining high (and relatively constant) body temperature. There are three types of feathers. *Contour feathers* are the large complex feathers that cover the body, wing and tail. *Down feathers* are the simple juvenile and adult feathers, and the very simple *filoplumes* are the "pinfeathers."

The contour feathers are not randomly distributed over the body of the bird, but occur in tracts with bare spaces in between. Down feathers and filoplumes, however, may be located in these spaces. They help to insulate the bird and aid in warming the eggs during incubation. Feathers are shed periodically. Molting, under the control of hormones, usually occurs in the fall or spring. All the feathers are not lost at one time; they are shed gradually. The feather follicle, which remains intact in the integument, replaces the feather. Feathers arise in the embryo in much the same way as a reptilian scale. Mesodermal cells condense under the epidermal ectoderm and induce the ectoderm to thicken and protrude as a cone-shaped structure from the surface of the body.

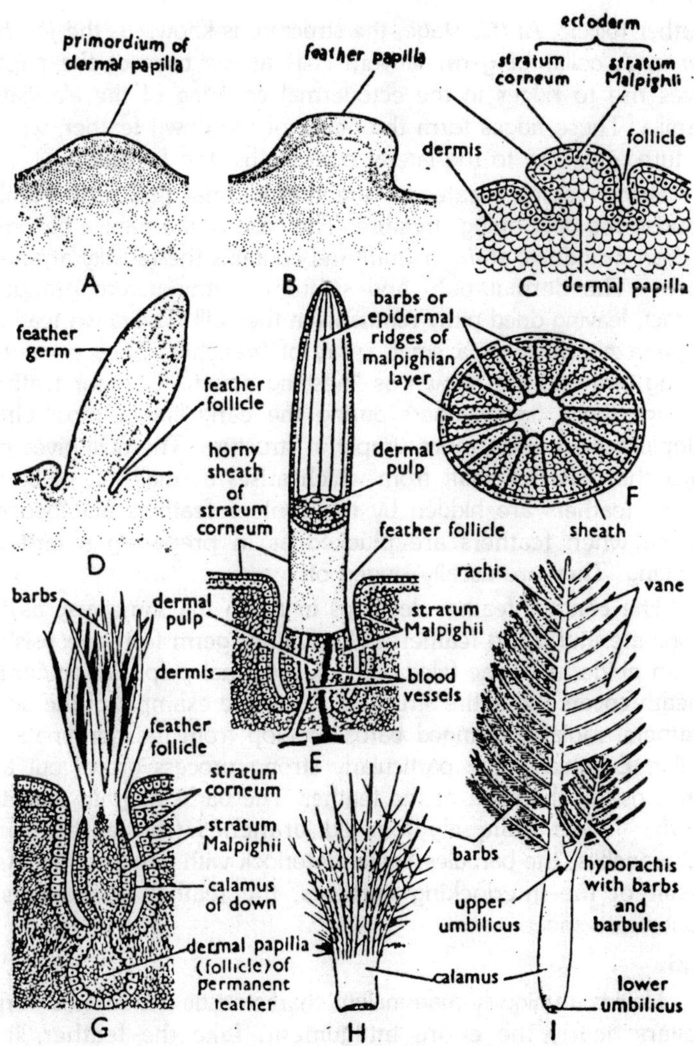

Fig. 8.5. Stages in the development of a feather. A—Beginning of feather papilla. B—Feather papilla rising above skin surface. C—Feather papilla in vertical section. D—Elongation of feather germ. E—Feather germ in vertical section. F—Feather germ in transverse section. G—Down feather in its follicle. H—Mature down feather of a newly hatched bird. I—Young contour feather.

Although the early development of a feather resembles that of a scale, its further development differs greatly. A groove appears around the base of the papilla, indicating the beginning of the feather follicle. At this stage, the structure is known as the *feather germ*. A collar of germinativum cells at the base of the papilla gives rise to ridges in the ectodermal covering of the elongated papilla. These ridges form the *barbs* of the down feather, which, in turn, give rise to the lateral outgrowths, the *barbules*.

Blood vessels located in the mesodermal pulp (the papilla) nourish the developing structure. The base of the papilla becomes a short cylindrical *stalk* or *quill* and contains the radially arranged barbs. The dermal pulp and stratum germinativum gradually retract, leaving dried pithy remnants in the quill. The down feathers appear externally as a small spray of branches. They cover the young bird and are known as "nestling down." Contour feathers replace these first feathers during the early life of most birds. Filoplumes are even more simple in structure. The quill gives rise to a short slender shaft from which arises a small tuft of barbs. These feathers are hidden by the contour feathers and become visible when feathers are plucked as in preparing a bird for cooking. They are usually singed off.

The contour feather develops in much the same way as the filoplume and down feathers. The feather germ forms, consisting of an epidermal cone filled with mesodermal pulp. An epidermal sheath covers the entire structure. As in the example of the down feathers, radially arranged barbs develop from the germinativum collar. Eventually one particularly strong process grows out and gives rise to the shaft of the feather. The barbs migrate onto the shaft. These obliquely located branches (the barbs) form subbranches, the barbules, which interlock with one another. As a result of the interlocking barbules, the feather appears as a continuous sheet.

Hair

Hair is a uniquely mammalian characteristic and in most forms covers nearly the entire integument. Like the feather, it is composed entirely of epidermal tissue but unlike the feather it is believed to have evolved from tactile sensory pits rather than from reptilian scales. The bodies of all mammalian embryos at sometime during development are covered with fine hair known as the *lanugo*. It is well-developed in the human embryo by the seventh month but is usually shed by the time of birth. The shedding and replacement of hair, like feathers, occur either periodically or constantly. If the stratum germinativum composing the hair follicle is lost through injury, no regeneration is possible.

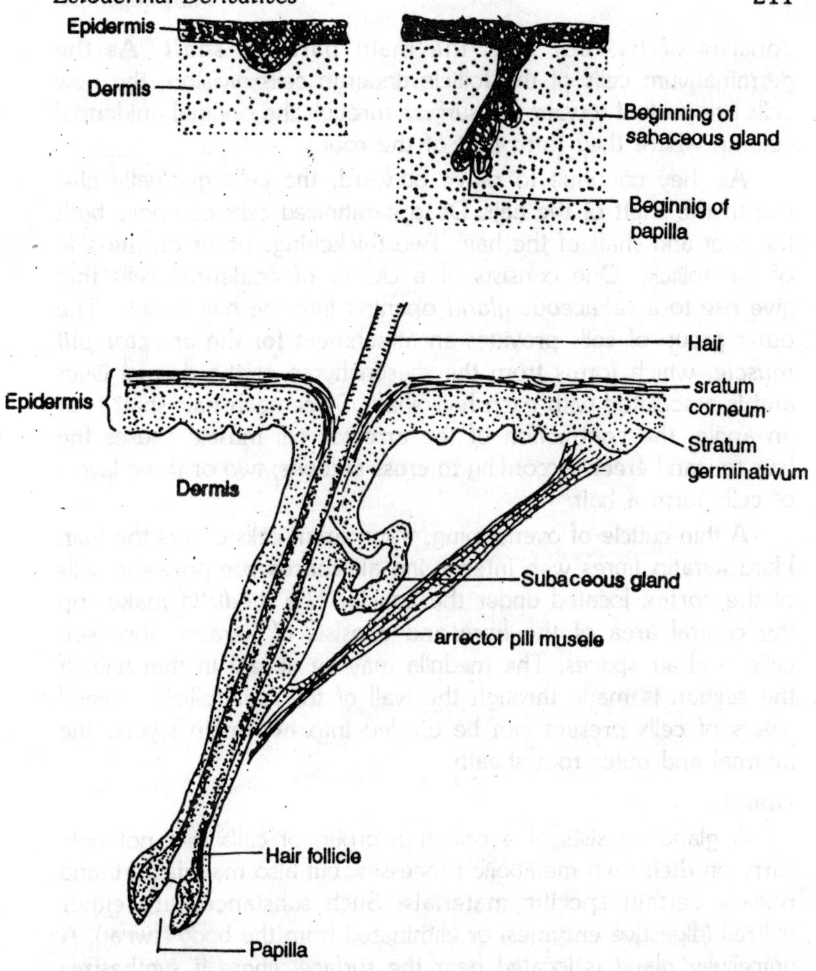

Fig. 8.6. Diagrams of longitudinal sections showing successive stages in the development of mammalian hair.

The density of hair varies in different mammals as well as over the body of the same animal. Manes, eyelashes, and eyebrows all differ from that covering the rest of the body. The development of hair is initiated when the epidermal ectodermal grows obliquely down into the dermis. The base of the column becomes bulb-shaped and the dermal mesenchyme pushes into the bulb to form a papilla. Blood vessels form in the papilla and supply nutritive material to bulb cells, the only living part of the hair. The papilla and its covering of epidermal cells make up a hair follicle. A hair consists of its *root* and, the main part, its *shaft*. As the germinativum cells of the follicle undergo cells division, the new cells are pushed toward the surface through the original epidermal column where they form part of the root.

As they continue to move outward, the cells gradually give rise to the shaft of the hair. Dead keratinized cells compose both the root and shaft of the hair. Two thickenings occur on the side of the follicle. One consists of a cluster of epidermal cells that give rise to a *sebaceous gland* opening into the hair follicle. The other group of cells provides an attachment for the *arrector pili muscle*, which forms from the mesenchyme of the dermal layer and is associated with each hair shaft. Since each hair shaft is at an angle, the contraction of the arrector pili muscle causes the hair to stand erect. According to cross sections, two or three layers of cells form a hair.

A thin cuticle of overlapping, transparent cells covers the hair. Hard keratin fibres with intervening air spaces compose the cells of the *cortex* located under the cuticle. The *medulla* makes up the central area of the fibre and consists of keratin, shrunken cells, and air spaces. The medulla may be absent in thin hair. If the section is made through the wall of the hair follicle, several layers of cells present can be divided into two main layers, the internal and outer root sheath.

Glands

A gland consists of a cell or a group of cells that not only carry on their own metabolic processes, but also manufacture and release certain specific materials. Such substances are either utilized (digestive enzymes) or eliminated from the body (sweat). A unicellular gland is located near the surface where it synthesizes and releases its specific secretion. Several cells (multicellular) formed from invaginations of the germinative layer compose most glands. These invaginations differentiate into a secretory portion and a non-secretory portion, or duct, that carries the secretion to the surface. The term, *exocrine*, refers to this type of gland.

Endocrine glands, on the other hand, lose their connection with the surface epithelium and are ductless. Exocrine glands exist in a variety of shapes. If the secretory cells are arranged in a simple tubelike form that opens into an unbranched duct, the gland is said to be *simple*. However, if the duct system is branched, the gland is *compound*. The secretory portions of either simple or compound glands may be arranged as tubules, in which case it is known as a *tubular gland*. If the secretory portion expands to form a sac-like structure, the gland is either a simple or compound *alveolar gland*. Glands can be classified not only by their shape but also according to the type of material secreted.

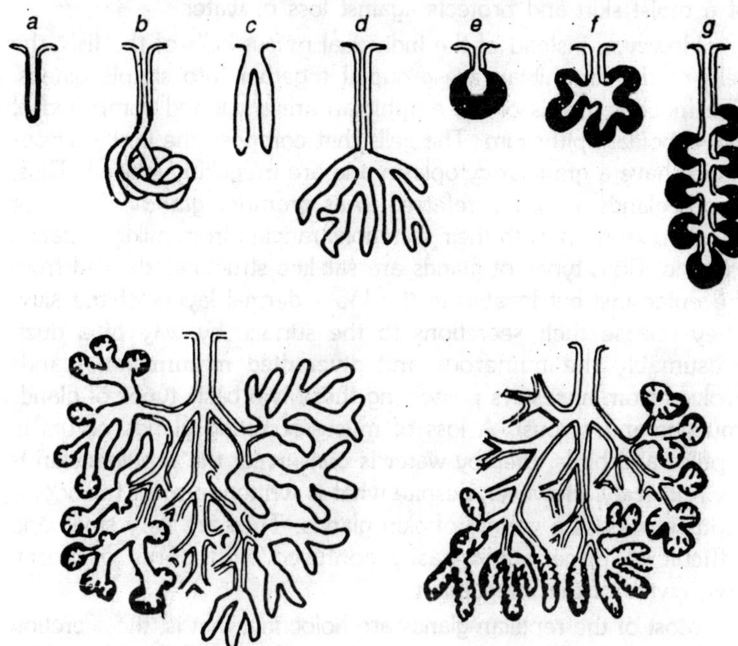

Fig. 8.7. Diagrams of various exocrine gland types. Above (a-g), simple glands; a, tubular; b, coiled tubular; c, d, branched tubular; e, alveolar; f, g, branched alveolar. Below, examples of compound glands. In all diagrams, ducts are in double lines, secretory portions in solid black.

Mucous glands secrete a viscous, slimy material; the secretion of *serous glands* is more watery. Some *mixed* glands produce both types of secretion. The cells composing the glands release their secretory products in different ways. A *merocrine gland* releases its products through the cell membrane with the cell remaining intact. If some of the apical cytoplasm is lost along with the material secreted, the gland is referred to as *apocrine*. In a *holocrine gland*, the entire cell disintegrates, forming a part of the secretion. Glands develop in the epidermis of all vertebrates. In general, the glands of fish are unicellular, although some multicellular glands exist. They differentiate as small cells from the stratum germinativum and migrate toward the surface.

Most are mucus-secreting glands; some are serous glands. The mucus forms a layer on the surface of the fish and makes it more impermeable to water. In some species, multicellular poison glands are associated with the base of the spine of the fins. The aquatic amphibians continue to secrete mucus that aids in the maintenance of a moist skin and protects against loss of water.

However, instead of the individual mucus cells of the fish, the cells of the amphibian are grouped together into simple glands. The mucous glands of the amphibian are small and composed of low cuboidal epithelium. The cells that compose the larger serous glands have a granular cytoplasm and are irregularly spaced. Thus, serous glands are often referred to as *granular glands*. They are usually poisonous, with their secretions ranging from mildly irritating to toxic. Both types of glands are sac-like structures derived from the epidermis but located in the loose dermal layers of the skin. They release their secretions to the surface by way of a duct. Presumably the numerous and diversified mammalian glands evolved from ancestors possessing these two basic types of glands (mucous and serous). A loss of mucus-secreting glands occurs in reptiles and birds, thereby water is conserved; the reptilian skin is dry and scaly. However, despite what is written in most textbooks, reptiles possess a variety of skin glands. They are very small and difficult to discern and easily confused with pores or horny outgrowths of the integument.

Most of the reptilian glands are holocrine, that is, the secretion consists of the entire secretory cells sloughing off into the cavity of the gland and degenerating into an amorphous fluid. Most of the glands are specialized scent glands and reach a greater size and activity in the male during the breeding season. The paired uropygial or preen glands at the base of the tail of birds secrete oil, which the bird dispenses with the help of its beak. This secretion is apparently necessary to maintain the structural integrity and water-repellent characteristic of the feathers, these glands are especially well-developed in aquatic birds. There are also traces of sebaceous-like glands in the ear and trunk regions of some birds.

The mammalian skin contains many glands, but basically they are either modified *sebaceous glands* or *sweat glands*. The sebaceous glands are spherical or ovoid in shape and are simple or branched alveolar glands. They are believed to be homologous with the amphibian alveolar mucous glands.. In the mammal the sebaceous gland secretes oil and is associated mainly with the hair follicles. They develop first as thickenings of the outer sheath cells. Their ducts open into the neck of the follicle and their oily secretion appears to be the direct product of cellular disintegration.

Several types of sebaceous glands are distributed over the body where hair is absent. They are located, for example, at the corners of the mouth and lips, and on the nipple, the eyelids, and glans penis. Sweat glands may have evolved from the amphibian serous glands. They are found only in mammals. They appear in the embryo as solid, cylindrical in growths of the stratum germinativum. The deeper part of the ingrowth coils on itself and forms the body of the gland. The central cells eventually degenerate and produce a tubular structure that is simple coiled or branched. The coiled portion rests in the sub-cutis region of the skin and a long duct extends through the dermis, entering the epidermis between two papillae. The duct opens to the surface by way of a sweat pore. Some of the surrounding ectodermal cells give rise to myoepithelial cells and, presumably, are able to contract. The sweat glands are apocrine glands since the distal ends of the cells are discharged as part of the secretion.

The sweat glands are important excretory organs. Sweat is a watery secretion containing certain salts, urea, and other wastes; it is a significant method of eliminating metabolic wastes of various kinds. It also helps maintain the body temperature, since the evaporation of sweat on a hot day produces a cooling effect. Mammary glands, ceruminous (wax) glands of the ear, and certain scent glands also have an apocrine type of secretion and probably evolved from sweat glands.

Mammary glands, present only in mammals, are responsible for the name of the class. Mammary glands assume various forms. In the monotremes they consist of two glands in the abdominal wall that secrete a sticky material onto the surface ; the secretion is licked off this mammary patch by the young. The glands are located in the pouch of the marsupial. At birth the young crawl into the pouch, and each attaches to a nipple. In ungulates, several ducts empty into a common collecting chamber and a single duct opens to the outside. The primate's mammary gland basically consists of several lobes (15 to 20) each, of which is considered to be a compound gland with a duct opening into the tip of the nipple. Mammary glands first form as linear ectodermal thickenings, known as the milk, or mammary, ridges running down the ventrolateral region of the embryo of both sexes. Various areas of this region continue to develop; the precise number of glands usually corresponds with the number of offspring produced.

The development of the mammary ridge normally occurs only in the pectoral region in man and the elephant and only in the inguinal region in the sheep cow and horse, but it is extensively developed down the middle of the ridge in such forms as the cat, dog, or pig. Thickenings of the epidermis represent the beginnings of the nipples. The ectodermal cells undergo proliferation and become bulbous; they grow in toward the mesenchyme layer.

Several buds (about 15 to 20 in the human embryo) push out into the dermis from the epidermal bulbous area. These buds are the primordia of the mammary gland ducts. Near the end of embryonic life, they undergo further branching, but then until puberty the mammary glands of both sexes remain in an infantile state. The secretion of estrogen by the female at sexual maturity stimulates duct growth; progesterone stimulates the growth of the alveolar secretory cells. During pregnancy when the estrogen and progesterone levels remain high because of production by both ovaries and placenta, the ducts and alveolae develop extensively, establishing a presecretory state. The lactogenic hormone (luteotrophic hormone, LTH) from the pituitary gland appears necessary in certain species for secretion to take place.

Chromatophores

The various colour patterns of vertebrate animals function in various ways and are determined by heredity. They are involved in camouflage, protection, sexual attraction, and warning. Colour has definite survival value. The pigment melanin, for example, may function in some cyclostomes as a screen, protecting the central nervous system against the sun. Since melanin is associated with both locomotion and spine response in the invertebrates (sea urchin), some authors suggest that pigment may play a role in regulating the vertebrate central nervous system.

Integumental pigmentation has undoubtedly been important in evolution, but we do not understand its specific role. Pigment-containing cells also penetrate more deeply than the integument. They are often found along the blood vessels, along the walls of the coelom, and in the liver, all areas where it is difficult to determine a functional advantage for the pigment. Neural crest cells invade the epidermis and dermis and differentiate into colour-forming cells known by the general term *chromatophores*. Since they have been traced to the neural crest cells, they are considered ectodermal in origin.

Soon after the completion of neurulation some of the neural crest cells (melanoblasts) migrate out to the various locations they will occupy in the postembryonic body. The specific colour pattern of vertebrates depend on the distribution and activities of these cells. They allow the animal to respond to many environment stimuli by changing colour. Since the chromatophores are stellate and have long slender processes, they are sometimes referred to as *dendritic cells*. Pigment granules may either be concentrated in the center of the cell in one mass, in which case little colour appears, or else the pigment may be distributed throughout the cell including the long dendritic processes.

When dispersal of the pigment occurs, the colour effect is maximum. The colour-bearing cells are also able to shift their relative positions in the body of the animal by ameboid movement. The colour changes displayed by such animals as the flounder and chameleon are attributable to both these mechanisms—the migration of the cells and the amount of dispersal of the granules within the individual cells. Chromatophores are under hormonal control and are influenced by pituitary thyroid, gonad, pineal, and adrenal gland secretions. For example, the central nervous system influences the intermediate lobe of the pituitary to release its melanotrophic hormone (intermedin) into the bloodstream of fish amphibians, and some reptiles. Dermal capillaries carry the hormones to the pigment-synthesizing cells. Colour changes controlled by intermedin take minutes or hours to occur. In poikilotherms (cyclostomes, elasmobranchs teleosts, amphibians, and reptiles), the chromatophores are also effector cells and respond to nervous stimuli. Consequently their response is rapid.

Colour pattern in the environment influences colour change in the animal by pathways leading from the eyes to the brain of the organism. If a single cutaneous nerve in a fish is destroyed, it is unable to change colour in the area of the cut. Nerves produce their effect by liberating a chemical a *neurohumor*, at their terminal ends that diffuses to the chromatophore and brings about the change. Epinephrine-like substances released by the nerve terminals cause the melanin to concentrate in the center of the cell and acetylcholine-like materials induce melanin granules to disperse into the dendritic arms. Birds and mammals apparently lack a pigmentary effector system, since colour changes associated with these classes of animals involve relatively slow alterations in the amount of pigment. The cells respond to seasonal changes, allowing the skin to tan in the summer or the coats of certain animals to change from brown to white.

In some cases these colour variations are caused by the amount of pigment present and in other cases to the type of pigment. There are various kinds of pigments in the integument of all vertebrates. The specific type present depends on the synthetic process of the chromatophores. The chromatophore that synthesizes *melanin* the predominant vertebrate pigment, is referred to as a *melanophore* and is located in both the epidermis and dermis of at least some member of each class of vertebrates. The skin colour of man depends on the presence and amount of melanin deposited in the epidermal layers. Melanin granules vary in colour from yellow through orange and reddish brown to dark brown, depending on their state of oxidation.

Melanin also may be present as reduced leukoform, a bleached form. Thus it acts as an oxidation-reduction indicator; in its oxidized form melanin is dark, in the reduced state it is bleached. Although the *epidermis* contains only melanophores, other types of chromatophores are present in the *dermis*. These cells may originate from a common cell type and differentiate into the various chromatophores according to the specific environmental stimuli they receive. *Iridophores* are described as reflecting cells that contain guanine, hypoxanthine, adenine, or a combination thereof. They are present in anamniotes and the iris of some birds, but are usually absent from the integument of birds and mammals.

Xanthophores responsible for the secretion of yellow pigment, and *erythrophores*, containing red pigment, are usually found only in the dermis of elasmobranchs, teleosts, amphibians and reptiles. Vertebrate colour changes result when these different chromatophores interact. The modern history of pigmentation shows that colouration in vertebrates is the result of interaction not only between pigment-containing cells, but also between these cells and other cells of the skin, such as the cells responsible for forming keratin. This interrelationship can be demonstrated in the human skin.

The neural crest cells destined to form melanocytes take up a position at the interface between epidermis and dermis. They extend their branches between the basal cells of the germinativum layer and discharge their pigments into the keratinocytes. Pigmentation of the mammalian epidermis, therefore, is not caused by the melanocyte acting alone, but is the result of an interrelationship between the melanocyte and as associated group of keratinocytes, forming the *epidermal melanin unit*. These units, established during development, are the result of tissue interaction. According to Quevedo (1972) the melanocytes in man exist in a ratio of one to 36 keratinocytes. The maintenance of such a relationship suggests a precise control over mitotic activity. The melanocytes contain tyrosinase, an enzyme that produces melanin by oxidation of the amino acid, tyrosine.

Melanocytes are the only cells capable of synthesizing this material. The pigment is deposited in the cytoplasm within a membrane-bound organelle, the melanosome. Melanin is then transferred to the keratinocytes when these cells phagocytize pieces of the melanocyte dendrites containing the melanosomes. The pigment-containing organelle is transported, and ultimately shed, along with the cornified cells, or catabolized along the way by the keratinocytes. There is significant variation in the number of melanocytes with regard to sex or race. Skin colour, therefore, appears to depend upon the activity of the pigment cell rather than the number of cells present. This activity, in turn, is genetically determined. Similar units located in the dermis, called *dermal chromatophore units*, are described for poikilothermic animals.

The basic types of chromatophores, previously described for poikilothermic vertebrates, are arranged in such a way in the amphibian that the xanthophores are situated in a top layer, the melanophores form a basal layer, and the iridophores are located in one or more rows in the middle. The dendritic arms of the melanocyte entwine about and between the layers above them. When the frog adapts to a dark background, melanosomes migrate to these dendrites of the melanophores, thus obscuring the other chromatophore types and causing the skin to darken. Migration of the melanosomes out of the dendrites to a more central location exposes the xanthophores and iridophores, causing the animal to lighten.

Nervous System

In vertebrates there are two integrated systems—the nervous system and the endocrine system. Nervous controls the activities mechanically and endocrine system metabolically. The nervous system is divided into the *central nervous system* (CNS), composed of the *brain* and *spinal cord*, and the *peripheral nervous system*, consisting of sensory and motor nerves. This division is mainly one of convenience, since both act together as one unit.

The central nervous system develops from the neural plate of the late gastrula stage. The peripheral nervous system derives partly from the neural plate, partly from neural crest cells, and partly from the thickened ectodermal areas we have identified as placodes. The basic structural unit of the nervous system is the nerve cells, or *neuron*, consisting of a *cell body* and its processes, the *dendrites* and *axons*. An intricate network of neuron communications carries out co-ordinating functions. Individual nerve cells (neurons) generate messages, which they conduct from one part of the body to another by way of their cell processes. These messages are then chemically transmitted to other nerve cells or to other cells of the body. The analysis and interpretation of these communications and the attempts at describing them in terms of neural mechanisms mark the frontiers of biological investigation in this decade.

Many physiologists believe that the complexities of the nervous system will never be comprehended. They believe that man is at the limits of his understanding and that the inadequacies of his brain will prevent him from completely analyzing and explaining its physiology. Only fragments of behaviour have been correlated with their neural mechanisms. Perhaps the simplest reflex, present in all vertebrate animals, is most readily understood. A stimulus produces a quick withdrawal movement and involves what is described as a *reflex arc*. This simplest of reflexes, considered to be a monosynaptic (one neuron in contact with one neuron) reflex action, may actually produce many more side effects in the central nervous system than we previously believed. Some of these effects are only beginning to be identified and described. Although such a simple reflex arc is possible, nervous reactions seldom involve only two neuron chains but are considerably more complex than this. Laterally, billions of nerve cells make contact with one another, forming a maze of interacting fibres.

In an attempt to understand basic nerve circuits, the investigator must seek common properties of neural structure and function. To accomplish these ends, the crayfish, the frog, and the cat have been used extensively in the past as investigative animals. However, at the present time, the research carried out on mammals, including man, exceeds that for all other animals.

DEVELOPMENT OF CENTRAL NERVOUS SYSTEM

Development of Neural Tube

By the end of neurulation the neural ectoderm is in the form of a dorsal, hollow neural tube that extends under the surface epithelium of the embryo from its anterior to its posterior end. From the beginning, the neural tube expands in the anterior regions of the body. During gastrulation the notochord assumes a position under the neural plate. As a result of this migration the prechordal plate region lies immediately beneath the presumptive brain areas and the notochordal and somatic mesoderm under the middle and posterior regions. The brain of the organism develops from the expanded anterior region of the neural tube, and the spinal cord develops from the more posterior parts.

As a result of further differentiation, induced by the prechordal plate and notochordal and somatic mesoderm, the primary rudiment (the neural tube) gives rise to the central nervous system of the adult. After the neural tube is established, it undergoes a general organization. At early stages one can distinguish certain

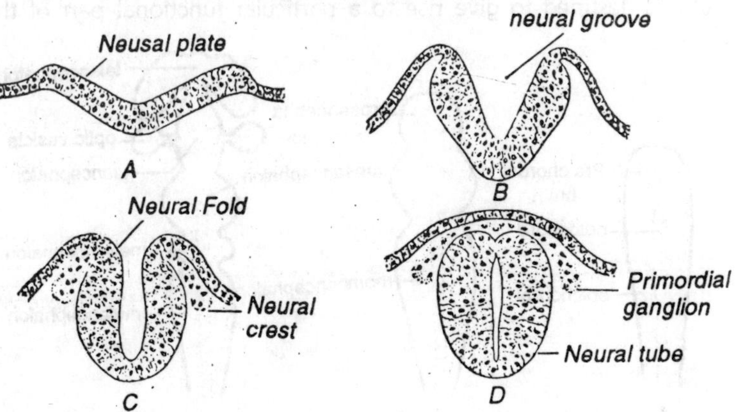

Fig. 8.8. Origin of the neural tube and neural crest, illustrated by transverse sections from early human embryos.

concentric layers and longitudinal bands in the walls of both the presumptive brain and neural tube regions. The three concentric layers are (1) an inner layer lining the central canal of the tube, (2) a middle cellular region, the *mantle layer*, and (3) an outer *marginal layer* consisting of fibres from the cells located in the first two layers. As development continues, the cell bodies located in the mantle layer give a characteristic configuration to the tube.

In transverse section, the cellular (mantle) region appears darker because of the closely packed nuclei and is now referred to as the *gray matter* of the tube. The marginal layer lacks cell bodies and contains fibres wrapped with a white, fat-like material, myelin. This layer becomes the *white matter* of the neural tube. The tube grows in diameter by the proliferation of cells from the inner layer and their migration into the mantle layer, and also by the further deposition of myelin around the fibres. The primitive neural tube can be divided into six longitudinal bands. A relatively thin *roof* (roof plate) and *floor* (floor plate) form the dorsal and ventral areas of the tube.

Each lateral wall consists of a dorsal *alar plate* and a more ventral and thicker *basal plate*. Once this basic organization is established, the brain and spinal areas of the cord organize further, both internally and externally, and give rise to the central nervous system unique to the specific organism. Evidently the neural epithelium at the time of cellular proliferation becomes determined, forming a rather rigid mosaic. Each region of the early neural tube is destined to give rise to a particular functional part of the

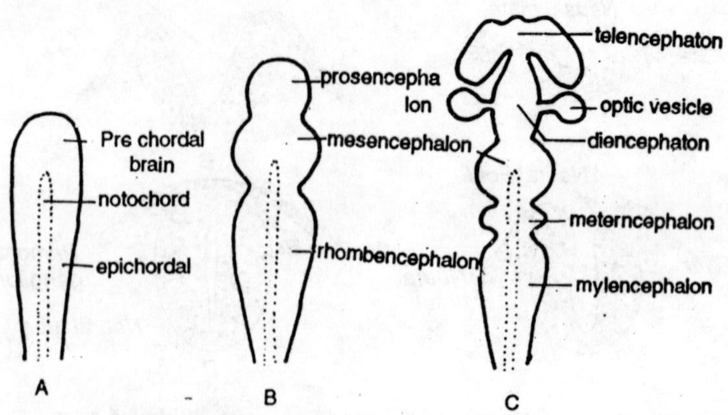

Fig. 8.9. Differentiation of various parts of brain.

future cord. For example, removal of the right dorsal half of the neural tube from the chick embryo at 44 hours of incubation results in a normal left side of the spinal cord but the right half is missing.

Regionalization of Neural Cord

The anterior region of the neural tube soon divides into three major regions usually referred to as the *primary brain vesicles*; the forebrain, midbrain, and hindbrain, better known by their more technical terms as the *prosencephalon*, *mesencephalon*, and *rhombencephalon*, respectively. These primary vesicles are associated with the three major sense organs. The prosencephalon receives impulses from the olfactory organs, the mesencephalon from the eye, and the rhombencephalon from the ear and lateral-line organs. The walls of the brain undergo further differentiation and the prosencephalon soon divides into an anterior *telencephalon* from which the olfactory lobe and cerebral cortex form, followed by a *diencephalon*, from which the eyes and thalamus develop. The mesencephalon does not divide further. The rhombencephalon differentiates into the *metencephalon*, which gives rise to the cerebellum and the pons, and the *myelencephalon*, from which the brainstem and upper part of the spinal cord develop. The rest of the tube remains of uniform diameter along its length and is destined to become the spinal cord. As the neural tube develops, it bends and gives rise to the *cephalic* and *cervical flexures*.

Cavities of Brain and Spinal Cord

The cavity of the neural tube, the *neurocoel*, remains in the adult animal and fills with cerebrospinal fluid. As the brain develops, the original small neurocoel in the telencephalon expands to form the *lateral ventricles* of the future cerebral cortex, *first* and *second ventricles*. The *third ventricle* is a median cavity located in the diencephalon and is connected to the first two ventricles by the *interventricular foramen*, or *foramen* of *Monro*. In amniotes a thin passage-way in the midbrain connects the third ventricle with the fourth, located in the medulla oblongata. In lower animals the cavity of the mesencephalon remains large, forming the mesocoel. The ventricles of the brain continue into the spinal cord as the small central canal.

Meninges

The skull and neural arches of the vertebrae are laid down around the brain and spinal cord in the embryo. Connective tissue membranes, the *meninges*, wrap intimately around the nervous tissue and offer additional mechanical protection and support for the blood vessels of the central nervous system. The outer membrane is derived from the surrounding mesenchyme; the inner membrane appears to be composed, at least in part, of neural crest cells. In fish, only one membrane, the *meninx primitiva*, covers the brain and spinal cord. Fine collagen fibres pass through the meninx to the lining of the brain case. Two membranes are present in amphibians, reptiles, and birds: an outer, tough, connective tissue membrane, the *dura mater*, and an inner, more fragile, vascular, *pia-arachnoid layer*. The *subdural space* separates the two membranes; a second space exists between the dura mater and lining of the brain-case, or neural arch. In mammals, the pia-arachnoid layer divides, and as a result three membranes are present.

There is the outer, tough dura mater, a middle *arachnoid layer*, and an *inner pia mater*. The pia meter is intimately applied to the brain and follows its contours. A space, the *subarachnoid* space, separates the pia mater from the arachnoid and is traversed by delicate fibers that tie the two membranes together. The middle layer, unlike the pia mater, does not dip down into the sulci and fissures of the brain. The outer layer, the dura mater of mammals is closely applied to the lining of the skull, obliterating any space, although in the spinal cord a space continues to exist between this membrane and the lining of the vertebral canal. The dura mater involutes and forms a septum (falx cerebri) that helps to divide the right and left cerebral hemispheres longitudinally. It also contributes to a transverse septum (the tentorium) separating the cerebrum from the cerebellum. Cerebrospinal fluid fills the ventricles of the brain and central canal of the cord, as well as the spaces traversed by the connective tissue fibers.

The fluid has two functions. It carries the nutrients derived from the blood vessels to the tissue of the central nervous system, and it acts as a cushion to the brain and spinal cord. The *choroid plexuses*, vascular networks located in the roof of the diencephalon and medulla oblongata, secrete the fluid into the third and fourth ventricles. It flows from the ventricles into the subarachnoid space. Here it makes its way into the lymphoid vessels, or arachnoid villi, that connect with certain large venous sinuses under the dura mater. Since secretion of the cerebral fluid occurs continuously, its removal must also go on constantly in order to maintain a stable intracranial pressure.

Basic Embryonic Plan

This basic plan of five cranial divisions plus the spinal cord, along with their cavities and connective tissue membranes, make up the primordium of the central nervous system. It forms early in the development of the organism and is similar in all vertebrate embryos at this early stage. It is far more advanced in the simplest vertebrate, the cyclostome, than in any member of the invertebrate phylum. Although the nervous system is basically the same in the embryo of all vertebrates and it functions in a similar manner in the adult, the increase in complexity from the lowly cyclostome to the mammal is truly outstanding. The growth of the cerebral cortex brings about an increase in co-ordination and association centers. It is this growth in the cerebral cortex that offers man his great potentialities. According to Romer, "The evolution of the cerebral hemispheres is the most spectacular story in comparative anatomy."

Histogenesis of Nervous Tissue

Although the presumptive neural tissue at the blastula and gastrula stages is only one cell thick, by the time the neural tube is established its walls consist of a pseudostratified epithelium composed of columnar cells and rounded cells. These cells attach not only to one another, but also at the lumen of the neural tube by special devices known as terminal bars. Watterson (1965) suggests that this attachment at their terminal ends imposes a radial orientation on the neuroblasts and helps to collect similar cells together. All the cells composing the early neural tube are capable of division. They appear similar to one another, but differentiate along two lines. They give rise to the *neuroblasts* that are destined to form the neurons, or else they form the supporting tissue, the *neuroglia*.

Stem-Cells of the Neural Wall

At the beginning of its development, the neural plate is composed of undifferentiated, proliferative epithelium. Its daughter cells enter into two lines of specialization. One path leads toward *nerve cells*, in which irritability and conductivity have become predominant functions; the other course is toward *ependymal* and *neuroglial cells*, which constitutes the distinctive supporting tissue of the nervous system. The embryonic nerve-cell is a *neuroblast*;

it passes through a bipolar stage, with a process at each end, before reaching a multipolar stage, or immediate precursor of the typical neuron of the central nervous system. The *spongioblast* is the forerunner both of the *ependymal cells* and of neuroglial cells known as *astrocytes*. Some spongioblasts are migratory in nature; these differentiate into *oligodendroglia* and into additional astrocytes, as well. It has been claimed that some migratory spongioblasts convert into neuroblasts, but this interpretation is open to doubt.

Layers of the Neural Wall

The neural plate was originally a single layer of columnar cells, but rapidly becomes many-layered (C): in doing this the component cells lose their sharp outlines and seemingly resolve into a compact syncytium (D). Some authorities, however, maintain that the resemblance to a syncytium is an artefact and the constituent cells are always distinct. In the sixth week the neural wall is bounded on the outer and inner surfaces by an *external* and *internal limiting membrane*, respectively, while the cellular elements of the wall are arranged radially (D). At this stage the neural tube is sufficiently organized so that three concentric zones may be distinguished: (1) an inner *ependymal layer*, with its cell bodies abutting on the internal limiting membrane (next the neural canal) and their processes extending peripherally; (2) a middle nucleated *mantle layer*, derived by proliferation of the innermost cells; and (3) an outer, non-cellular *marginal layer*, into which many nerve processes (nerve fibers) grow. The ependymal layer, originally the uppermost stratum of the neural-plate stage, not only contains the nucleated bodies of inertly-supporting ependymal cells but also mitotic cells. These are perhaps not special stem-cells, as commonly described; rather, it is now urged that whenever cells of the mantle zone are about to divide they retract and round up, so that they come to lie temporarily in the ependymal zone.

The mantle layers makes up the future *gray substance* of the central nervous system; it is predominantly cellular in structure and contains the cell bodies of the neurons and many neuroglial cells. The marginal layer is a 'fibres' mesh which lacks cells in the early months. It provides a basis into which the processes of nerve cells grow and reach their destinations; thereby neuron is linked with neuron and center with center. It becomes the *white substance* of both the brain and spinal cord. The details of the transformation of neuroblasts into neurons and spongioblasts into ependyma and neuroglia demand further treatment and will occupy the description that follow.

The Differentiation of Neuroblasts

The process of which neural epithelium specializes into neurons is named *neurogenesis*. In accomplishing this end each neuroblast (and its descendants) sooner or later loses the power of cell division, develops cell processes and converts into a definitive *neuron*. A neuron is a discrete structural and functional unit of nervous tissue; it consists a nerve cell and all of its processes. The only relation of one neuron to another is that of contact between processes, or between a process and a cell body; this touching constitutes a functional contact known as a *synapse*. Mitosis among neuroblasts ceases during the first year of postnatal life.

Thereafter the nervous system matures and enlarges, but the ability to produce new neurons is forever lost. The total number of neurons developed for the use of the nervous system is remarkably constant, and this is true regardless of the size of the individual. The origin of the nerve fibers as extensions from the neuroblasts is easiest understood in the development of the root fibers of the spinal nerves. The dorsal roots contain sensory, or *afferent fibers* which are processes from ganglion cells located in

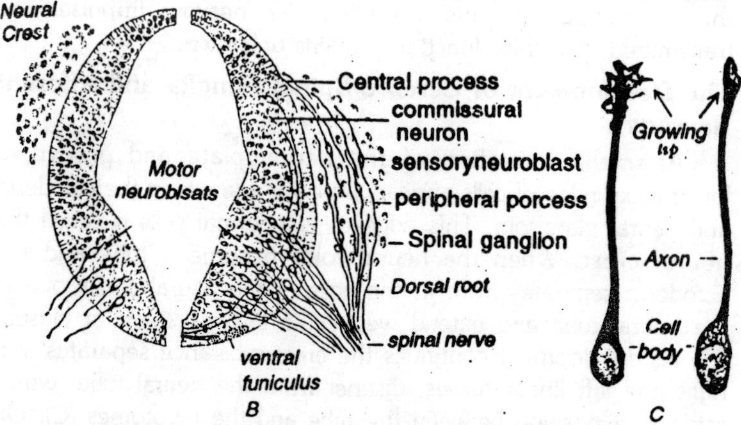

Fig. 8.10. The differentiation and growth of human neuroblasts. A, Spinal cord, in transverse hemisection. B, Spinal cord, in transverse hemisection. C, Two neuroblasts, demonstrating neurofibrils and the enlarged, growing tip.

ganglionic enlargements in these roots. The ventral roots, on the other hand, contain motor, or *efferent fibers* arising from nerve cells located in the neural tube itself. The efferent fibers are the earlier ones to appear.

The Development of Efferent Neurons

Toward the end of the first month neuroblasts separate from the general '*syncytium*' in the mantle layer of the ventrolateral wall of the neural tube. The young, bipolar neuroblasts become pear-shaped, and from the slender end of each cell a thin primary process grows out. This extension becomes the chief process, or *axon*, of the nerve cell and is destined to conduct efferent (motor) impulses away from the cell body. Axons penetrate the marginal layer of the spinal cord and, on emerging, straightway combine as a *ventral root* of a spinal nerve. In a similar manner the efferent fibers of the cranial nerves grow out from neuroblasts of the brain wall. The cell bodies of all efferent neurons soon become multipolar by the development of branched secondary processes, the *dendrons* (or *dendrites*). Within the cytoplasm of even young nerve cells and their secondary processes, fine *neurofibrillae* can readily be demonstrated by selective staining methods. Since neurofibrils have been observed in living cells and after instantaneous fixation that eliminates chemical treatment, they are apparently not artefacts. There is no proof that neurofibrils are the conducting elements through which nervous impulses are transmitted, and their function remains unknown.

The Development of Cerebro-Spinal Ganglia and Afferent Neurons

After the formation of the neural plate and groove, a longitudinal ridge of cells appears on each side where the ectoderm and neural plate join. This ridge of ectodermal cells is called the *neural crest*. When the neural folds become a tube and the ectoderm separates from it, the cells of the neural crests overlie the neural tube and extend wedge-like into its seam of closure (B). As development continues the crest substance separates into right and left linear halves, distinct from the neural tube, which settle to a position between the tube and the myotomes (C). On its arrival in this location the neural crest is a cellular band extending the full length of the spinal cord and far rostrad along the brain wall. At regular intervals, agreeing with the position of somites, the proliferating cells of the crest give rise to bead-likc enlargements, the *spinal ganglia* (D).

The serially repeated ganglia of each side are interconnected for a short time by parts of the originally continuous crest substance, but these bridges soon disappear. In the hind-brain region certain ganglia of the cranial nerves develop also from the crest; they are spaced in relation to the branchial arches. The cells of the ganglion primordial differentiate into *ganglion cells* and *supporting cells*, groups that are comparable to the neuroblasts and spongioblasts of the neural tube. Each neuroblastic forerunner of a ganglion cell elongates into a spindle-shaped element and, by developing a primary process at each end, transforms into a neuron of the bipolar type. All of the growing processes that are directed toward the neural tube converge into a distinct bundle and thus constitute a *dorsal root*. Each component fiber of the root penetrates the dorsolateral wall of the neural tube, courses in the marginal layer of the spinal cord and, by means of end-twigs, comes in contact with a neuron of the mantle layer.

The peripheral processes of the bipolar ganglion cells complete a dorsal spinal root by passing outward and joining the corresponding ventral root; the common bundle, thus formed, constitutes the trunk of a *spinal nerve*. Although bipolar at first, ganglion cells become unipolar in a way not surely understood. Presumably a part of the cell body draws out into a common stem that bears the two processes at its tip. In Figure B there can be traced intermediate stages between typical bipolar and unipolar cells.

The Development of Autonomic Ganglia

In addition to forming the cerebro-spinal ganglion cells, certain neuroblasts migrate ventrad and differentiate into cells of the autonomic ganglia. The source of these neuroblasts in disputed, but the evidence seems to favour an origin from the neural crest. The course of differentiation of autonomic ganglion cells differs from that in the cerebro-spinal series in as much as the final product is multipolar cells whose chief process is an axon. Functionally, therefore, these ganglion cells are efferent.

Differentiation of the Supporting Elements

Both the primitive neural tube and the early ganglia furnish cells that become non-nervous elements. These are permanent constituents of the brain and spinal cord and of the peripheral nervous system, respectively.

Supporting Elements of the Neural Tube

The brain and spinal cord are given stability by ectodermal, interstitial tissue in the form of ependymal cells, which bound the spinal canal and extend outward toward the periphery, and by neuroglial cells which are more irregularly distributed. A preceding paragraph has described how the spongioblasts originate from the undifferentiated cells of the neural-plate tissue and become more or less altered. The degree and direction of this specialization determine whether they result in ependyma or neuroglia. For a while the spongioblastic elements are radially arranged, like columnar epithelium. One end, which also contains the nucleus, borders the cavity of the neural canal and projects cilia into it; in the other direction the slender cell extends even to the periphery of the neural tube. Those spongioblasts that retain their primitive shape and bordering relation to the neural canal are known as *ependymal cells* (C, 'Floor plate'). The majority of spongioblasts, however, differentiate further. These elements lose their relation with the neural canal, migrate outward and convert into *neuroglial cells* (C,); some cells, so displaced, retain a peripheral attachment, but most abandon both central and distal connections.

It is of interest to note that the several developmental stages encountered in mammals recapitulate the progressive neuroglial conditions found within the chordate group. In its final state the ependymal tissue consists of elements whose nuclei lie next the cavity of the brain or spinal cord, and whose cell bodies radiate outward like columnar epithelium. Primitive ependymal relations are clearly retained only at the midplane of the spinal cord and medulla in other regions the distal processes of ependymal cells extend merely for varying distances beyond the cell body.

Elsewhere in the brain and spinal cord the supporting elements are neuroglial cells, distributed throughout the mantle and marginal layers. They are of two morphological types: (1) *astrocytes*, stellate in shape and with long processes and (2) *oligodendroglia*, with a smaller cell body and fewer, finer processes (C). A third type *microglia* (D), should be mentioned although it would appear that they do not belong developmentally, structurally or functionally with the true neuroglia. In spite of counterclaims, these elements, which are potential ameboid phagocytes, appear late in the neural tube and probably originate from mesenchymal cells. They could, therefore, be appropriately named *mesoglia*. The astrocytes are derived from full-length primitive spongioblasts, from spongioblasts that (because of the thickness of the tube) never connect with the periphery, and from wandering spongioblasts. Astrocytes appear first in the third month. Those occupying the gray substance are named *protoplasmic astrocytes*, another type, *fibrous astrocytes*, develop fibrils within their cytoplasm and are typical of the white substance (B). The oligodendroglia, derived solely from migratory spongioblasts, arise at a later period than astrocytes (C).

Supporting Elements of the Ganglia

The supporting cells of the ganglia at first make up an apparent syncytium, in the meshes of which are found the neuroblasts. The interstitial elements differentiate both into flattened *capsule cells*, which invest the ganglion cells, and into *sheath cells*, which migrate peripherally along the developing nerve fibers and envelop the axons. Still other migratory non nervous cells, of early neural-crest origin, mingle with the general mesenchyme. Their subsequent history has been explained, at least in part, by properly designed experiments.

The Neurolemmal Sheath

Peripheral nerve fibers are enclosed within a cellular sheath (of Schwann). The component cells, although indistinguishable, have two sources of origin. As already mentioned, some differentiate from the tissue of the early neural crest. Slightly later others emerge from the neural tube by way of the ventral roots. The young sheath cells are spindle-shaped and wrap about bundles of nerve fibers. Multiplication on the part of the sheath cells then separates the bundles into single fibers, each with its own *neurolemma*.

The Myelin Sheath

Between the fourth month of the fetal life and the third month following birth a fatty *myelin sheath*, also called a medullary sheath, begins to appear about many nerve fibers. It surrounds the chief process (axon) and, in turn, is enclosed by the neurolemma. The origin of myelin is in doubt. It does not appear until neurolemmal cells are present on a peripheral fiber, and is laid down as units agreement with the extent of those cells. Yet myelin is closely related to the axon and possibly is 'secreted' by it. Probably the origin should be assigned to the co-operative effort of neurolemma and axon. In the central nervous system there are no characteristic neurolemmal sheaths investing the fibers, yet many acquire a myelin sheath. Nevertheless, scattered *'sheath cells'* are said not only to be present but also most numerous during the period when myelin is differentiating. Some trace their origin merely to the spongioblastic supporting cells of the neural tube, while others identify them more specifically with oligodendroglia.

Myelination

The *myelinated fibers* (i.e., those with a myelin sheath) have a glistening, white appearance which gives the characteristic colour to the *white substance* of the brain and spinal cord and to most peripheral nerves. Many of the fibers of the central nervous system remain *unmyelinated*; this is true of certain fibers coursing in the white substance, while the portions of all fibers lying within the *gray substance* never acquire myelin. Many fibers in the peripheral nerves are also unmyelinated, yet every peripheral fiber is supplied with a neurolemmal sheath. Myelin is deposited first near the cell body of a neuron and then spreads progressively along the fiber. The myelin sheath also appears at widely different times in the

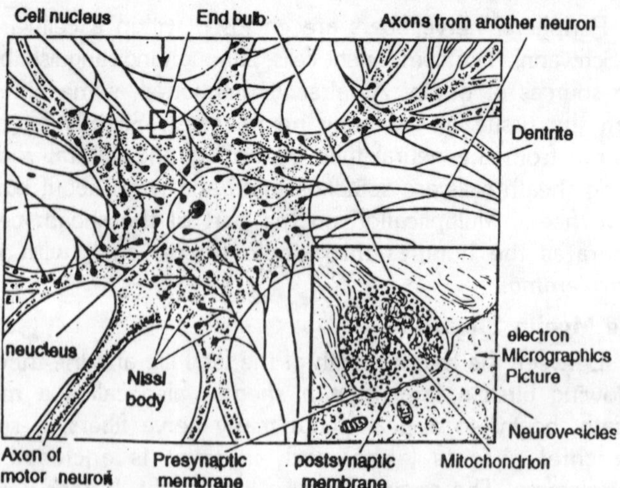

Fig. 8.11. Drawing of the cell body of a motor neuron to show its own dendrites and single axon and the numerous end-bulbs or synaptic knobs from other (afferent) nerve fibers pressing against the cell body.

various fiber-systems. The oldest tracts historically, which are also the earliest to attain function, are myelinated soonest.

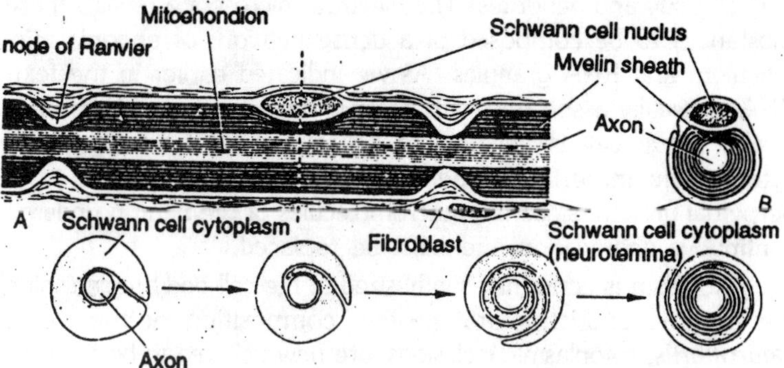

Fig. 8.12. Diagram from an electron photomicrograph of an axon including a node of Ranvier. In the lower series, note how the Schwann cell wraps itself like a sheet around the axon, thus making the myelin sheath. B is a cross section at the dotted line in A.

In general, tracts acquire myelin at about the time they become capable of functioning. Since myelin is deposited in the various fiber tracts at different developmental periods, this condition has been of great help in tracing the origin, course and extent of the various fiber-groups within the nervous system. The development of myelin in the spinal cord begins at the middle of fetal life and is not completed in some fibers until adolescence. It occurs first in the cervical cord and then extends progressively to lower levels. Fibers of the ventral roots acquire myelin before those of the dorsal roots. Tardiest of all are certain descending motor tracts (cortico- spinal; tecto-spinal) which myelinate largely during the first and second postnatal years. The brain begins to myelinate in the sixth fetal month, but progress is slow and only the fibers of the basal ganglia and those that continue the structure of the spinal cord upward possess myelin sheaths at the time of birth. In fact, the brain of a newborn is still largely unmyelinated, so that deposition goes on chiefly from birth through puberty. First to acquire sheaths in this period are the primary sensory-motor fields- that is, the olfactory, optic and auditory cortical fields and the motor cortex. The projectional and commissural fibers myelinate last.

Typical Neurons

The cell body, or *perikaryon*, of a typical neuron is usually round or oval, but it may take various shapes. Its large, lightly basophilic nucleus contains a single prominent nucleolus. Basal condensations. *Nissl substance*, are present in the cytoplasm of the cell body and dendrites. The electron microscope reveals these substances to be composed of a dense network of endoplasmic-reticulum and RNA granules. As we indicated earlier in the text, RNA is usually associated with protein synthesis and its presence in the nerve cell is believed to be related to the turnover of cytoplasmic material in the axon. The mature neuron is in perpetual growth, since the macromolecules of the neuron undergo continuous degradation and must be replaced.

Axoplasm is constantly synthesized in the cell body replenishing the enzyme and structural protein composition of the axon. *Neurofibrils*, cytoplasmic inclusions, are now believed to be artefacts produced in preparing the tissue for histologic sectioning. However, the *neurofilaments* and *microtubules*, seen in the electron micrographs, are considered part of the cytoplasm, although we do not completely understand their function. Two types of processes extend from the cell body, the *dendritic* and *axonal* processes.

In the past, their identification was based on what was traditionally considered a typical nerve cell, the motor neuron. However, anatomists named the processes one way, neurophysiologists, another way. Today, we realize that no cells in the body differ from one another more than the cells of nervous tissue and that there is no such thing as a typical nerve cell. In an attempt to take into account current functional-anatomical concepts, the conventional terminology was re-examined and a modified nomenclature of vertebrate neuron structure was proposed. Since the perikaryon (cell body) is considered primarily a growth center, its position on the fiber has little to do with electrochemical functions. Therefore, terminology based on the direction of travel of the nerve impulse, that is, toward or away from the cell body, is meaningless.

The new nomenclature refers to a *dendritic zone* of a neuron. In this zone the receptor membrane consists of cytoplasmic branches modified either to receive synaptic endings of other neurons (such as an association neuron), or to respond to stimuli by generating a nerve impulse (as in a sensory neuron). The many dendrites (*dendron*, '*tree*') branch in the motor neuron and end in knoblike structures called *gemmules*. On Purkinje cell of the

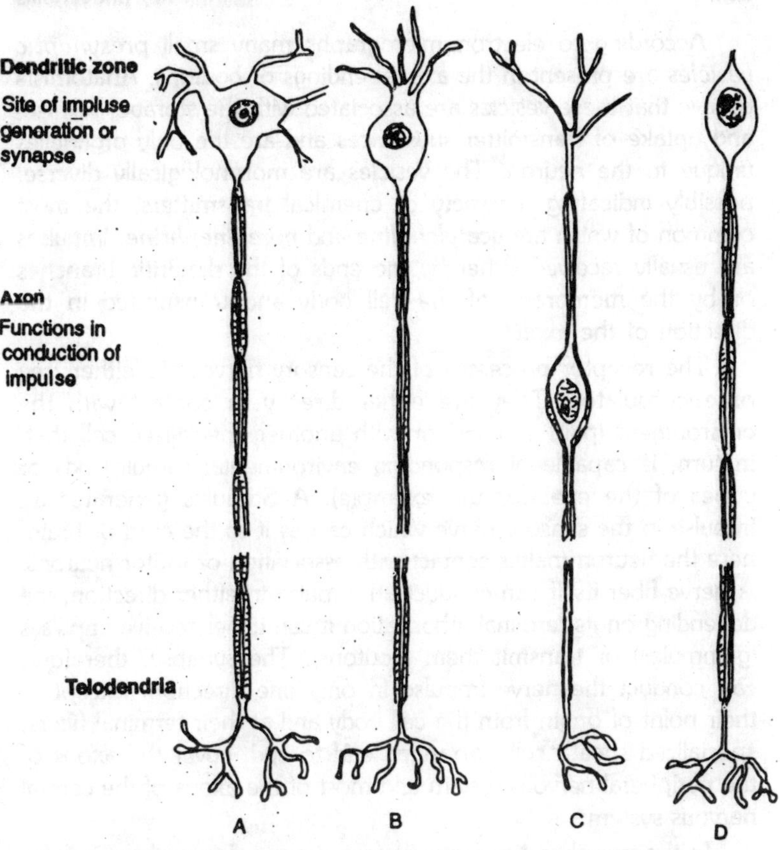

Fig. 8.13. Diagrams of a typical motor neuron, A, and three different types of receptor neurons, B to D.

cerebellum, for example, may have as many as 200,000 terminals. The *axon* is a long extension of cytoplasmic material differentiated to carry nerve impulses away from the dendritic zone. It is synonymous with the general terms "long peripheral fiber" or "axis cylinder". It may be branched and is of uniform diameter. A *neurolemma* (specialized sheath cells) usually wraps around the axon. *Axon telodendria* are the endings of axon functioning in synaptic transmission and neuro-secretory activity. They may terminate in button-like enlargements, *boutons*. These are in contact with the processes or cell membrane of another neuron, or they terminate at a gland, a special unfolded area of a muscle, or at a blood vessel.

According to electron micrographs many small *presynaptic vesicles* are present in the axonal endings or boutons. Anatomists believe that these vesicles are associated with the storage, release, and uptake of transmitter substances and are the only organelles unique to the neuron. The vesicles are morphologically diverse, possibly indicating a variety of chemical transmitters, the most common of which are *acetylcholine* and *norepinephrine*. Impulses are usually received either by the ends of the dendritic branches or by the membrane of the cell body and transmitted in the direction of the axon.

The receptor processes of the sensory nerves are either free or encapsulated. They are either directly in contact with the environment (pain endings) or with another specialized cell that, in turn, is capable of responding environmental stimuli (rods or cones of the eye, for the example). A Stimulus generates an impulse in the sensory nerve which carries it to the cord or brain; here the neuron makes contact with association or motor neurons. A nerve fiber itself can conduct an impulse in either direction, but depending on its terminal arborization it can either receive impulses (gemmules) or transmit them (boutons). The synapse, therefore, can conduct the nerve impulse in only one direction. Except at their point of origin from the cell body and at their terminal fibers, specialized sheath cells, arranged end to end, cover the axons of the peripheral nervous system and most of the axons of the central nervous system.

In the peripheral nervous system these cells are derived from the neural crest cells and are known as the *neurolemma*, or *Schwann cells*. Most of the axons also have a sheath of fatty material. *Myelin*, inserted between the outer membrane of the Schwann cells and the membrane of the axonal fibers. Those nerve fibers that contain this fatty sheath are referred to as *myelinated fibers*; those fibers that lack the sheath are *non-myelinated*. The myelinated fibers are usually larger than the non-myelinated and conduct impulses at a faster rate. The post-ganglionic fibers of the autonomic system, those that pass from the autonomic ganglion to a muscle or gland, are examples of fibers that lack myelin sheaths. The axons of these fibers are usually grouped together with as many as 20 axons in a bundle.

Each axon sinks into an invagination of the cell membrane of a neurolemma cell, the so-called *mesaxon A*. As a result, one sheath cell encloses a bundle of axon fibers. On the other hand, as shown by electron microscopy, a single Schwann cell wraps a myelinated fiber, often as many as 50 times or more, as illustrated. This is sometimes referred to as the "*jelly roll*" formation of myelin. A cross section of a myelinated nerve fiber reveals a sheath composed of several concentric layers of the Schwann cell's plasma membrane. As the Schwann cell wraps itself around the axonal fiber, the cytoplasm is squeezed out in some manner and the plasma membranes of the Schwann cells are left. These membranes consist of a lipid layer of molecules sandwiched between two protein layers.

The protein layers of two adjacent membranes, fuse, freeing the lipid layer. It is this lipid component of the Schwann cell membrane that constitutes the myelin. Myelination begins to develop in the human fetus about the fourth month and may continue until the second or third year after birth. A constriction, the *node of Ranvier* marks the point where one Schwann cell ends and another begins. Although the myelin sheath is absent at these nodes, the axonal fiber is not naked at this point. The ends of one Schwann cell overlap another at the nodes and form a protective membrane.

Since the cytoplasm is squeezed out from between the layers of the membrane as the Schwann cells rotate around the fiber, most of the cytoplasm accumulates along with its nucleus in the outermost membrane. It is this outer portion of the Schwann cell that is customarily referred to as the neurolemma. The cells of the central nervous system lack Schwann cells (neurolemma). Instead, one type of glial cell (oligodendroglia cells) provides the myelin around the fibers. The myelinated fibers of the central nervous system, therefore, structurally resemble the myelinated fibers of the peripheral nervous system. We do not understand the exact function of the myelin sheath, but assume it acts as an insulating material that prevents nerve impulses from jumping from one axon to another. The thickness of the myelin sheath apparently influences the speed of conduction, since the thickest fiber is the fastest conductor. The myelin may also function in a nutritive role for the enclosed axon.

Neurosecretory Cells

Although the neuron functions conventionally as the conductor of nerve impulses, some neurons are modified to secrete. These neurons, referred to as *neurosecretory cells*, are found in the hypothalamus and other parts of the nervous systems of both vertebrates and invertebrates. Sensory neurons, relaying all possible types of information from other parts of the body, synapse with the dendrites of the neurosecretory cells. Unlike ordinary neurons, the axons of these cells end near a capillary bed. They do not synapse with other neurons nor innervate effector organs. A stimulated neurosecretory cell synthesizes its product and transports it is the form of visible droplets down the axon toward the blood vessel. Here it is either temporarily stored or discharged into the bloodstream, as suggested.

The circulatory system carries the product of the neurosecretory cell to some distal site on the body where it produces prolonged effects. The neurosecretory cells link the nervous system with the endocrine system, the two co-ordinating systems of the body. Transmissions of information to these cells enables them to convert sensory information into a hormonal message that can produce a physiologic response in the organism. Such a circuit is referred to as a *neuroendocrine reflex*.

Synapse

New information resulting from electron microscopic studies and microelectrode and pharmacologic analysis forced the neuroanatomists to change some of the their concepts of the organization and structure of nervous tissue. The nature of the *synapse* morphology is one of these areas. Anatomists no longer talk of the *Synapse* but realize that there is a variety of mechanisms involved in cell-to-cell communication. A synapse is traditionally described as the end-to end coupling of the axon telodendria of one neuron with the dendritic receptor sites of another neuron. New data reveal, however, that this conventional picture is only one of several relationships that exist between neurons demonstrates some of the diversity of neural linkages that have been described and that do not fall readily into the suggested nomenclature system proposed earlier.

The coupling of one cell body with another dendrite to dendrite, and axon to axon occur frequently. However, cell body to dendrites and cell body to axons have also been described. Impulses travel down nerve processes until they come to the point where one cell process contacts the cell process or membrane of the cell body or another. This physical junction between the

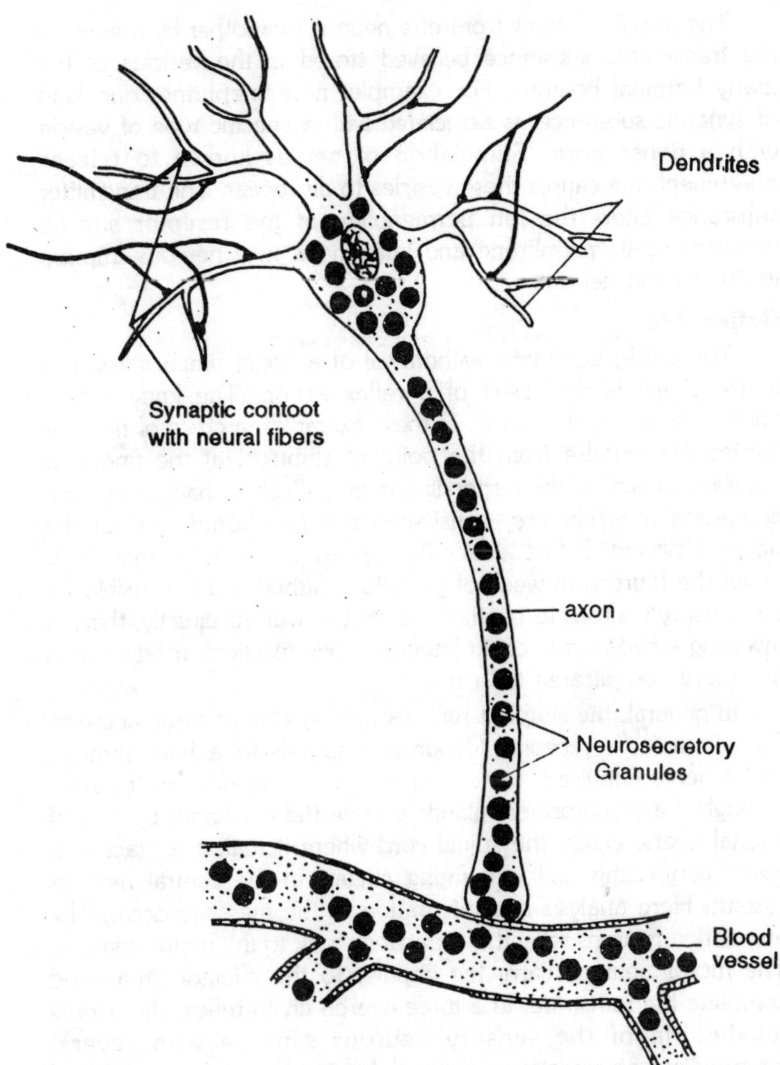

Fig. 8.14. Sensory neural fibers synapse with the dendrites of a neurosecretory cell, stimulating it to synthesize its product. These neurosecretory materials are then transported in the form of visible droplets down the axon toward the blood vessel.

transmitter site of one neuron and the receptor site of another is known as the *synapse*. The two membranes, closely apposed to one another, have only a small but variably sized *synaptic cleft* between them.

The impulse passes from one neuron to another by the aid of the transmitted substance believed stored in the vesicles of the many terminal boutons. For example, norepinephrine, one kind of synaptic substance, is associated with a specific type of vesicle with a dense core. Stimulation of nerves known to release norepinephrine causes these vesicles to disappear. The transmitter substance alters the ion permeability of the receptor site by depolarizing its membrane and initiating a new nervous impulse in the second neuron.

Reflex Arc

The quick, automatic withdrawal of a finger when it touches a hot object is the result of a reflex action. The knee jerk is another common illustration. In these examples, a chain of neurons carries the impulse from the point of stimulus, at the finger or patellar ligament to the responding muscle. Such a chain of neurons composes a *reflex arc*, considered the *functional unit* of the nervous system. It first makes its appearance in the human fetus about the fourteenth week of gestation. Although it is possible for the sensory neurons to contract the motor neurons directly, thereby involving a two-neuron chain (monosynaptic reaction), most reflexes are more complicated than this.

In general, the simplest reflexes involve at least three neurons. For example, a receptor (dendrite) responds to a heat stimulus and a nerve impulse is generated in the sensory neuron. It passes through the cerebrospinal glands outside the cord and, by way of axonal fibers, enters the spinal cord where it makes contact with many connecting and integrating circuits in the central nervous system. Here analysis and integration of the message occur. The association neurons relay the commands back to the motor neurons. The motor neurons carry the impulse to the effector organ and complete the reflex arc. In a three-neuron chain reflex, the axonal telodendria of the sensory neurons synapse with several intermediate or association neurons located in the gray matter of the spinal cord. These synaptic connections increase considerably the number of possible pathways. Each association neuron, in turn, receives impulses from several different sensory neurons. The spray of branches at the end of the axonal fiber then allows the association neurons to relay the impulses to several motor neurons.

Obviously such interrelationships result in a kind of nerve net. Sometimes the response is visible such as an overt behaviour

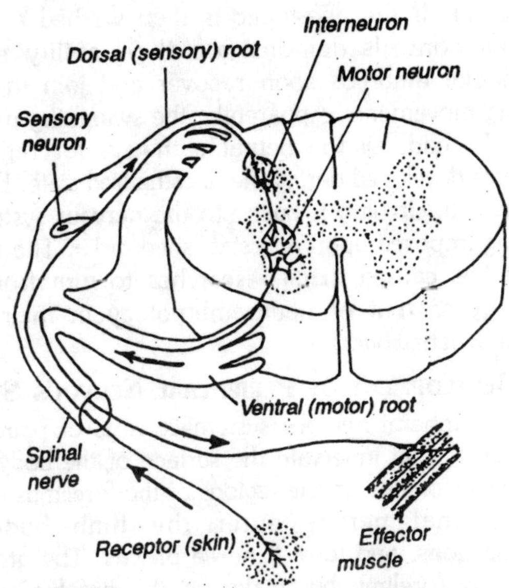

Fig. 8.15. Cross section of spinal cord to show a very simple reflex pathway.

reaction (running away) or the knee jerk mentioned above, sometimes the response is invisible and involves regulative changes in internal organs (for example, secretion by glands or changes in circulation). It is those *automatic reactions*, visible or invisible, that are defined as *reflex actions*.

A reflex may go on at one level of the central nervous system or it may involve other parts of the nervous system as well. For example, in the withdrawal of the toe from the stimulus of a sharp object, the individual may move his arms and utter an exclamation of pain as he removes his foot. If both arms are raised, fibers of the intermediate neurons must have crossed to the other side of the cord and relayed the impulse to motor neurons on this side, the opposite side to that in which the stimulus had occurred. Intermediate neurons that cross to the other side of the cord are referred to as *commissural neurons*. There is evidence that reflex circuitry is laid down early in the embryo and that the nervous system develops orderly pattern of behaviour long before they are actually used. One can demonstrate this by anesthetizing a young embryo at its learning stage.

A young amphibian embryo anesthetized before its muscles have functioned remains quiescent during the completion of their development. If the anesthetic is then washed away when the untreated controls demonstrate their ability to swim, the experimental embryos soon recover and join in the vigorous swimming movements. Apparently, the swimming pattern was not learned. Instead, all the neural pathways for reflex swimming movements developed during the anesthetized state. Evidently much of behavioural circuitry is built into the nervous system and is not something imposed upon an established order. The recognition of this fact has caused many researches to turn from a study of psychology to that of neuroembryology in their attempts to understand behaviour.

Development of Peripheral Nervous System

The peripheral nervous system consists of paired spinal and cranial nerves that innervate the surface of the body or penetrate to the deeper organs. In the regions of the forelimbs and hindlimbs, several spinal nerves enter the limb buds, establish interconnections, and form a *nerve plexus*. The group of nerves entering the forelimb but make up the *brachial plexus*; those that innervate the hindlimb bud compose the *lumbosacral plexus*.

Formation of Spinal Nerves

The location of the somites on either side of the neural tube in the young neurula determines the position that some of the neural crest cells (future sensory nerve cells) take as they migrate away from the newly closed neural tube. One can demonstrate the importance of the somites by surgically removing them or transplanting additional somites along one side of the body. The opposite unoperated side remains as a control.

A corresponding decrease in neural crest cell clusters occurs when somites are removed. Conversely, additional somites induce an increase in numbers of cells clusters when compared with the unoperated side. There are 31 pairs of spinal nerves in the human embryo, but only 10 pairs in the frog. Each spinal nerve is connected to the spinal cord by way of two roots, *dorsal* and *ventral*, that have different embryonic origins. In amphioxus and some cyclostomes the two- roots remain separated, but in all gnathostomes they unite to form a *spinal nerve*. Both sensory and motor elements must be present in the nerves if reflex arcs, the functional units of the nervous system, develop. The axons of the neuroblasts located in the motor nuclei grow laterally and penetrate the outer wall of the neural tube, the *marginal zone*.

Here they emerge as the *ventral*, or *motor*, *root* of a spinal nerve, A. These motor fibers are classified as either *somatic* or *visceral fibers*.

Somatic motor fibers carry impulses to the voluntary muscles of the body, those usually derived from myotomes. Visceral motor fibers, on the other hand, pass deeply into the body of the organism and innervate such internal organs as the heart, stomach and intestine. They maintain those routine functions of the body that are under more or less automatic control. The motor fibers of this visceral system are involuntary in nature and are usually referred to as the *autonomic nervous system*, as compared to the conscious, or willful, nature of the somatic system. The neuroblasts located in the cerebrospinal ganglia develop into *sensory* neurons, either *somatic* or *visceral*, depending on the organs they innervate. In either case, processes extend from the neuroblast (formerly the neural crest cell) in both directions, out to the receptor and in toward the spinal cord. In the gray matter of the spinal cord they synapse with association neurons. The fibers of the nerve cells collect into bundles outside the spinal cord and give rise to the *dorsal*, or *sensory*, *root* of the spinal nerve. The dorsal root unites with the ventral root to form a *spinal nerve*, as can be seen in the section of a 10 mm. pig. A spinal nerve is usually a *mixed nerve*; it can be composed of four different types of nerve processes. It may contain somatic sensory (afferent) and visceral sensory (afferent) fibers, as well as somatic motor (efferent) and visceral motor (efferent) processes.

Growth of Nerves

The factors responsible for directing or guiding certain neurons to specific structures or other neurons have been the subject of much investigation. According to Weiss (1955), there are three phases in the development of nerves; the pioneering phase, the application phase, and the towing phase. In the pioneering phase the cytoplasm protrudes from the side of the nerve as a short thread. The thread lengthens because of the synthesis of protoplasm by the cell body and its flow to the tip of the sprout where ameboid processes develop. During the application phase the free pioneering tip of the fiber attaches itself to a receptor or to an effector cell, located nearby at this early stage.

As the embryo grows, the peripheral cell shifts and during the migration "*tows*" (third phase), the nerve fiber with it. If these peripheral cells migrate outward and become, for example, the big toe of a hindlimb the distance the axonal fiber is towed can be considerable. We do not understand why the pioneering fiber attaches to a particular cell. The problem of innervation obviously concerns the orientation of the pioneering fibers. Once the application phase is established, additional new fibers from other cells follow the course set by the pioneers, obviously using it as a solid substrate. There is general agreement that nerve fibers follow oriented structural or ultrastructural elements.

One hypothesis suggests that the tips of nerves grow into actively growing limb buds by following the stress lines set up as the rapidly dividing mesenchyme cells absorb water. The tendency of nerve fibers to grow along solid objects is referred to as *stereotropism*. The growth of the optic nerve offers a more obvious illustration. The fibers from the inner layer of ganglion cells in the retina converge, pierce the choroid and sclera, and follow the optic stalk as a pathway into the brain. As the fibers grow back into the brain substance, they group together and give rise to the optic nerve. There is general agreement that growing nerves are subject to contact guidance.

Whether additional factors (chemical or electrical) are involved remains a controversial issue. Some investigators explain the migration of nerve fibers toward the source of their stimuli by postulating a difference in electric potential. However, the application of actual electrical currents does not affect either the rate or direction of growth of the fibers. There is little evidence that a fiber can electrically select a particular pathway. According to Sperry (1965), the modern concept of inherent chemical affinities of the neurons can best explain the selectivity of a growing nerve fiber for a particular pathway. Each of the billions of cells within the nervous system has a specific chemical identification tag dictated by the genetic code.

According to this thesis, as the neurons mature and begin to form associations with other neurons, they are attracted to chemically compatible cells or fibers from the millions, or billions, of nerve cells with different chemical identities. As a result numerous neurons having similar chemical identities establish brain pathways. However, chemical affinity most not be total answer. There are too many different connections that must be made in the total organism for such a simple explanation to apply. Like most embryonic cells, the developing peripheral nerves appear to be unspecialized as they grow out toward and end organ. There is evidence that a motor nerve makes connection with any muscle cell in its path and becomes specialized after it makes this contact.

Motor fibers connect to muscles and glands, and sensory fibers connect to receptors. Evidently these end organs bring about specific chemical changes in the nerves they contact. Once attained, the specificity of a neuron is irreversible. Sperry (1951) obtained evidence for this specificity when he crossed the cutaneous nerves of the left hind foot to the right hind foot in rats 14 to 26 days old. After the rats recovered from the operation, he found that stimulation of the right foot caused the animal to lick its left foot caused the animal to lick its left foot. No response occurred on the stimulated right side and the proper reactions to the stimuli could not be induced by training. These reactions suggested with the end organs of their receptors and this reaction determined the specificity of these nerves. This information was recorded in the brain. As development proceeds, the brain irreversibly registers a copy of the non-neural portion of the body.

A condensed, point-by-point map of all the body's surfaces, both internal and external organs and muscles, is imprinted on the nervous system. By the time the operation was carried out in the above experiment, the nerves passing to the left side of the animal had contacted the end organs, become specialized as the left sensory; or motor leg nerves, and been registered as such in the central nervous system. Therefore, when the left nerve was transplanted to the right side in the postnatal animal and the right side was stimulated, the animal lifted its left foot. This lack of plasticity offers a possible explanation as to why; an amputee continues to feel pain in his leg when the stump is stimulated.

Formation of Cranial Nerves

The origin of the cranial nerves is much the same as spinal nerves, in that motor fibers arise from nuclei located in the gray matter and sensory neurons form from neuroblasts derived either from neural crest cells or ectodermal placodes. There is one difference, however. Although cranial nerves may be mixed nerves, in many cases they are one or the other, either sensory or motor in function. Some biologists believe that cranial nerves at one time contained two roots that did not unite, like the spinal nerves in amphioxus and the cyclostomes. In some cases, the dorsal root became the predominant component and the ventral root degenerated leaving a purely sensory nerve. In other instances, the dorsal root was lost and the ventral root remained as the motor nerve. Segmentation is also lost in the cranial region since some of the cranial nerves represent roots from more than one segment. As an illustration, the hypoglossal nerve (twelfth nerve) forms from the union of ventral roots from several occipital nerves.

Structure of a Nerve

A spinal nerve consists of axons from the somatic as well as the visceral nervous systems. A cross section of the nerve as seen best reveals its structure. Each nerve axon is surrounded by a sheath of Schwann cells (*neurolemma*), but may or may not be myelinated. The axons, therefore, are of various sizes ranging from large axons with a thick myelin sheath, fibers of smaller diameter and thinner sheath and thin fibers with no sheath. The large myelinated fibers have a more rapid rate of conduction than do the smaller myelinated fibers. The fibers are arranged in parallel groups and collected into bundles, or *fascicles*, by a connective tissue sheath, the *endoneurium*. When we refer to a nerve, we are speaking of a group of fascicles bound together by the *perineurium*. This connective tissue sheath not only protects the nerve bundles, but also functions as a diffusion barrier. Several nerves make up a *nerve trunk*. A connective tissue membrane, epineurium, ensheathes a nerve trunk. This membrane may extend into the nerve trunk as a septum helping to further separate the groups of nerves.

Spinal Cord

General Description

At first the spinal cord is a thick-walled tube with a central canal. This canal gradually diminishes in size as development continues; in the human adult it is usually obliterated. The spinal cord extends down most of the length of the body and undergoes considerably less modification than that of the brain. The adult cord is usually oval in shape and flattened more on its ventral side than on its dorsal side. There is no special boundary between it and the brain at the anterior end, other than the cervical flexure present in amniotes. At the posterior end, the spinal cord tapers to an end, the *filum terminale*.

In tetrapods, two prominent bulges appear, a cervical enlargement at the level of the forelimbs and a lumbar enlargement opposite the hindlimbs. They are most pronounced in amniotes and reached their greatest size in dinosaurs, where they are believed to have been much larger than the brain itself. The enlargements are the result of an increase in numbers of cell bodies at the level of the limbs and mark the point of departure from the cord of neurons that pass laterally into these structures. Early in their development, the neural tube and vertebral column are approximately the same length. However, the vertebral column (cartilage or bone) develops at a faster rate than does the cord. The coccygeal end (posterior) of the cord terminates at the lumbar region in most vertebrates, and the cord, except for the filum terminale, is absent from the posterior regions of the vertebral column. In man the spinal cord terminates at the level of the third lumbar vertebra; in other animals it may be longer but does not usually reach the full length of the vertebral column.

As a result of this unequal growth, the spinal nerves in the cervical regions pass laterally into the organs they innervate, but in the posterior part of the cord they are arranged in a more craniocaudad direction. A *ventral fissure* is present on the undersurface of the cord of higher vertebrates. A longitudinal *dorsal septum* extends inward from a *dorsal sulcus* (depression). This septum, together with the dorsal sulcus and ventral fissure, partially divides the cord into two symmetrical halves as may be. The two halves are connected with one another by nerve fibers (*commissures*).

Gray Matter of Cord

A cross section of the adult cord of amniotes shows that the mantle area of the original neural tube has undergone considerable expansion during development. It forms a "butterfly", or "H-shaped", structure that fills the central part of the section. Anatomists speak of the *dorsal* and *ventral horns* of the gray matter. These are actually cross sections of *columns*, areas of common function, that run up and down the cord. In all vertebrates *dorsal* and *ventral gray commissures* located on either side of the central canal connect the two sides of the cord. The presence of sulci and fissures, as well as dorsal and ventral columns, is much more pronounced in higher vertebrates than in lower forms. For example, the spinal cord of cyclostomes lacks sulci and fissures, and there is no definite demarcation between gray and white matter. The cell bodies of the *association neurons* are located in the dorsal columns of the gray matter. These multipolar neurons receive the impulses from the axons of the afferent (sensory) neurons. They co-ordinate the impulses brought in by the sensory neurons to those sent out by the motor neurons. The fibers of the association neurons may remain on the same side of the cord or cross to the other side. They may make contact ventrally (synapse) with the dendrites of the motor neurons, or they may send their axons into the white matter where they ascend to the brain.

The cell bodies of the motor neurons, located in the ventral columns, are distinguishable by their large size and large amount of Nissl material. Their axons pass out laterally, forming the ventral root fibers of the spinal nerves. The gray matter, therefore, consists of interstitial cells, association neurons, cell bodies and dendrites of motor neurons, portions of myelinated and non-myelinated fibers, and synapses. Neural anatomists distinguish functional areas (somatic or visceral) within the dorsal and ventral columns of the gray matter. They identify four such areas on each side of the cord: two areas in the dorsal column where somatic and visceral sensory neurons synapse with association neurons, and two in the ventral area where the cell bodies of the visceral and somatic motor neurons are located. From dorsal to ventral, the areas may be listed sequentially as the somatic afferent and visceral afferent, located in the dorsal column of the gray matter, and visceral efferent and somatic efferent, in the ventral column A.

The somatic areas are usually larger than the visceral ones. These general regions persist and can be identified as far forward as the medulla oblongata in higher animals. Anterior to the medulla, this structural plan changes. However, in the amphibian, for example, the entire central nervous system (brain and cord) resembles the cord of higher animals.

White Matter of Cord

Longitudinal tracts, referred to as *funiculi* (to differentiate them from the columns of the gray matter) compose the white matter of the cord. Myelinated and some non-myelinated fibers of sensory, motor, and association neurons make up the funiculi. Some transverse fibers are also present running to and from the gray matter. In general, there are no cell bodies and no dendrites in the white matter. The shape of the gray matter divides the surrounding white matter into three general areas on each side of the cord. They are the *dorsal funiculi* located between the dorsal sulcus and dorsal column, the *lateral funiculi* situated between the dorsal and ventral columns, and the *ventral funiculi* between the ventral column and ventral fissure. A white commissure connects the two halves of the cord. The three funiculi can be further divided into specific *fiber tracts* (fasciculi). A detailed discussion of the fiber tracts of the cord is beyond the scope of this book.

One should realize, however, that these tracts connect one part of the cord with another and join the brain with the peripheral system. Their importance increased as the brain evolved and assumed more of the co-ordinating and association functions of the body. The fiber tracts developed to a greater degree in higher animals than in the fist where the trunk is more or less semi-autonomous. They are two-way conduction paths, consisting of ascending axons carrying impulses up the cord and to the brain and descending axons relaying message back to the cord and to the peripheral neurons. Bundles of axons compose the tracts and

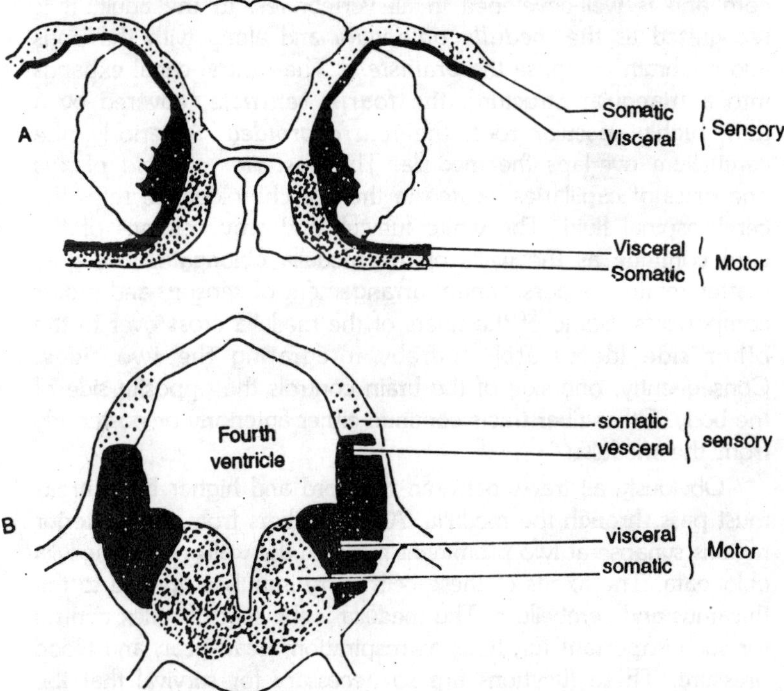

Fig. 8.16. Diagram showing similarity in distribution of sensory and motor columns in spinal cord, A, and embryonic medulla oblongata, B.

many tracts make up the funiculi. In general, ascending fibers from some of the sensory neurons compose each dorsal funiculus. The fibers enter the cord through the dorsal root and then ascend to a higher level instead of synapsing with an association neuron.

Ascending axons from association neurons are also present in this region of the cord. The ventral funiculi carry impulses down the cord and from the brain centers. The cell bodies of these neurons are located in the higher centers in the brain, and the descending fibers synapse either directly with a motor neuron in the ventral column of the cord or with an association neuron, which then contacts the motor neuron. The lateral funiculi consists of both ascending and descending fibers that appear to connect one area of the cord with another.

BRAIN

Myelencephalon

The myelencephalon is the anterior continuation of the spinal cord and is well-developed in all vertebrates. In the adult, it is recognized as the *medulla oblongata* and along with the pons and midbrain compose the *brainstem*. The central canal expands into a triangular structure, the *fourth ventricle*, covered by a thin, highly vascular roof, the *tela choroidea*. Anteriorly, the cerebellum overlaps the medulla. The *posterior choroid plexus* (the mass of capillaries located in the tela choroidea) secretes the cerebrospinal fluid. The white funiculi and gray columns of the cord continue as the walls of the medulla oblongata. The gray matter retains its dorsoventral arrangement of sensory and motor components. Some of the fibers of the medulla cross over to the other side (decussate), thereby integrating the two sides. Consequently, one side of the brain controls the opposite side of the body. Other fiber tracts continue either anteriorly or posteriorly from the medulla.

Obviously all tracts between the cord and higher brain areas must pass through the medulla. Afferent fibers from the posterior regions synapse at two prominent nuclei in the walls of the medulla oblongata. The axons of these cells carry impulses upward to the thalamus and cerebellum. The medulla is the seat of reflex control for such important functions as respiration, heart-beat, and blood pressure. These functions are so necessary for survival that the nuclei of the medulla are referred to as *vital centers*. An injury to this area of the brain often results in death. The reflex centers for certain non-vital functions are also located here including centers for vomiting, coughing, sneezing, swallowing hiccuping and salivation. The majority of the cranial nerves are associated with the medulla oblongata. The fifth to twelfth cranial nerves leave or enter the medulla.

Metencephalon

The metencephalon gives rise to the anterior regions of the medulla oblongata, the cerebellum, and the pons. The dorsal part of the metencephalon becomes the elevated and thickened *cerebellum*, which seems to be directly related to the locomotor activity of the organism. The cerebellum does not itself direct activities, but it moderates and co-ordinates those of higher brain centers. It is poorly developed in sedentary forms of vertebrates and most highly developed in birds and mammals. It is a receiving center for sensory and motor information and functions by controlling skeletal muscles.

The cerebellum is poorly developed in the cyclostomes, some fishes amphibian and reptiles. In these sluggish and poikilothermic forms, it consists of little more than a shelf of tissue anterior to the fourth ventricle. The posterior dorsal area of the metencephalon receives information from the vestibular portion of the eighth cranial nerve carrying impulses from the semicircular canals. Also nerves returning from the lateral-line organs give information about vibrations, water currents, and movement of nearby objects. After receiving these sensory data, the cerebellum sorts, correlates, and regulates the body movements.

In more active animals (Elasmobranchii, Dipneusti, etc.), irregular-shaped lobes called *auricular lobes*, or *restiform bodies*, project laterally from each side. Their cavities are continuous with the fourth ventricle and they represent the most ancient part of the cerebellum. These lobes are intimately connected with the inner ear and function primarily as organs of equilibrium. In amniotes, the lateral line components drop out, and as the muscles associated with terrestrial living develop, the proprioceptors increase in importance. The cerebellum of birds is more advanced than in most of the lower animals. It is divided into a middle "*worm-shaped*" portion, called the vermis (consisting of many horizontal ridges), between two lateral lobes or hemispheres. Its development is associated with the co-ordination of the intricate flight muscles.

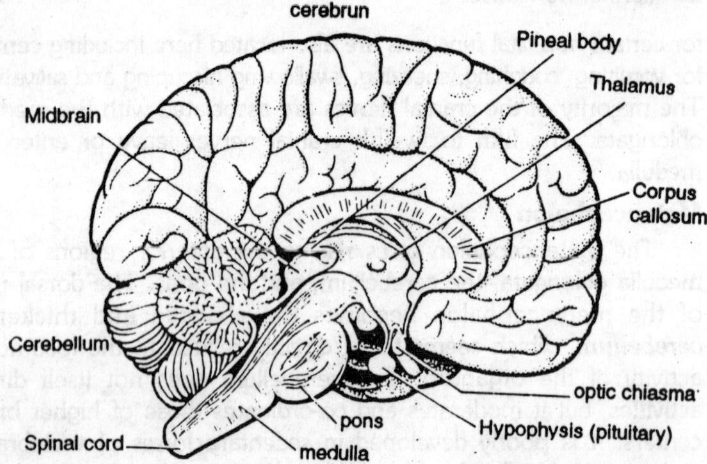

Fig. 8.17. Left half of the human brain, median aspect. The corpus callosum connects the two hemispheres.

In the mammal the cerebellum reaches its greatest degree of development. It also receives information about the position of body parts from the organ of equilibrium and on the state of muscle tension from muscle and tendon spindles types of receptors. The cerebellum sends and receives fibers to and from the higher brain centers. In the mammal the highly developed cerebral cortex takes over much of the control from the cerebellum. Conspicuous

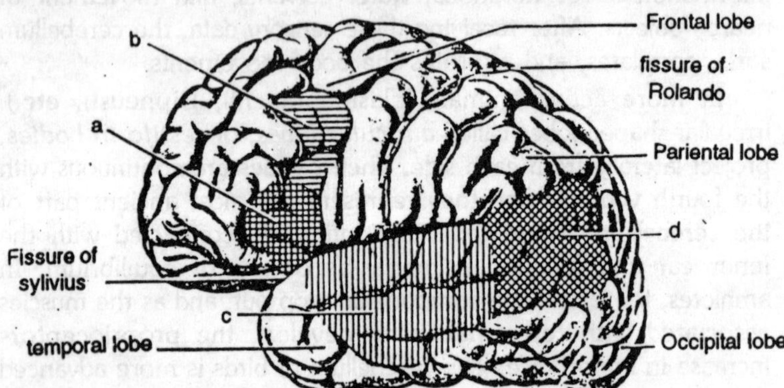

Fig. 8.18. Left hemisphere of the human brain, lateral view. Shaded areas are connected with speech. Destruction of region a results in loss of ability to speak, of b to write words, of c to recognize spoken words, and of d to recognize written words.

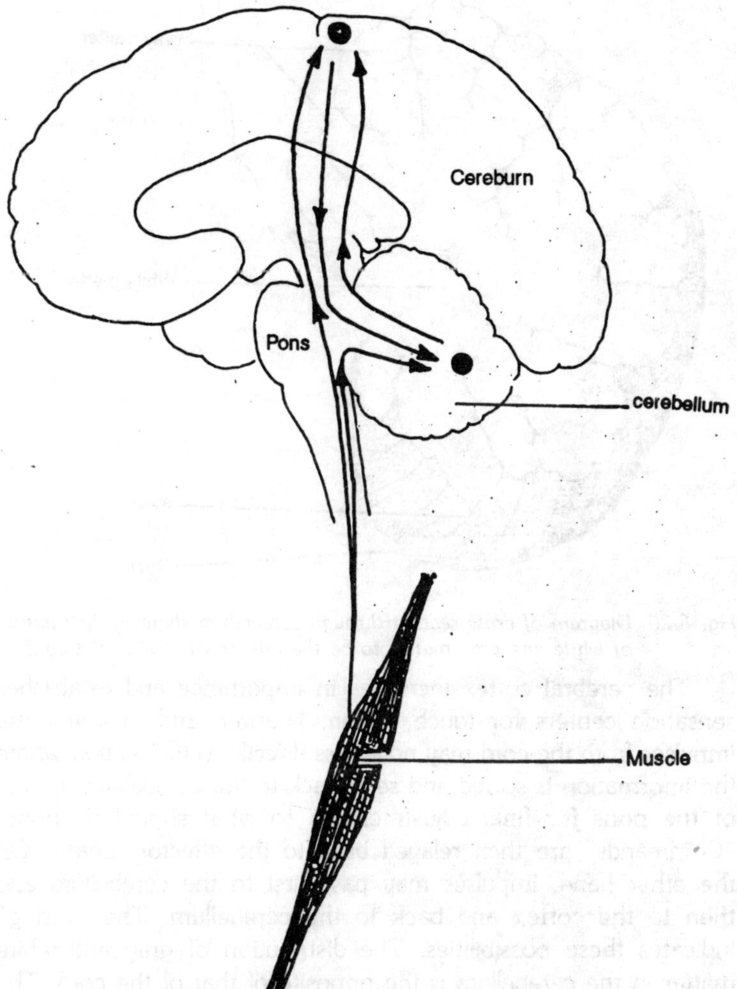

Fig. 8.19. "Wiring diagram" indicating that impulses from spinal cord may pass either-directly to cerebrum or first cerebellum and then to the higher center. The cerebrum sorts the information it receives and then sends instructions back to the cerebellum as to what should be done.

fiber tracts connect the spinal cord and medulla with the cerebellum, and the cerebellum with the cortex. These tracts are located on the ventral part of the myelenecephalon and metencephalon and externally appear as a bulge, the pons (Latin word meaning 'bridge').

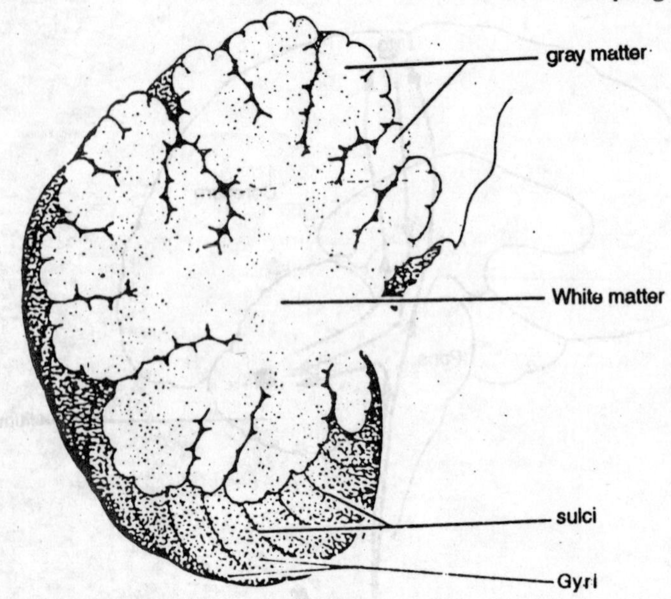

Fig. 8.20. Diagram of cross section through cerebellum showing distribution of white and gray matter to be the reverse of that of the cord.

The cerebral cortex increases in importance and establishes sensation centers for touch, vision, hearing, and muscle tone. Impulses from the cord may now pass directly to the cortex, where the information is sorted and send back to the cerebellum by way of the pons for final adjustment as to what should be done. "Commands" are then relayed back to the effector organs. On the other hand, impulses may pass first to the cerebellum and then to the cortex and back to the cerebellum. The "wiring" indicates these possibilities. The distribution of gray and white matter in the cerebellum is the opposite of that of the cord. The white matter is concentrated in the center of the cerebellum, and the gray matter, which has increased in amount, forms a convoluted covering over the white matter. The surface of the gray matter has numerous grooves (sulci) and convolutions (gyri). A sagittal section through the cerebellum shows that the white fiber tracts spray out into the gray matter of the cerebellar cortex.

Mesencephalon

Unlike the other parts of the embryonic brain, the midbrain does not divide further. In lower animals it is linked to the visual sense and is an important association center. In higher forms many of its activities are taken over by the cortex, and it remains primarily a tract to and from these higher centers. In cross section, the mesencephalon consists of the thickened dorsal region of gray matter, the *tectum*, and the thinner walls and floor of gray matter referred to as the *tegmentum*. The cavity of the mesencephalon decreases during development, forming the narrow *aqueduct*. It communicates posteriorly with the fourth ventricle and anteriorly with the third ventricle.

In 'lower animals the tectum, of root, an important visual center, appears as a pair of dorsolateral *optic lobes* (*corpora bigemina*). The fibers of the optic nerve enter the diencephalon and then pass into the roof of the mesencephalon. Fibers from the spinal cord and medulla also terminate in these lobes. The optic lobes are thus important correlating centers for optic and exteroceptive impulses. The tectum is most important in fishes and amphibians, a little less important in reptiles and birds, and least important as a visual center in the mammal. If the cerebral hemispheres are removed from a higher animal, it can no longer function. A frog, on the other hand, after recovering from such an operation, can continue functioning in an apparently normal fashion as long as the brainstem remains intact. The tectum of mammals is reduced to four prominent swellings known as the *corpora quadrigemina*. The larger anterior pairs are the *superior colliculi* (containing visual receptive centers); the posterior pairs are the *inferior colliculi* (integrating centers for auditory sense).

Although optic and auditory fibers are still sent to the superior and inferior colliculi, the ascending spinal tracts, in general bypass the mesencephalon of mammals and continue anteriorly to the diencephalon, or higher centers. The highly evolved cerebral cortex of mammals takes over the major visual and auditory control, leaving the tectum as essentially a reflex center for light and sound. Nuclei of the third and fourth cranial nerves are located in the gray matter of the midbrain. Ascending and descending fiber tracts from the rear and from the anterior cerebral cortex form the tegmentum. These are more prominent in mammals than in some of the lower animals, as might be expected, and constitute the major bulk of the mesencephalon. The majority of these fibers originate in the cerebrum and pass by way of the midbrain to the pons, medulla, or spinal cord.

Diencephalon

The diencephalon, or the tween-brain represents an ancient reflex center. It can be divided into a thin roof, the *epithalamus*; the thickened sides, the *thalamus*; and the floor, the *hypothalamus*, each part carrying out *specialized* functions. The third ventricle forms the cavity of the diencephalon. Several dorsal and ventral outgrowths arise from the diencephalon.

Epithalamus

The anterior portion of the roof of the diencephalon resembles the covering of the fourth ventricle in the medulla. It is a thin, non-nervous, vascular membrane that extends into the third ventricle and contains the *anterior choroid plexus*. Along with the posterior choroid plexus, it forms the cerebrospinal fluid, which fills the cavities of the nervous system. In some forms, a fold of the membrane gives rise to the paraphysis, a structure of unknown function. The posterior portion of the roof of the diencephalon evaginates to form a pair of median stalklike structures, the *anterior parapineal body* and a *posterior epiphysis*, or *pineal structure*.

The function of the pineal has long intrigued the biological investigator, but Turner and Bagnara (1971) state that the "time has probably arrived when it is justifiable to consider the pineal as an endocrine gland". There appears to be a decided seasonal interaction between the pineal and the reproductive organs, but other functions are not consistently present among the classes of vertebrates. According to a symposium presented by comparative endocrinologists of the American Society of Zoologists (1970), the pineal may control some basic homoiokinetic mechanism that allows the organism to adjust seasonally to its environment.

Thalamus

The thalamus is an integrating center for fibers that pass to and from the cortex, and it contain several important areas (nuclei). The walls of the diencephalon can be divided into a dorsal and a ventral half. In general, the cells located in the dorsal half are involved with sensory impulses and, next to the cerebrum, reflect the greatest evolutionary change. All sensory pathway, with the exception of the olfactory one, pass to the cerebral cortex of amniotes by way of these tracts. In lower animals the dorsal thalamus is of less importance. The ventral half of the thalamus is concerned with motor impulses.

Hypothalamus

The hypothalamus is of great importance, since it controls involuntary actions relating to sleep, temperature regulation, rate of breathing, genital impulses and metabolism. In the human being it weighs a little more than 7 grams. It is possible to identify four areas. The *optic chiasma* is located in the anterior parts of the hypothalamus. It is here that the optic nerve fibers cross to opposite sides of the brain before they proceed to terminate in the tectum of the midbrain of the lower animal. In higher animals, they pass to special areas in the gray matter of the cortex.

The *tuber cinereum*, believed to be the center of the parasympathetic system, is located behind the optic chiasma. Posterior to this center are the *mammillary bodies*, centers of olfactory function. The fourth area, the *infundibulum*, grows ventrally between the optic chiasma and mammillary body to meet the ectodermal material from the stomodeal region. Together they give rise to the pituitary gland of the individual. The infundibulum contributes to the neurohypophysis (posterior pituitary and median eminence), and the stomodeal ectoderm contributes to the adenohypophysis, consisting of the anterior and intermediate lobes of the pituitary. The hypothalamus regulates the function of the anterior pituitary gland, not by way of nerve fibers but by releasing into the portal blood vessels chemicals that stimulate or inhibit its activity.

According to present information, the hypothalamus secretes nine regulating or releasing hormones (RH) that affect the pituitary gland. Most of them stimulate the pituitary to both synthesize and release particular hormones. Some, however, inhibit the production of pituitary hormones. A dual system of hypothalamic control regulates three of the pituitary hormones: the growth hormone, prolactin, and melanocyte-stimulating hormone. One factor stimulates the synthesis and release of the hormone and one inhibits its production. These hormones need a dual regulatory system since their target organs release no negative- feedback products. On the other hand, the thyroid, gonads, and adrenals all produce hormones. The circulating levels of these hormones inhibit the pituitary and hypothalamus secretions by a negative-feedback mechanism. Therefore, only a hypothalamic-releasing factor controls each of these glands. The hypothalamus also synthesizes oxytocin and vasopressin. The long axonal fibers transport these hormones to the posterior lobe of the pituitary where they are stored until needed.

Telencephalon

Of all the parts of the brain, the telencephalon, or forebrain, evolved the most. Originally a single anterior expansion of the neural tube, it divides into the olfactory lobe and the *cerebral hemisphere*. In tetrapods, the hemisphere increases in size and separates into two definite lateral swellings, the cerebral hemispheres. The olfactory sense is the most important sense influencing behaviour in fish, amphibians, and most reptiles. In primates, on the other hand, the sense of smell is of less importance. The story of the evolution of the telencephalon involves the structural and functional expansion of its cerebral part. This progressive enlargement of the cerebrum in higher groups is diagrammed.

The function of the telencephalon shifts from a primarily olfactory one in primitive forms to that of the dominate association and co-ordinating center for all the senses in the mammals. The cerebral hemispheres are responsible for the complex thought, intelligence, and acute sensation associated with man. The evolutionary history of the telencephalon involves a shift in the location of gray and white matter as compared with that of the spinal cord. The cells of the gray matter move progressively outward as the phylogenetic scale is ascended.

The migration of the gray matter outside the white matter occurs slowly and is associated with the growth of a new area, the neopallium. As a result of this reversal of position, the cells of the gray matter occupy a peripheral position in the telencephalon of mammals where there is room for greater surface expansion. The white fiber tracts occupy the central position. When viewed in cross section, the forebrain of vertebrate embryos consists of two general areas surrounding the *telocoel*, or ventricle; the dorsal roof, or *pallium*; and a thickened floor, the *basal area*. The pallium becomes greatly modified during evolution; it is the primordium of the cerebral cortex. The basal area develops into the *basal nuclei*, essentially equal to the *corpus striatum* of mammals. It forms a part of the brain most concerned with basic instinctive behaviour patterns. Certain cells representing additional sense centers appear in the basal nuclei of the more active fish. The pallium is thin in the lower vertebrates. Only a small amount of gray matter line the ventricle; the rest of the pallium consists of an outside covering of white fibers.

In the primitive vertebrate and in the early embryo of all types, the telencephalon consist of a single cavity. Bilaterally located olfactory bulbs project forward from the forebrain toward the nasal sac. In the cyclostomes the pallium is related entirely to the sense of smell and the basal nuclei is poorly developed. Even in the cyclostomes, the olfactory bulb of fishes separates from the forebrain and extend toward the olfactory sacs. In the elasmobranchs, for example, a slight longitudinal depression indistinctly divides the single telencephalon into two anteroventral *olfactory lobes* and two posterodorsal *cerebral hemispheres*.

The olfactory lobes have only an olfactory function; the hemispheres are probably mostly olfactory, but they may contain a few other centers. Attached to the olfactory lobes are two stalklike anterior *olfactory tracts*, which pass into the rostrum of the fish, terminating in the *olfactory bulbs*. The bulbs are the true anterior ends of the brain. The receptor cells located in the epithelium of the adjacent olfactory sac send their fibers into the olfactory bulb. These fibers form the olfactory, or *first cranial nerve*. The paired amphibian hemispheres are larger than those of the fish, but it is difficult to differentiate between olfactory lobes and cerebral hemispheres. In these forms the pallium divides into two general regions, a dorsomedial *archipallium* and a lateral *paleopallium*. Although the archipallium continues to be predominantly an olfactory center, the new association center of other sensory impulse first appear here in the amphibians; the paleopallium and the basal nuclei remain olfactory center. The changes that occur in the amphibian correlate with the transition to a terrestrial life. Dependence on the sense of smell begins to diminish, the sensitivity to light, sound, and cutaneous stimulation increases. The shift of additional sense centers from a more posterior area into cerebral hemispheres represents another trend in the evolution of the mammalian brain. It is also in the amphibians that a few cells from the gray matter begin to move away from the ventricle and into the walls of the pallium. Most of gray matter of amphibians, however, is internal. All of the areas receive fibers from the olfactory center and the thalamus.

Fibers that connect the archipallium with the paleopallium and with the basal nuclei also develop, and so these regions are able to correlate olfactory function with a few other simple impulses, possibly gustation. The basal nuclei relays impulses back to the thalamus and to the midbrain. It begins to form an important correlation center that increases in complexity in more advanced groups. A fourth region, the *neopallium*, differentiates laterally to the archipallium in some reptiles and all mammals. According to Romer, "the evolutionary history of the mammalian brain is essentially a story of neopallial expansion and elaboration". It represents a complex association center that provides unlimited intellectual capacity. From the beginning it received sensory impulses from the brainstem, correlated them with other information, and relayed commands back to the brainstem.

The neopallium consists of only a small area of superficial gray matter in reptiles and a somewhat larger region in the primitive mammals; in advanced mammals its surface area is considerable. It reaches its greatest development in man and in mainly responsible for the large size of the cerebral hemispheres. The neopallium assumes more and more of the functions that in the lower vertebrates were centered in the brainstem and in the basal nuclei. As it increase in size, it covers the other areas, pushing them to the inner part of each hemisphere. In the majority of mammals it overlaps the midbrain and part of the cerebellum.

The paleopallium is restricted to a small ventral area, the *piriform lobe*. The well-developed corpus striatum or basal nuclei moves toward the interior of the hemispheres, and the archipallium is forced medially, where it is known as the *hippocampus*. Automatic reactions remain centered in the corpus striatum. Beginning with the marsupials, a band of white myelinated fibers, the *corpus callosum* connects the two hemispheres. As expansion continues, the cells of the gray matter composing the neopallium migrate nearer the surface. Here they become concentrated and are known in mammals as the *cerebral cortex*. The walls of the cortex (neopallium) measure in man about ½ to $^1/_{12}$ inch in thickness. As many as six different layers of neurons, with millions of neurons in each layer make up the gray matter of the placental animals cortex.

The cells of the various layers of this continues blanket enter in to complex connections with one another by means of their axons and dendrites. It is impossible to comprehend the exact number of correlations that can be made. The roof of the cerebral

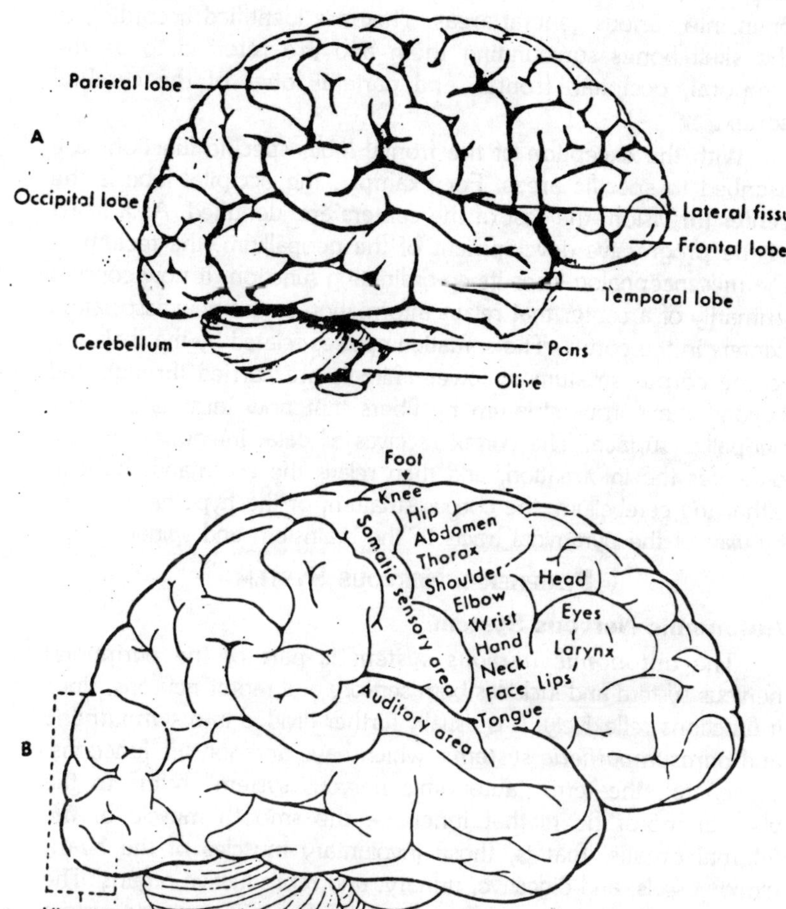

Fig. 8.21. Convolutions of mammalian cerebral cortex divide it into temporal, occipital, frontal, and prietal lobes, A. Except for the frontal area, specific functions have been ascribed to specific areas as indicated in B.

hemispheres is smooth in many of the lower vertebrates. In mammals, however, the more rapid increase in number of neurons at the surface of the neopallium as compared to the deeper lying parts, as well as the increase in underlying fiber volume causes the surface to be thrown into convolutions. The folds are referred to as *gyri*; the indentations are *sulci*. The convolutions divide the brain into various general areas. They are identified according to the skull bones surrounding them and are referred to as the temporal, occipital, frontal, and parietal lobes of the cerebral cortex.

With the exception of the frontal lobe, specific functions are ascribed to specific areas. For example, the occipital lobe is the center for vision; the rest of the centers are identified. As a result of the progressive development of the neopallium, the tectum of the mesencephalon loses its co-ordinating function. It now consists primarily of a center that relays information to the new association centers in the cortex. The somatic impulses related by the thalamus to the corpus striatum of lower animals are carried through and beyond the corpus striatum by fibers that now terminate in the neopallial surface. The cortex receives all data, integrates and co-ordinates the information, and then relays the commands back to either the cerebellum, the corpus striatum, or the hypothalamus, or by way of the pyramidal tracts to the brainstem and spinal cord.

Peripheral Nervous System

Autonomic Nervous System

The *autonomic nervous system* is part of the peripheral nervous system and includes both sensory and motor neurons, since it functions reflexively. It is usually further divided into *sympathetic* and *parasympathetic systems,* which have antagonistic functions. In general, the term "autonomic nervous system" refers to the visceral motor fibers that innervate the smooth muscle of the internal organs, that is, those involuntary muscles of the heart, blood vessels, and digestive, urinary, and reproductive organs. The visceral sensory system differs from the somatic sensory system only in that it carries impulses to the cord from the viscera rather than from the body wall. However, the input of any somatic sensory nerve can also produce an automatic motor response. For example, something that an individual sees and finds disturbing can cause his heart beat rate to increase. The visceral motor system differs distinctly from the somatic motor system; in fact, most authors define the autonomic system as consisting of only theses visceral efferent fibers. Sometimes the student has the impression that the autonomic system is an independent system, but it should be considered as a functional division of the whole. Its control is located in the central nervous system and it is intimately bound together with the rest of the nervous system.

Comparison of a Somatic and Autonomic Nervous Systems

Earlier in the chapter we described a typical spinal nerve as being composed of somatic and visceral fibers. Certain functional and structural characteristics help to differentiate the autonomic (visceral) system from the somatic. When an impulse leaves the somatic motor nuclei in the cord, it passes down an axon to a muscle or gland. Should that muscle be located in the big toe, the axon extends from its cell body in the posterior regions of the cord to the big toe, a distance of perhaps 3 feet. Only one neuron is involved between the cord and the muscle. In the autonomic system, however, the impulse is conveyed from the cord to the effector organ via two neurons.

The cell body of the first neuron is typically located in the visceral motor area of the gray matter, previously described. Its axon may extend to the *lateral ganglia*, lined up in a chain outside the central nervous system on either side of the vertebral column, or it may extend to the *collateral ganglia* located farther away, or even to the organ itself. The ganglia are those described as originating from the neural crest cells. In the ganglion the first neuron may synapse with a second neuron, the axon of which passes to the organ it innervates. In some cases, as will soon be described for the parasympathetic system, this ganglion is actually embedded in the tissues of the organ. *Preganglionic fibers*, as their name implies, are axons that transfer the impulse from the cord to the lateral ganglion. The second neuron, whose axons terminate in the muscle or gland, is referred to as the *postganglionic* neuron. Unlike the preganglionic fibers, they lack myelin sheaths. When the ventral and dorsal roots of a spinal nerve unite outside the cord, they form a nerve trunk consisting of many fibers. This trunk soon divides into three branches, or *rami*.

A *dorsal ramus*, composed primarily of somatic afferent and efferent fibers, passes to the dorsal epaxial region of the organism; a *ventral ramus* extends to the somatic hypaxial region of the organism; and a *visceral ramus* sends branches to the internal organs. The visceral rami can be further divided into two components, the *white* and *gray rami*, best understood by referring. The preganglionic fibers of the sympathetic system leave the cord in the ventral root along with somatic motor fibers, but then pass to the lateral ganglia by way of the visceral rami.

The myelinated preganglionic fibers make up the white ramus. In the ganglion some of the preganglionic fibers synapse with postganglionic neurons, and these non-medullated (non-myelinated) postganglionic fibers give rise to the gray ramus. As stated above, all sympathetic fibers do not synapse at the lateral ganglion; some may pass through the ganglia and extend anteriorly or posteriorly in the sympathetic chain before synapsing at levels other than where they originated. A third possibility exists for the preganglionic fiber to pass through the lateral chain ganglia to certain large ganglia located some distance away known as the *collateral ganglia*. The celiac, anterior mesenteric, and posterior mesenteric ganglia compose these *collateral ganglia*. Regardless of the location of the synapse, any one preganglionic neuron synapses with several postganglionic neurons. The latter, in turn, may then carry the impulse to several different organs. As a result sympathetic responses are widespread and involve many different structures.

Sympathetic and Parasympathetic Divisions

On the basis of function, structure, and location, the autonomic system of higher animals is further divided into two main divisions, the *sympathetic system* and the *parasympathetic system*. Both types of fibers innervate most internal organs. The two systems can be differentiated in several ways. They differ in their physical relationship to the central nervous system.

The preganglionic fibers of the sympathetic system leave the cord in the thoracolumbar regions. For this reason they are sometimes referred to as the *thoracolumbar autonomic system*. However, the preganglionic neurons of the parasympathetic fibers are mostly located in the medulla oblongata. Included are a few fibers to the iris and ciliary body that travel with the third cranial nerve, as well as a part of the facial (seventh), the glossopharyngeal (ninth), and the vagus (tenth) nerves. Also, nerves from the posterior parts of the cord belong to this system and send fibers to the large intestine, bladder, and reproductive tract. As a result, the parasympathetic portion of the autonomic system is referred to as the *craniosacral system*. The locations of the synapses between preganglionic and postganglionic fibers is a second difference between the sympathetic and para-sympathetic systems.

The preganglionic fibers of the parasympathetic system do not synapse with the postganglionic fibers in the lateral or collateral ganglia as described above for the sympathetic system. Instead, they pass directly to the organs they innervate. Here, either in or near the organ itself, synapses occur with the second neuron. Since synapsis occurs some distance from the cord, the axons of the preganglionic neurons of the parasympathetic system are long compared to those of the sympathetic system that synapse just outside the cord. Of course, the postganglionic fibers are short in the parasympathetic system and long in the sympathetic system. The two systems have opposite effects on the organs that they innervate.

On the whole, the sympathetic system alerts the body for activity. It prepares it to meet an environmental challenge by speeding up its heartbeat and respiration rate and stimulating the integumentary glands, hair muscles, and skin capillaries. It depresses those activities not conducive to body activity, such as digestion, excretion, etc. The parasympathetic system, on the other hand, is sometimes referred to as the vegetative system, since it produces a reverse effect. Neither system is entirely stimulative or repressive, but produce both excitatory or depressant effects depending on the structure under consideration.

Under normal circumstances, the organism co-ordinates their effects and maintains a homeostatic state, with either system being above to take over temporarily if the occasion demands it. The systems produce their effect by liberating certain chemicals, acetylcholine or norepinephrine, at their end organs. These were referred to as a transmitter substances earlier in the chapter and are given the general term of *neurohumors*. The ends of all somatic neurons liberate acetylcholine. It is also released by the preganglionic fibers of both sympathetic and para-sympathetic axons and by the axons of the parasympathetic postganglionic neurons. Such neurons are referred to as *cholinergic nerves*. Epinephrine-like (adrenaline-like) compounds are liberated by the postganglionic endings of the sympathetic system and consequently this system is referred to as *adrenergic*.

Cranial Nerves

General Characteristics

The nerves that arise from the ventral surface of the brain and emerge through special openings (foramina) of the skull are referred to as *cranial nerves*. Together with the spinal nerves, they complete the peripheral nervous system. There are 10 pairs present in most anamniotes and 12 pair in reptiles, birds and mammals. The nerves are numbered starting at the anterior end. Roman numerals are often used. The name, number, origin, and distribution of the nerves from fish to man are similar, although, of course, where structures have been added or deleted through evolution, modifications occur. For example, nerves pass to the gills of aquatic animals. When amniotes undergo development, these nerves are either lost or they innervate the structures that evolved in the gill region. Although absent in man, a terminal nerve is present in most vertebrates, anterior to the first cranial nerve. Since it was discovered after the other nerves were numbered, it is usually labeled 0. Earlier in the chapter, we described four functional components of the spinal nerves. These same general components are present in the cranial nerves. The cell bodies of the cranial motor nerves are located in nuclei in the brainstem.

The cell bodies of the sensory neurons collect in the ganglia near the brain. Somatic sensory fibers send twigs to the skin of the head and other somatic structures. Visceral sensory fibers relay information from the membranes of the pharynx, the anterior part of the gut. Somatic motor fibers terminate in structures derived from myotomes, for example, the eye muscles, and visceral motor fibers pass to the muscles of the gut. However, a special group of nerves, distinguished from the general nerves described above form an additional category of cranial nerves. They are the *special somatic sensory*, *special visceral sensory*, and *special visceral motor nerves*. The terms "special" and "special visceral" need further clarifications. The nose, eye, and acoustolateralis organs are defined as special sense organs, distinguished from the simpler receptors. The sensory nerves that pass from these organs to the brain retain this term and are known as the special somatic sensory nerves.

A special visceral sensory nerve consists of fibers from the taste organ. The muscles that work the gill bars of fish are a special kind of muscle. They are not derived from myotomes and, although they are formed from lateral plate mesoderm, they differ from the smooth muscle of the gut (of splanchnic mesodermal origin). These gill muscles, referred to as visceral muscles, are striated, and in this respect are similar to skeletal muscles. They are innervated by cranial nerves known as special visceral motor nerves.

Classification of Cranial Nerves

In order to minimize rote memorization of names and functions of the cranial nerves, we advise grouping them in different ways. For example, two of the nerves, cranial nerve 1 (olfactory) and cranial nerve II (optic), are considered atypical nerves, since their embryonic origins do not conform to the usual pattern. These sensory nerves do not possess ganglia. The cell bodies of most sensory nerves are located near the spinal cord or brain, but cranial nerve I does not actual develop from neural tissue. Instead, the olfactory receptor cells, derived from the olfactory placode, send their projections into the olfactory lobe of the brain. These fibers combine to form the olfactory nerve. Cranial nerve II is actually a brain tract, since the retina is an evagination of the diencephalon and is considered part of the brain.

The optic nerve forms by a combination of fibers from the ganglionic cells located in the wall of the retina. These cells send their fibers back into other regions of the brain. It is also helpful to remember that the forebrain arose in relation to the olfactory sense, the midbrain with the visual sense, and the hindbrain with the acousticolateralis sense. Cranial nerves can be grouped therefore. According to the region of the brain with which they are associated. The fibers of the olfactory nerve extend to the telencephalon and the optic fibers extend to the diencephalon and then on to the tectum of the mesencephalon; cranial nerves III and IV arise from the mesencephalon and nerves V to XII from the medulla oblongata.

The cranial nerves can also be divided into three general groups. Cranial nerves 0 (terminal nerve), I (olfactory), II (optic), and VIII (auditory) are all sensory nerves. Motor nerves consist of cranial nerves III (oculomotor), IV (trochlear), VI (abducens), and XII (hypoglossal). The remaining nerves, cranial nerves V (trigeminal), VI (fascial), IX (glossopharyngeal), X (vagus), and XI (spinal accessory), are all mixed nerves.

Specific Cranial Nerves

The following cranial nerves are grouped according to their functional components:

Somatic sensory

0, *Terminal nerve.* This is a small nerve of unknown function that runs parallel to the olfactory nerve but enters the diencephalon. It is apparently sensory in function, since it bears one or more ganglia. It may represent the remnants of an anterior nerve existing in ancestral vertebrates.

Special sensory

These nerves transmit impulses from the special sense organs.

I, *Olfactory nerve*. This is a sensory nerve that carries impulses from the olfactory sac to the telencephalon. It has no ganglion, but develops from the cells of the epithelial lining of the olfactory sac.

II, *Optic nerve*. This is a sensory nerve that carries impulses from the eye to the higher regions of the brain. The fibers from each side of the head unite in the floor of the diencephalon where some of them cross to the other side and give rise to the optic chiasma. The fibers then continue in the brain as optic tracts passing to the mesencephalon (in lower animals) or to the cerebral cortex (in higher forms).

VIII. *Acoustic nerve*. The ganglion giving rise to this nerve divides into two parts in higher animals: the spiral ganglion with fibers passing to the cochlear duct, and the vestibular ganglion that sends fibers to the semi-circular canals and the sacs of the inner ear. The lateral line nerve is closely related to the acoustic nerve. It enters the medulla of fish with nerve VIII, but is usually considered to be more closely combined with nerve VII.

Somatic motor. These nerves carry impulses to the muscles that move the eyes and to the tongue muscles. Apparently they lost their dorsal roots; the remaining ventral roots compose the nerve.

III. Oculomotor nerve. This nerve arises from the mesencephalon and supplies the eye muscles that are derived from the first cranial myotome: the inferior oblique and the superior, inferior, and medial rectus muscles. A few fibers from nerve V usually accompany nerve III as they pass to the smooth muscles of the iris and ciliary body. However, these fibers are usually not considered a part of nerve III.

VI. Trochlear nerve. This nerve arises from neuroblasts in the floor of the mesencephalon just posterior to nerve III; it innervates the eye muscle derived from the second myotome. Its fibers pass upward in the wall of the brain where they then pass dorsally and cross to the opposite side. They emerge dorsally from the brain and pass to the superior oblique muscle of the eye.

VI Abducens nerve. The abducens nerve originates from the motor nucleus located in the pontine region of the mesencephalon.

It innervates the external rectus muscle of the eye, derived from the third myotome.

XII. Hypoglossal nerve. The hypoglossal nerve has several origins. The ventral roots of three or four nerves from the occipital region (back of head and neck) leave the medulla oblongata and merge into one nerve. This nerve innervates the occipital myotomes, which in the fish embryo migrate beneath the gill region to form the hypobranchial muscles. In higher forms, when the gills are lost, this region is remodeled. These muscles are lost in amphibians but contribute to the tongue musculature of amniotes. The hypoglossal nerve therefore innervates the tongue.

Mixed nerves

Mixed nerves all arise from the medulla oblongata and each is related to a gill arch of the fish. As a result, they are sometimes referred to as the *branchial nerves*. When amniotes lose their gills, the nerves innervate the structures derived from the gill area. Each nerve usually includes somatic sensory as well as special visceral sensory fibers and special visceral motor fibers. Cranial nerves V, VII, IX and X are all considered mixed nerves. Originally, each nerve consisted of three main branches, a pharyngeal branch passing inward to the pharynx, a pretrematic branch running in front of the gill slit, and a posttrematic branch running in back of the gill slit.

V. Trigeminal nerve. This is a stout nerve that has a large semilunar (or gasserian) ganglion located outside the hindbrain. It is usually divided into three main branches: the ophthalmic, maxillary, and mandibular nerves. The fibers are primarily sensory from the face, nose and mouth; they enter the brain at the somatic sensory column of the gray matter of the medulla. Motor fibers pass to the muscles of mastication. Since these muscles are derived from the muscles of the original first branchial arch, the nerves are considered special visceral motor nerves.

VII. Facial nerve. In fish, the facial nerve passes to the second arch and the spiracular gill slit (considered the second gill slit). It contains both somatic and visceral sensory, as well as visceral and special visceral motor fibers. Some of these fibers become intimately associated with fibers from nerve V and pass along with them to the anterior regions of the head. This nerve is called the *facial nerve* because in mammals the muscles of the second arch become the muscles of expression and retain the same innervation. There is a sensory ganglion, the geniculate ganglion, housing the cell bodies of the sensory portion of the nerve. The visceral sensory fibers pass to the taste buds and to the lining of the pharynx. The large somatic sensory portion of the nerve passes to a lateral line system.

IX. Glossopharyngeal nerve. In fishes this nerve passes to the first functional gill and therefore contain both special visceral sensory and special visceral motor fibers. Its sensory cell bodies are located in two ganglia, one near the myelencephalon, the superior ganglion, and the other more laterally located, the petrosal ganglion. In fishes it contain both a pretrematic and a posttrematic branch; in mammals this nerve is lost to a great extent. Its visceral motor fibers innervate some of the muscles of the pharyngeal areas and parotid salivary gland; its visceral sensory fibers pass to the taste buds.

X. Vagus, and XI, spinal accessory nerve. The vagus is one of the most important nerves. It is a composite nerve formed by the union of several branches that pass to the posterior gill arches, those numbered four to six. When the amniotes lose their gill arches, some of these branchial fibers degenerate, but a few remain and innervate those structures derived from this region, such as the pharyngeal musculature. The main branch of the vagus is visceral and extends backward into the body along the gut where it gives off branches to the stomach, heart, lungs and anterior intestine. It carries the visceral impulses to the medulla oblongata. It also relays visceral motor impulses back to these organs by way of the preganglionic branch of the parasympathetic system. Since the vagus represents a union of several nerves, it has several sensory ganglia as might be expected. A motor portion of the vagus of lower animals divides into a separate nerve. In mammals, it is known as nerve XI, or the spinal accessory nerve.

Receptors

Animals are living in a very fluctuating environment. If an animal is to survive successfully, it must be responsive to the condition of both its internal and external environments and to any changes that occur in them. In the primitive, one-celled animal, this information is easily made known to the active part of the organism, since all cells posses irritability. However, communication between various parts of the multicellular animal becomes more of a problem. As the result of neurulation, the central nervous system (the brain and spinal cord) moves from the surface of the animal to the interior, and the neural crest cells that form the ganglia of the sensory nerves end up in proximity to the brain and cord. The epidermal ectoderm develops into an integument that covers, insulates, and protects the individual.

The central nervous system, therefore, usually cannot be directly stimulated by any change that goes on in or around the organism. A sensory receptor system evolved in the multicellular vertebrate that detects changes in the internal or external environment and transmits the information to the central co-ordinating areas. These areas, in turn, send appropriate instructions for action and organs. If a prompt reaction is necessary to remedy a temporary situation, the message to react is usually sent through the proper nerve to the muscle. On the other hand, if the body needs a slower more generalized adjustment to some situation, the endocrine system functions in producing the response.

In general, sense organs are associated with the nervous system. *Sensory receptors* are hypersensitive, intermediary structures, located between the environment and central nervous system, that receive and respond to stimuli. They are usually specialized cells that are sensitive to specific chemical or physical stimuli and are associated with nerves that relay the information to the central nervous system. Each receptor responds to one particular kind of stimulus, and depending on the nature of the stimulus receptors are classified as *photoreceptors* (those that respond to visible radiation), *chemoreceptors*, (those sensitive to dissolved chemical materials), *mechanoreceptors* (those stimulated by touch or pressure), *thermoreceptors* (those that react to temperature), *statoreceptors* (those stimulated by changes in body position), and *phonoreceptors* (those stimulated by vibrations). The sensory receptors develop from ectoderm and their origin may be traced to epidermal ectoderm, neural ectoderm, or neural crest cells. In some instances, a combination of ectodermal types compose the sense organ; often mesoderm, in the form of connective tissue capsules, combines with the ectodermal cell.

The receptors are associated with the somatic sensory and visceral sensory nerves. The somatic sensory nerves carry impulses from the cutaneous surface and muscles of the body wall to the central nervous system; visceral sensory nerves relay impulses from the viscera. The higher centers of the brain interpret the impulses as *sensations* and the individual experiences cold, pain, warmth, sight, sound, etc. The receptor do not perceive anything; they merely refer information to the nervous system where it is sorted and integrated. Obviously most of our insight into sensation is based on man's *reaction to stimuli*, since there is no way of knowing whether the toad of the dog possesses similar sensations. There may be many sensations experienced by lower animals that man is unable to comprehend and, therefore, he has no way of knowing about them.

The fact that similar structures (the eye, for example) originate in all vertebrate embryos in a similar manner and that they give rise to organ that are anatomically similar in the adult lead biologists to believe that some sensations, at least, are shared by both lower and higher animals. Those receptor located over the general body are referred to as the *general sensory organs*, although in some cases bare fibers serve as the receptors and the term "organ" is applied from general use. Other receptors concentrate only in special areas of the body and react to very specific stimuli. They are known as the *organs of special senses*: the organs olfaction, sight, hearing and equilibrium. We classify sense organs according to their morphology and usually recognize at least three levels of organization. These levels also reflect their evolutionary status.

At the simplest levels, the epithelial cell, located at the surface of the animal, serves as both receptor of stimuli and conductor of nerve impulses. At the second level, the cell body migrates inward and the modified cell processes react to the stimulus and transmit the information. In the third type of organization, a second cell, inserted in the pathway and located at the surface of the animal, reacts to the stimulus. The generated impulse passes to the nerve cell, which, in turn, transmits it to the brain. Some organs are classified according to their functions. *Exteroreceptors* are those receptors that receive information about the external environment; *proprioceptors* are those located in the striated muscles and tendons. *Interoceptors* are associated with internal organs.

General Sensory Organs

The receptors composing the general sense organs are less sensitive to stimuli than those of special senses. They react to touch, pressure, pain, temperature, position, movement, and the viscera. A variety of these stimuli apparently produce the sensation of pain. The most primitive type of sensory cell, found primarily in invertebrates, consists of a modified ectoderm cell located in the superficial layers with long processes extending inward. The vertebrate's olfactory organs (which we shall a simple arrangement. In general, the cell bodies of the general sensory organs do not remain in the superficial ectoderm of the organism but move inward and become associated with the brain and spinal cord. They belong to the second level of organization mentioned above. These well differentiate from neural crest cells that were brought in at the time the neural tube formed.

The cell bodies collect in clusters (ganglia) outside the brain and cord and send processes into the central nervous system as well as to the integument of the organism. These distal processes, modified during evolution, provide a variety of nerve endings usually classified as free (non-encapsulated) or *encapsulated*. The free type of nerve ending occurs when the peripheral process of the sensory nerve merely ends in a spray of terminal, non-myelinated, slender twigs or knoblike swellings. This type of sensory cell is distributed abundantly over the entire body, and its fibrils entwine or embed in epithelial cells, connective tissue, muscle, or serous membranes.

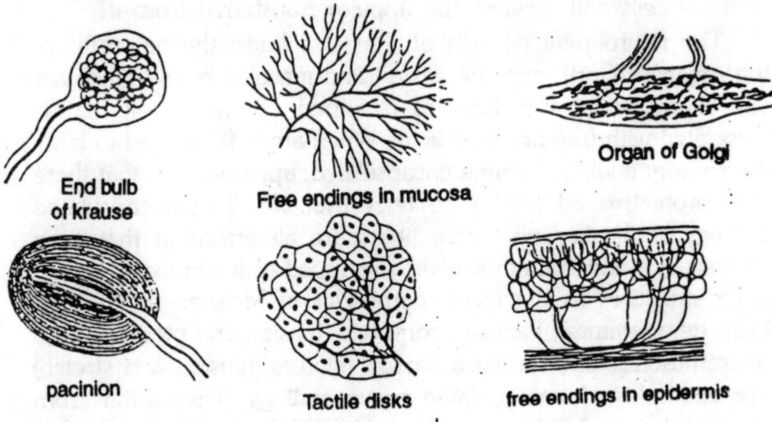

Fig. 8.22. Examples of encapsulated and (non-encapsulated) free nerve endings.

The fibril terminate in the epithelium of the cornea, oral cavity, respiratory passages, and kin of all vertebrates. They also appear among the fibers of the connective tissue of the dermis, periosteum, blood vessels, and individual muscle fibers, to name a few example.

Properly stained histologic sections of the integument show this type of three ending to penetrate the skin as far as the stratum germinativum layer. Sections through the bronchial epithelium, on the other hand, reveal branches of the nerve wrapped around the cell, with the ends of the fiber actually pushing to the surface. Pain endings are of this type. The sense of touch appears to depend on stimulation of similar endings. For example, nerve endings in hair follicles are important tactile organs.

The sensory fibers divide and form a meshlike ending around the hair shaft. Any pressure sufficient to move the hair shaft also stimulates the nerve and adds to the sensitivity of touch. The epithelial cells associated with the nerve endings are generally similar to one another although they differ in some cases. The tactile cell is an example of an epithelial cell that varies in staining properties from its neighbour. This specialized epithelial cell, together with the arborization of its nerve endings, is known as a *tactile corpuscle of Merkel* and is located in the deeper epithelial layers of the skin. Other modified epithelial cells, the *neuroepithelial cells* or *neuromasts*, receive the termination of a special sensory nerve. The nerve fiber forms a net around one of these cells and receives the impulse transferred from it.

The neuroepithelial cells of the taste buds, the hair cells of the organ of Corti, and the cristae and maculae of the inner ear are examples of this type of modified epithelial cell that is associated with free nerve endings. When amniotes moved to land, the sensory endings became encapsulated, apparently so that there is a protective adaptation to terrestrial life. The encapsulated endings of the sensory nerve fibers are all similar in that each consists of a terminal nerve ending wrapped with connective tissue. Such structures as the tactile corpuscles of Meisner (touch), end bulbs (temperature), Pacinian corpuscle (tension and pressure), and neuromuscular and muscular-tendon bundles (tension and stretch) are examples of encapsulated nerve endings. They differ from one another in the degree of branching of the nerve, the number of supporting cells present (connective tissue), and the size and shape of the structure.

The terminal nerve endings that pass to skeletal muscles and tendons, for example, entwine around a muscle fiber that is encased in a connective tissue capsule. When the muscle fiber stretches, it stimulates the nerve fiber, revealing the position of the body and its parts of the organism. In some instances the connective tissue wrapping around the ends of the nerve is so extensive that the structure can be seen with the naked eye (Pacinian corpuscle).

Organs of Special Senses

Organs of Taste

The *taste buds* develop from modified epithelial cells, the neuromasts, mentioned earlier in the chapter. In fish, these cells are distributed over the surface of the body and are innervated by extensions of the cranial nerves. Taste buds are also found buried in the epithelium of the gill rakers and gill arches as well as in the mouth and pharynx of fishes. Materials dissolved in the water stimulate the exposed ends of the cells; taste buds therefore are examples of chemo-receptors that are stimulated by substances at close range rather than at a distance. During their evolution, taste buds lost their generalized distribution over the body of the animal and became localized in special areas. In terrestrial animals they are usually restricted to the tongue, although they may also be present in the palate and pharynx. The taste buds are usually associated with small elevations of the tongue epithelium, known as *papillae*. Trenchlike depressions separate one *papilla* from another, and the taste buds are located in the walls of these trenches.

The taste buds of man can discriminate between only four taste sensations: acid, bitter, salty and sweet. All other "tastes" are actually olfactory sensations. Fish are believed to discriminate between these same four tastes but amphibians show no response to substances classified as sweet or bitter. The horny tongues of birds lack taste buds. Histologically, the epithelial cell differentiates into two types of cells, a *gustatory receptor* cell (neuromast) and a *sustentacular*, or supporting cell. A chemically sensitive hairlike process projects from the apical surface of the elongated neuromast. In mammals, clusters of neuromast cells group together with their supporting cells to form a taste bud. The number of neuromast cells composing a taste bud varies. The supporting cells are interspersed among the neuromast cells and also surrounded them like staves of a barrel.

The hairlike process of each neuromast is in contact with the surface on the epithelium by way of a taste pore. Nerve fibers that break up into a basketlike arrangement of telodendria (nerve endings) surround the deep ends of the neuromast cells. The cell membrane of the nerve ending is in direct contact with the cell membrane of the neuromast cell. Consequently, when the neuromast cell responds to chemical stimulation, the physiologic changes activate the nerve endings. The nerve then relays the impulse to the central nervous system. Taste buds of all vertebrates are supplied with branches of the eighth, ninth, and tenth cranial nerves. Since these nerves contain visceral sensory fibers, the gustatory organs are considered special visceral receptors. Presumably as a result of the chemical stimulation, the primitive animal moved toward or away from such objects as food, enemies, or a mate.

Organs Derived from Placodes

Although most of the epidermal (ectoderm) covering of the embryo develops into the epidermis and epidermal derivatives, a number of other structures are derived, from it. Many of these structures begin their development in the form of *placodes*, plate-shaped thickenings in the epidermal ectoderm. In the early neurula stage the placodes are arranged in an almost continuous horseshoe-shaped swelling around the open cranial neural folds. In most cases the initiation and further development of the placodes depends on induction by the tissues beneath it. By the time the neural fold closes, the generalized swelling becomes identified as contributing to a specific sense organ. The thickened epidermal ectoderm located anterior to the neural plate (nasal placodes) is stimulated to form olfactory sacs. The optic vesicle contacts the epidermis and induces it to thicken (*lens placode*) and develop into the lens of the eye. An *auditory placode* located laterally to the hind brain gives rise to the ear. A *lateral-line placode* associated with the auditory placode becomes the lateral line sense organs of aquatic animals. Several placodes give rise to the cranial ganglia. These contribute to the fifth and seventh to tenth cranial nerves.

Olfactory Organ

The olfactory or nasal placodes of all vertebrates first appears oval thickenings on the ventrolateral surfaces of the head of the embryo, anterior to that part of the closed neural tube destined to become the brain. With the exception of the cyclostome, which has a single nasal pouch, the olfactory placodes are bilaterally arranged They appear in the human embryo as early as the fourth week of gestation. The cells composing the central part of the placode sink inward to form the *olfactory pit*; the walls of the pit soon thicken to become the *olfactory sac*. The opening of the sac to the outside is called the *external naris*. Some of the cells in the wall of the olfactory sac differentiate into sensory cells. These take the form of rodlike projections ending in sensitive brushes extending to the surface.

Processes develop on the proximal end of each cell (end closest to the brain) and grow into that part of the forebrain destined to become the *olfactory bulb*. In the bulb the processes (axons) synapse with the mitral cells, which carry the impulse via the olfactory tracts to the appropriate center in the brain. These processes of the nasal epithelial cells leading from the sac to the telencephalon make up the first cranial nerve, the *olfactory nerve*. Most nerves develop either from the brain itself or from ganglia that are associated with the brain or cord. The olfactory nerve is an exception. It does not develop from nervous tissue at all but from the epithelium of the nasal sacs. It is, therefore, not considered a typical cranial nerve. In most fishes the olfactory sacs of the embryo develop into similar adult structures. Their primary function is an important one since they act as distance receptors informing the organism of the presence of food some distance away. The olfactory receptors are good examples of extero-receptors. The sacs remain as a pair of pockets in the fish without any communication with the mouth. Each pocket is usually divided into two openings. The water with its dissolved chemical substances flows in one side of the nasal pocket and out the other side.

The mucosa of the sac consists of many folds of columnars epithelium thereby increasing its surface area. As the water passes over the sensory end brushes, the dissolved chemicals stimulate the nasal epithelial cells. The proximal processes of the cells conduct the generated impulse to the brain. The *olfactory organ* increases in complexity in the Crossopterygii and in the tetrapods. The olfactory sacs in these animals are no longer blind pouches but communicate with the oral cavity by way of an opening, the *internal naris*, or *choana*. When is a cross section of a 10 mm. passing through the head region as from frog it shows the internal naris opening into the mouth cavity. In these animals the nose is also involved in breathing. In the amphibians the nasal sacs is a simple elongated structure that opens into the anterior region of the roof of the mouth and functions both on land in the water. On land it becomes filled with air and special mucous glands keep the epithelial cells moist and capable of functioning.

The *sensory epithelium* of the fish was thrown into folds. In contrast, the olfactory pouch of the amphibian is smooth, and only part of its surface contains the sensory cells. The nasal organs of reptiles and birds are more complex. Although birds retain the reptilian plan of nasal structure, their olfactory organs are of modest size. Birds rely more on their eyes than on their sense or smell to learn about their environment. The olfactory sac divides into two main regions- a small anterior *vestibule* into which the air passes from the external naris, and a larger olfactory chamber opening into the oral cavity by an internal naris more posteriorly placed than in the amphibian.

Only the dorsal area of the olfactory chamber contains sensory cells. This sensory epithelium covers the curved, lateral ridges known as the *conchae*, which increase the surface of this area. The walls of most of the chamber, however, function as an air passage and are free of sensory epithelium. The nasal structures of most mammals increase in size and may occupy as much as half of the skull space (dog). As in the reptile, the olfactory area divides into a dorsal sensory area and a non-sensory air passage. The development of a secondary roof to the oral area, the secondary and soft palates, moves the openings of the internal nares more posteriorly in the mouth. The chonchae, formed from coiled scrolls of the ethmoid, maxilla, and nasal bones, are highly

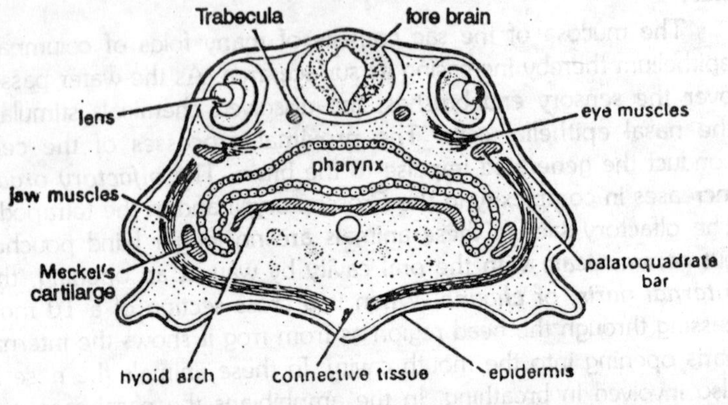

Fig. 8.23. Diagram of structure of neuromast.

developed and covered with sensory epithelium. The incoming air passes through the scrolls and is warmed and filtered in the process.

The nasal cavities of eutherian mammals also communicate with cavities (sinuses) of adjacent bones, lightening the skull. Except for some mammals, turtles, and crocodiles, a second sensory area appears in tetrapods, the *vomeronasal organ*, or *organ of Jacobson*. It forms another part of the olfactory sac that retains its sensory function. It detects olfactory sensations from the food in the mouth. This "organ" is either a groove covered with olfactory epithelium, or else it exists as a separate pocket. In either case, it usually communicates with the oral cavity by way of the internal naris.

Jacobson's organ is particularly well-developed in reptiles. It may consist of a medioventral, club-shaped pouch whose opening into the oral cavity is associated with the internal naris, or it may exist as a completely separated pouch with its own opening. In lizards and snakes, the cleft, flicking tongue also serves as an olfactory organ. When drawn into the mouth, the tips of the tongue are inserted in to the vormeronasal pocket. Chemical particles adhering to their surface pass into solution and stimulate the sensory epithelium of the pocket. The pouch is absent in primates but is present in other mammal a functional structure.

Acousticolateralis Placodes

A lateral series of placodes, located on either side of the neural tissues of the head, develop into the lateral-line system in fishes and aquatic stages of amphibians. The cells composing this system in modern vertebrates are sensitive to mechanical stimulation. Apparently they evolved from the neuroepithelial cell (neuromast) considered common in ancestral types. A group of stereocilia and a long motile cilium extend from the apex of this nucleated, columnar cell. These processes are embedded in a knoblike mass of gelatinous material, the cupula which adheres of the cell and is secreted by it. The terminal endings of the associated neurons wrap around the neuromast and receive the impulse of the stimulated receptor cell. Although some neuromast cells respond to chemical and thermal stimuli, they are considered primarily mechanical receptors in the acousticolateralis system, sensitive to the shearing forces that affect the stereocilia and cilia. Both the lateral-line organs and inner ear contain these mechanoreceptors.

The inner ear, present in all vertebrates, presumably evolved from the acousticolateralis system of its ancestors.

Lateral-line System

The lateral-line system responds to pressure changes in the surrounding medium of fish and amphibians. Animals above them on the evolutionary scale, the amniotes, lack lateral lines. Even those reptiles and mammals that returned to the aquatic environment did not retain the system. The lateral-line receptors consist of clusters of neuromast cells located in pits or grooves that run in a definite line in the epidermis from head to tail. The pits open to the surface or communicate with one another under the surface through canals. The canals open to the outside by pores that pierce the scales. The cupula waves freely in the water of the pit or groove and stimulates the hairlike processes of the neuromast cells.

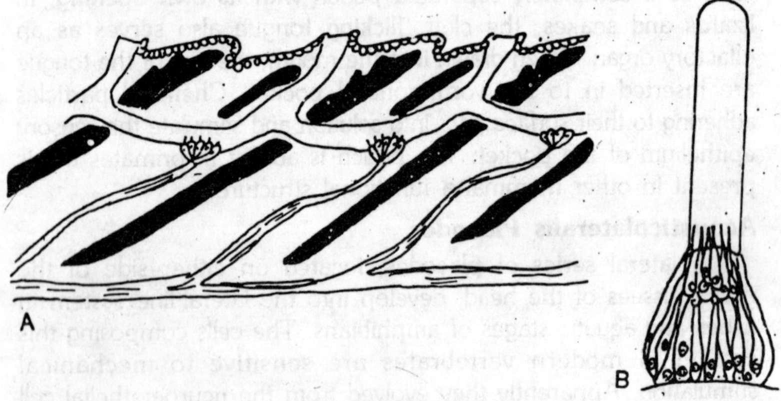

Fig. 8.24. Lateral line of fish. A, Part of the main lateral line canal of the perch (Perca) showing three pores to the surface and three neuromast organs. B, Diagram of a single neuromast organ showing the sensory processes and the gelatinous cupula.

The cilium is located on the apical end of the sensory cell at the periphery of a patch of stereocilia. Apparently the lateral-line organs enable the organism to distinguish the direction of mechanical stimulation, that is turbulence, vibration, or disturbances in the water. Lateral-line stimulation makes the fish aware of prey, enemies, or navigational obstructions. The lateral-line placodes are associated with three of the cranial nerves; the seventh (facial), ninth (glossopharyngeal), and tenth (vagus). Specific nerves innervate different areas of the lateral line. The head canals become associated with the seventh cranial nerve and the temporal region with the ninth nerve. The vagus (tenth) sends fibers to the occipital region and forms a stout trunk that accompanies the lateral line down the length of the animal's body.

Ear

The mammalian ear is composed of three regions; *external ear*, *middle ear*, and *inner ear*. All vertebrates have inner ears, but the middle ear and external ears are relatively recent additions. The middle ear first appears in the amphibians and is an adaptation to aerial living. As the animal changed from an aquatic to a terrestrial existence, one of the gill pouches became the middle ear chamber and one of the bones of the gill apparatus became adapted to transmit vibrations from the outside to the receptors of the inner ear. The external ear merely collects and directs sounds to the middle ear where they are magnified. Some reptiles have an external ear, but it is most developed in the mammals. Embryologic development of the inner ear suggests that it evolved from that part of the lateral line innervated by the eight cranial nerve.

The otic placodes located in the epidermis lateral to the hindbrain (rhombencephalon), on each side of the head, arise in much the same manner as other components of the lateral-line organs. The epidermis thickens to form a plate; it sinks more deeply into the loose mesenchyme cells of the head region than do the lateral-line placodes and becomes the otic sac. It may not retain its connection with the surface. The development of the inner ear of the bony fish is an exception. In these animals a solid plate of cells on the inner surface of the epidermis sinks downward and later becomes hollowed out. Although the exact mechanism varies with the species, the otic sac is converted into the *otic vesicle*, or *otocyst*. This completely closed structure differentiates into the inner ear, composed of the organ of equilibrium and the organ of hearing. In the fish the organ of hearing is under-developed and the ear functions primarily in equilibrium.

Since olfactory function was of primary importance in the primitive animal, auditory function in the ancestral vertebrates was perhaps of little value and may have been completely absent. However, the ability of the ear to function as an organ of equilibrium is more fundamental and remains relatively unchanged from fish to man.

Hearing becomes an important part of the sensory system at the tetrapod level and increases in complexity as the evolutionary scale is ascended. The saccule of the fish, for example, is only a small structure, whereas in man the lagena, or organ of hearing develops from its posterior end. Neuroblast cells differentiate from the otic vesicle and form a club-shaped mass on the medial wall. These cells are destined to become the acoustic, or auditory, ganglion. Processes develop on the cells of the ganglion and penetrate the rhombencephalon, giving rise to the root of the eighth nerve. In adult animals this nerve divides into two main branches, the vestibular nerve, which passes to the anterior regions of the organ of equilibrium, and the cochlear nerve, which services the posterior part of the organ of equilibrium and the cochlea. The fact that neuromast cells provide the sensory receptors for both the lateral-line organs and inner ear offers additional evidence of their close phylogenetic relationship.

Ear as organ of equilibrium

The otic vesicle develops into a pear-shaped structure with the pointed end oriented toward the dorsal surface of the head. The pointed area becomes the endolymphatic duct; the expanded portion of the otic vesicle give rise to the *membranous labyrinth*. The labyrinth undergoes irregular growth and soon divides into two chambers, a dorsal *utricule* connected to a more ventral *saccule* by way of a narrow *sacculoutricular duct*. The endolymphatic duct is attached either to this duct or to the saccule. Further expansion and constriction of the utricule give rise to the rudiments of the three *semicircular canals*, first present in the human embryo as early as the sixth week.

In the adult of all vertebrates, these rudiments become three distinct canals, each of which is oriented in a different plane (two vertical and one horizontal) and is open at both ends into the utricule. They are referred to as the anterior, posterior, and horizontal semicircular canals. In lower vertebrates, the saccule forms a pocketlike structure, the lagena, at its posterior end. This elongates in reptiles and birds and expands in mammals into the cochlear duct. The utricule, saccule, semicircular canals, and lagena compose the inner ear and should be differentiated from the auxiliary middle and outer-ear structures, which develop from different primordia. The thin walls of the membranous labyrinth consist of flat epithelial cells.

The medioventral area, however, thickens, and the cells develop into patches of supporting cells and sensory cells known as the *maculae*. In the adult organ they are distributed in a horizontal plane on the floor of the utricule and in a vertical plane on the inner wall of the saccule. These spots of sensory epithelium known as the utricular macula and saccular macula, respectively, are associated with the branches of the auditory nerve. The fibers of this nerve grow between the epithelial cells early in development and arborize about the bases of the sensory cells. Stimulation of these sensory cells offers information about the tilt of the head, and the individual is made aware of static position. An expansion at the end of each semicircular canal, referred to as the ampulla, also contains sensory areas, termed *cristae*. The hollow, irregularly shaped membranous labyrinth, consisting of the utricule, saccule, and semicircular canals, fills with fluid.

The *endolymph*, which is more viscous than water, resembles the interstitial fluid. Supporting cells and hair cells with long processes compose the maculae and cristae (the actual receptors for the sense of equilibrium). The hair cells are histologically similar to neuromast cells and respond to pressure or movements of the liquid in which they reside. The supporting cells are believed to secrete a jellylike substance, the *cupula*, in which the hairlike projections of the sensory cells of the cristae rest. Various concretions (otoliths) or grains of sand are associated with the gelatinous covering of the sensory cells of the macula. As the animal moves about, the weight of the otoliths, which rest on the sensory epithelium, stimulate different hair cells. This stimulation in turn causes an increase in the rate of impulses fired to the central nervous system by way of the vestibular branch of the eighth cranial nerve. The stimulation of these slightly elevated patches of cells furnishes the animal information about its turning movements.

The endolymph also surges about in the tiny semicircular canals as the animal changes positions, and the movement of the fluid stimulates the sensory hairs of the cristae, providing additional information about the activity of the animal. As mentioned above, the original placode (induced to form by the rhombencephalon) sinks into the mesenchyme of the head area and gives rise to the otic vesicles. This vesicle develops into the labyrinth. Depending on the species, the mesenchyme (derived from sclerotomes) is in turn induced by the otic vesicle to differentiate into skeletal tissue. It is of harder texture immediately surrounding the membranous labyrinth than elsewhere and forms a capsule about the ear known as the *cartilaginous* or *osseous labyrinth*. A space exists between the membranous labyrinth (derived from ectoderm) and the osseous labyrinth (composed of mesoderm). It fills with *perilymph*, a fluid very similar to the endolymph of the membranous labyrinth.

Ear as organ of hearing

Middle and *outer ear*. The organ of equilibrium remains relatively unchanged in tetrapods, but hearing becomes an important additional sense. How well fishes hear remains a controversial topic, but, in general, investigators believe that the sensory maculae act as possible phonoreceptors. Skeletal structures of the head transmit sound waves from the water to the inner ear of fishes. The air bladder also may function as a transmitter of sound vibrations. However, since air waves are relatively faint, there is a need for some kind of device to amplify the vibrations in land animals. The development of the *middle ear* from the anterior gill pouch (the spiracular gill cleft) occurs in terrestrial animals. This first pharyngeal pouch does not break through to the outside to form a typical gill cleft; instead, it remains as an expanded lateral cavity, the primordium of the *middle ear*, or *tympanic cavity*. The addition of mesoderm to the thin membrane left between the pouch and the anterior ectoderm converts it to the *tympanic membrane*, or eardrum.

Since the gill pouches are outpocketings of the pharynx, the internal lining of the tympanic membrane is endoderm. Its outer surface, however, is derived from superficial epidermal ectoderm. The resulting tympanic membrane thins and responds sensitively to air waves. The remnant of the attachment of the gill pouch to the pharynx becomes narrowed and remains as the *eustachian tube*. Mesenchyme condensations give rise to the primordium of the transmitting apparatus of the middle ear.

In amphibians and reptiles a piston-like bone, the columella (which evolved from the hyomandibular bone of fish to be discussed later), crosses this pouch from the eardrum to a membrane-covered opening in the ear capsule, the *oval window* (fenestra ovalis). One end of the columella (stapes) rests against this membrane. It picks up vibrations received by the ear-drum and transfers them internally to the fenestra ovalis. As a result, the liquid of the internal ear is set in motion on the other side of the membrane, stimulating the various sensory spots. A second membrane-covered opening, the *round window* (fenestra rotunda), is located at the other end of the perilymphatic duct. It functions as a release valve.

In mammals pressure waves, introduced to the perilymph by the action of the stapes, travel up the scala vestibuli and then across to the scala tympani (dorsal and ventral ducts of the original perilymphatic space) and back toward the middle ear. The membranous fenestra rotunda bulges into the middle ear cavity dissipating the pressure waves. In mammals, three *auditory ossicles* (malleus, incus and stapes) replace the columella and lead from the tympanic membrane to the fenestra ovalis. The mammal also adds a cartilaginous external ear (*pinna*) to help collect and concentrate air waves. In many animals, muscles attach to the pinna and move the ear about and orient it in the direction of sound. A short canal, the *external auditory meatus*, leads from the outside to the tympanic membrane. Wax glands are usually present in the walls of this canal.

Development of cochlear duct

The lagena, present in all vertebrates beginning with the fish, develops in higher animals into the basic structure of hearing, the *cochlear duct* with its *organ of Corti*. The evolution of the cochlea involves the *lagena*, the *perilymphatic duct*, and the *organ of Corti*. The lagena first appears as an endolymphatic papilla in the amphibian saccule. It expands into a long fingerlike extension in the crocodiles and birds, but in mammals it is so long that it must coil in order to fit within the ear capsule. In man, the cochlea makes 2 ¾turns. Since the lagena is an extension of the saccule, endolymph also fills its cavity. This duct, formed by the extended lagena, is the *cochlear duct*, or *scala media*. The *organ of Corti*, named after the Italian anatomist who first described it, is the true receptor organ for the sense of hearing. It extends down the length of the cochlear duct and its cells discriminate the sounds of different pitch. In the human embryo it develops as early as the third month as a local thickening of the floor of the cochlear duct.

Between the third and the fifth months the epithelial cells differentiate into the hair cells of the organ of Corti and establish the adults structure. The mature structure basically consists of a series of sensory hair cells (neuromast cells) intermingled with supporting cells. A gelatinous membrane, the *tectorial membrane*, extends over them and is in contact with the stereocilia of the hair cells. This membrane may be compared to the cupula of the neuromast cells previously described as composing the lateral-line sense organs. A basilar membrane, capable of vibrating, forms the floor of the organ. The roof of the cochlear duct (scala media) is known as the *vestibular*, or *Reissner's membrane*.

The *perilymphatic duct* is the name given to the space between the cartilaginous capsule and lagena portions of the membranous labyrinth. It is present in amphibians, reptiles, birds, and mammals. It consists of a well-defined channel that follows one side of the cochlear duct from its origin at the membranous labyrinth to the end of the coil and then turns and follows the other side back. A portion of the perilymphatic duct, therefore, lies above and below the cochlear duct. This relationship of the cochlear duct to the perilymphatic canals can best be seen in a cross section of the duct. As a result of its relationship with the perilymphatic ducts, three main longitudinal chambers referred to as *scalae* compose the cochlea proper. The upper chamber is the *scala vestibuli* and the lower one is called *scala tympani*; both form from the perilymphatic ducts and fill with perilymph. In between the two chambers is the *scala media*, or *cochlear duct*, filled with endolymph and the only part derived from the membranous labyrinth. The scala media ends blindly at the apex of the coil, but the scala vestibuli and scala *tympani* communicate with one another at the apex, the *helicotrema*, through a small junction. They are, therefore, continuous with one another.

The cochlear duct is attached to the sides of the bony labyrinth but is surrounded above and below by the perilymphatic ducts. Delicate strands of tissue cross these spaces and help anchor the fragile organ of Corti in place. In the mammal the external ear directs the sound toward the tympanic membrane. Sound vibrations that impinge on this membrane cause it to vibrate. These vibrations are magnified and transmitted inward by the three auditory ossicles. The stapes vibrates against the fenestra ovalis at the same frequency as the original sound. The ascending scala vestibuli is on the other side of the fenestra ovalis and a pressure-change is produced in the perilymph when the stapes moves the window in and out. Since the perilymph is enclosed by a rigid bony casing and cannot be compressed, as stated above, the membranous fenestra rotunda bungles outward into the middle ear cavity dissipating the pressure.

Pressure changes initiated in the scala vestibuli (when the stapes vibrates against the fenestra ovalis) are believed to travel up the scala vestibuli to its junction with the scala tympani at the helicotrema and then move down this canal. However, many vibrations in the perilymph are transmitted across the membrane separating the scala vestibuli from the cochlear duct. This causes in the cochlear duct a disturbance of the endolymph, which displaces the basilar membrane toward the scala tympani. This displacement (caused by waves in the perilymph) alters the contact of the hair cells with the tectorial membrane. This change in relationship acts as a stimulus for the hair cells. Terminal branches of the auditory nerve contact the hair cells. Presumably the stimulation of the hair cells is translated into a nerve impulse that passes to the medulla oblongata by way of the cochlear branch of the auditory nerve. Clinical and experimental evidence shows that the regionalization of response by the organ of Corti enables the individual to discern pitch. That part of the organ of Corti closest to the membranous labyrinth discriminates high tones; the receptors for low tones are located near the apex of the duct.

Eye

The eye is an exceedingly complex structure well-developed in even the lowest vertebrate group. There are few clues as to how it evolved. It is a sensitive detector of light waves and differs from the other sense organs in that it develops from the wall of brain. Its basic structure is the same from cyclostomes to mammals, although adaptive modifications allow it to function better in one ecologic niche than in another. It basically consists of a spherical *eyeball*, located in an *orbit*, protected by cartilage or bone, and connected to the brain by an *optic nerve*. The wall of the adult eyeball consists of three concentric coats: an outer *sclera*, a middle *uvea*, and an inner *sensory retina*. The eye also contains a crystalline lens and a transparent cornea (anterior portion of the sclera), which focus the light rays on the retina. Three cavities make up most of the eyeball: the *anterior* and *posterior chambers* and the *vitreal cavity*, which contain transparent viscous substances that allow the light rays on the retina. The eye results from a series of inductions. It is derived from three main groups of tissues in the embryo. The retina forms from neural ectoderm, the lens develops from epidermal ectoderm, and the tunic of the eye develop from the surrounding mesenchyme.

Early Development

Early in development, when the brain is still in the process of differentiation, lateral evaginations from the wall of the forebrain push toward the skin ectoderm. In the human embryo, the evaginations appear as early as the middle of the third week. These protrusions of the brain are the beginning of the *optic vesicles*. They become bulblike at their distal ends, where they are in contact with the surface of the head: each contains a cavity, the *opticoel*, continuous with the cavity of the brain (*prosocoel*). A constriction soon separates the proximal end of the vesicle from the brain; the resulting connection between the two structures becomes the *optic stalk*. The underlying chorda-mesodermal roof induces the optic vesicles to form.

As a result of this influence, the eye develops in a step-by-step manner, with specific tissue interactions necessary before the next stage of the development process can proceed. For example, mesenchyme cells must surround the evaginations of the brain for normal development to occur. Also, as the lateral evaginations from the diencephalon expand outward, they must contact the epidermal ectoderm for the retina to differentiate. If no contact is made, the presumptive neural retina becomes pigmented rather than sensory. When the optic vesicle contacts the epidermal ectoderm, the vesicle flattens and the distal wall invaginates, with obliteration of the opticoel and formation of a double-layered *optic cup*.

At first the wall of the optic vesicle is only one cell thick. However, the cells of the indented wall of the optic cup undergo mitosis and become the sensory *retina*, the photosensitive lining of the eye. Pigment granules invade the thin outer layer and form the *pigmented layer* of the retina. It absorbs any light that passes through the retina layer, thereby preventing scattering of the light and blurring of the image. The resulting structure is a double-layered cup attached to the diencephalon of the brain by a narrow, hollow, optic stalk. By the time the eye completes its differentiation, the retina and the pigmented layer fuse form one layer. The optic vesicle changes into the optic cup by an invagination that begins at the lower edge of the optic vesicle near the optic stalk and proceeds upward. As a result of the differential invagination, the ventral rim is not as well-developed as the rest of the cup and a groove remains, the *choroid fissure*.

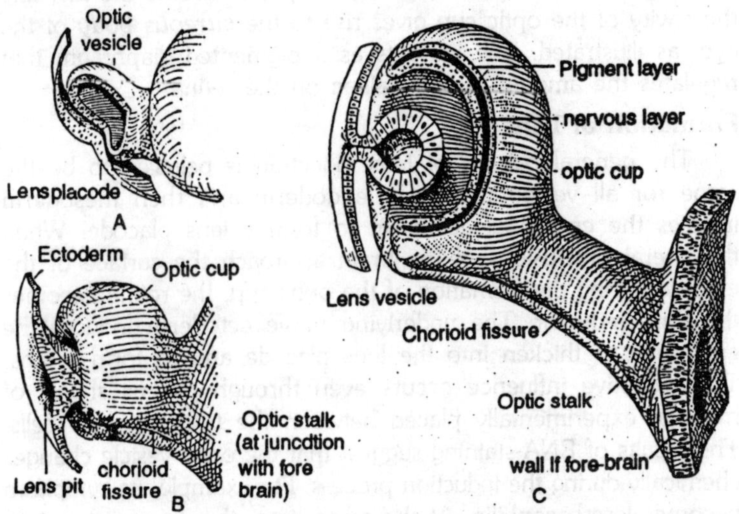

Fig. 8.25. Human optic primordia, shown as models in side view X 100. The lens is sectioned and the optic cup in A and C has been partly cut away. A, at 4.5 mm.; B, at 5.5 mm.; C, at 7.5 mm.

The indentation that gave rise to the cup continues along the ventral surface of the optic stalk. As a result, during early development of the eye a troughlike structure extends from the ventral rim of the cup toward the brain. The central artery follows the trough of the choroid fissure and enters the optic cup where it is known as the *hyaloid artery*. A network of capillaries forms from the artery around the developing lens. This step appears necessary in the early stages for differentiation of the lens. By birth the *hyaloid artery* regresses, leaving the *hyaloid canal* in its place in the vitreous body. The edges of the choroid fissure eventually fuse and a small tube forms within the already tubular optic stalk.

The central artery and vein remain in this canal and become the main blood vessels of the retina. When the optic nerve forms, the axons of the ganglionic cells converge to the walls of the optic stalk and follow its lumen into the brain. The walls of the optic stalk become the sheath of the optic nerve, enclosing the axons and the central artery and vein a well. At first the opening of the optic cup is very large, but as development continues, the edges of the cup converge, resulting in a small opening, the pupil. The edges of the cup surrounding the pupil become the iris, and the cavity of the optic cup gives rise to the *vitreous body* of the eye, as illustrated. The iris acts as a pigmented diaphragm that regulates the amount of light falling on the retina.

Formation of Lens

The general pattern of lens induction is believed to be the same for all vertebrates. First endoderm and then mesoderm induces the epidermal ectoderm to form a lens placode. When the lateral protrusions of the brain approach the surface of the embryo during the formation of the optic cup, the retina becomes the third inductor. The underlying nerve ectoderm induces the epidermal to thicken into the lens placode and to form a lens. The inductive influence occurs even through porous sheets of material experimentally placed between the two layers of cells. The results of RNA-staining suggest that the optic vesicle changes chemically during the induction process. For example, its cytoplasm becomes less basophilic. At the same time, the presumptive lens ectoderm shows a gain in basophilia, as if "something" were transferred to it. The influence of the optic vesicle on the epidermis can be further demonstrated by surgical removal of the lens placode.

The surrounding epidermal cells move into the empty space and develop into a second lens placode when they come in contact with the optic vesicle. The optic vesicle, therefore, induces other areas of the epidermis to form a lens, and determines the exact position of the lens in relation to the eye cup. If the optic cup is removed, either no lens forms at all or it is defective. This ability of the epidermal ectoderm to respond to the optic cup is a clear example of an induction process and has been extensively studied. The induced lens placode, under the continued influence of the retina, either bends inward or separates itself as a mass and reassembles to form a lens vesicle. In either case, the lens eventually occupies a position in the pupil of the iris almost filling the optic cup during early development.

Once the presumptive lens separates from the overlying ectoderm, it differentiates further. The cells of the inner portion

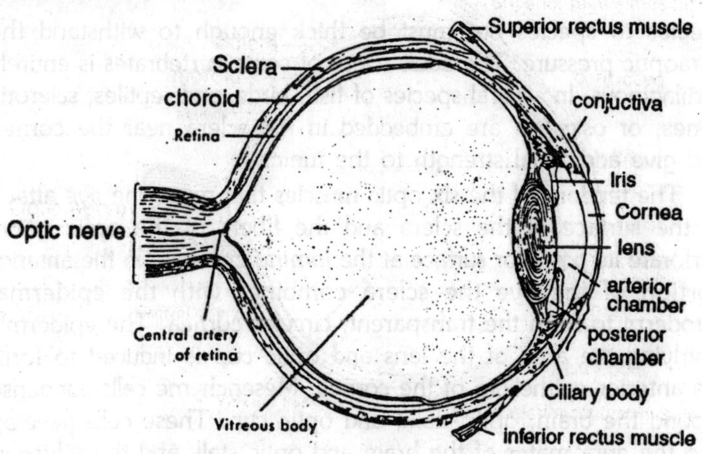

Fig. 8.26. Diagram of sagittal section of eyeball.

of the lens elongate, lose their nuclei, and transform into longer fibers, arranged in an orderly fashion. The outer layer of the lens becomes its epithelial capsule. The lens takes its place at the mouth of the cup as a small crystalline structure, capable of either moving forward and backward (in anamniotes) or changing shape (amniotes). This difference in the mechanism of lenticular adjustments is correlated with the kind of adaptations the animal must make to its environment. For example, the ability to change the shape of the lens gives the animal a greater range of focus necessary for aerial vision. In cyclostomes the lens is not attached but the pressure of the vitreous humor and the cornea keep it in its proper position. In other forms, a ring of fibers hold the periphery of the lens in place. The ability of the eyes to focus light on the retina is known as *accommodation*. It is accomplished in general, in land animals by the curved surface of the cornea focusing the image on the retina and the lens bringing it into sharp focus. This is not the case in fish. The fish lens has a spherical shape and does most of the work of focusing the image.

Tunics of Eye

While the basic structures of the eye differentiate, its accessory structures organize from mesenchyme. These structures are the fibrous and vascular tunics of the eye. The outer fibrous tunic (coat) gives rise to the *sclera* and *cornea*. The sclera is a whitish, tough, fibrous coat that forms a protective covering to the eye and is seen as the white of the eye. It varies in thickness from species to species but must be thick enough to withstand the intraoptic pressure. The adult sclera of some vertebrates is entirely cartilaginous. In several species of fish, birds, and reptiles, sclerotic bones, or *ossicles*, are embedded in the sclera near the cornea and give additional strength to the tunic.

The tendons of the six optic muscles that move the eye attach to the surface of the sclera and the fibers of the optic nerve perforate its posterior surface at the *lamina cribrosa*. In the anterior portion of the eye the sclera combines with the epidermal ectoderm to form the transparent, circular cornea. The epidermis overlying the area of the lens and optic cup is induced to form the anterior epithelium of the cornea. Mesenchyme cells condense around the brain, optic stalk, and optic cup. These cells develop into the dura mater of the brain and optic stalk and the sclera of the eyeball. Those mesoderm cells that lie under the epidermal ectoderm (anterior *epithelium* of the future cornea) become transparent and contribute to the cornea. The cornea, therefore, is a composite structure, derived from both ectoderm and mesoderm. It is at the corneal surface that much of the refraction of light occurs in terrestrial forms as pointed out above.

Irregularities in its curvature bring about an impairment of vision in the human eye known as *astigmatism*. The cornea is an avascular structure, nourished by the secretions of the eye and able to repair itself rapidly if injured. It must be kept moist at all times. The location of many pain receptors in the cornea makes it an extremely sensitive structure. The *uvea*, the inner tunic of the eye (the middle layer between retina and sclera), is derived from mesoderm and includes the *choroid coat*, the *ciliary body*, and the *iris*. The choroid coat is pigmented and extremely vascular, providing the nourishment for the eye. It contains arteries and veins of various sizes as well as capillaries. It also reflects light back through the retina in several species, a phenomenon that makes their eyes shine in the dark. The area of the choroid coat responsible for such eyeshine is known as *tapetum lucidum*. The uvea differentiates into the connective tissue of the ciliary body, the fibers of the ciliary muscle, and the connective tissue of the iris.

The iris (as well as the ciliary body) has a dual composition. This structure present in all vertebrates, receives contributions from both the choroid coat and the non-nervous part of the retina. The presence of pigments in this layer is responsible for the colour of the eyes. If no pigment is present at all, the blood vessels of the iris give the eye a pink colour, characteristic of albino eyes. Blue eyes occur when the pigment is present only in the inner portion of the retinal layer. The colourless tissue covering the iris interferes with light rays in such a manner that the iris appears blue. The presence of dark pigment in both the choroids and retinal layers produces brown or black eyes. However, the eyes of newborn children, even those of the black races, do not reflect their true colour and are blue. The addition of pigment later to the layers of the iris gives rise to its true adult colour. Muscles are present in the iris, arranged in circular and radial patterns. Contraction of these muscles results in decreased pupil size.

The iris, therefore, regulates the amount of light that passes to the lens and on to the retina. In dim light, the pupil opens wide, but in bright light it narrows. These muscles of the iris are derived from its retinal portion and are one of the very few examples of a muscle originating from neural ectoderm rather than the usual embryonic source of muscle, the mesoderm. The ciliary body also receives contributions from the optic cup. It is a ring-shaped-structure that encircles the equatorial zone of the lens and is located near the *juncture* of sclera and cornea. It functions in accommodation and may or may not actually contact the lens. When the muscles of the ciliary body push against the lens and cause it to thicken and shorten its focal length. The lens is a very elastic structure in mammals. It is anchored to the ciliary region by certain zonary fibers that hold the lens under tension and, therefore, flat.

Contraction of the zonary muscles brings the ciliary body closer to the lens, the fibers relax, and the elastic lens becomes more rounded. This allows the mammal to accommodate for closer objects. The eyeball consists of a series of cavities filled with transparent liquids of various consistencies. The *anterior chamber* of the eye is the space between the cornea and the iris; the *posterior chamber* is the small area between the iris and the lens. These two cavities are continuous with one another through the pupil. They are filled with a thin watery liquid known as the *aqueous humor* secreted by epithelial components of the ciliary body. This liquid is similar to the blood serum and nourishes the lens, iris, and cornea and produces the turgor of the eyeball.

Drainage of the fluid into series of sinuses that make up the canal of Schlemm, located in the angle of the iris, regulates the intraocular pressure. A thick gelatinous substance known as the *vitreous humor* fills the principal cavity of the eyeball, the space between the lens and the retina. The photosensitive cells of the retina secrete the material during the development of the eye.

Histogenesis of Retina

As demonstrated by the manner in which it is formed, the retina is an extension of the brain. The origin of the rods and cones, the two types of cells sensitive to light, can be traced to the ependymal cells of the neural tube and to the presumptive neural ectoderm on the surface of the blastula. A knowledge of their development explains why light must pass through several layers of the adult retina before reaching the neurosensory photoreceptors. These receptors are located in the outermost part of the retina, that layer closest to the choroid coat and farthest form the pupil. In the early embryo the ependymal layer lines the neural cavity. As the optic cup forms, this layer becomes adjacent to the pigmented layer of the cup. As a result, the rods and cones are located away from the surface exposed to the vitreous body and away from the source of light. Three layers of cells make up the retina of vertebrates.

Fig. 8.27. Early differentiation of the nervous layer of the human retina, shown in a ventral section at three months. At left, Cajal's analysis of the component element after silver impregnation; at right, the appearance with ordinary stains.

Listed in order, the processes of the rod and cone cells are located next to the pigmented cells; the cell bodies of these

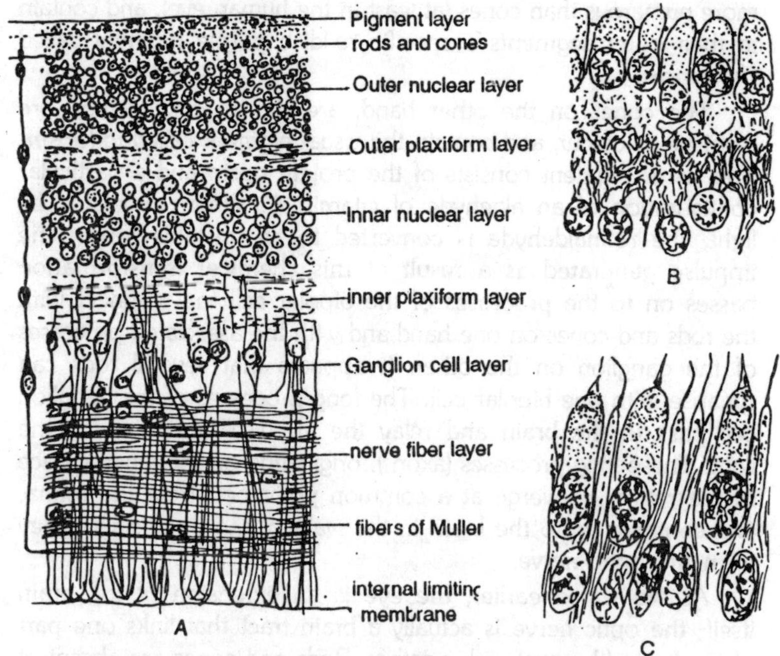

Fig. 8.28. Later differentiation of the human retina. A, At seven months, in vertical section at left, the chief neurons shown by silver technique; at right, appearance with ordinary stains. B, Early cone cells during the fifth month. C, Rods and cones during the seventh month.

photoreceptor cells make up the outer nuclear layer. This is followed by the *bipolar neurons*, with the layer of *ganglion* cells situated nearest the vitreous humor. The plexiform layers separate one cell layer from another and consist of cell processes synapsing with one another. The *outer plexiform layer* is the area of synapse between the rods and cones and the bipolar cells, the *inner plexiform layer* consists of the synapses between the bipolar cells and the ganglion cells. The distribution of rods and cones in the vertebrate retina differs among the various species. Nocturnal vertebrates, for example, usually have more rods than cones, and the retina of diurnal vertebrates contains more cones. A depression, known as the *fovea*, often marks the site of sharpest vision in certain teleosts, birds, and primates. The rods and cones differ from one another in shape, number, sensitivity to light, and specific visual pigment contained in the receptor. Vertebrate rods are slender, filamentous cells, sensitive to degrees of darkness, are more numerous than cones (at least in the human eye), and contain in their outer segments a specific reddish purple pigment called *rhodopsin*.

The cones on the other hand, are shorter and thicker, are sensitive to colour, and contain the visual pigment called *iodopsin*. The visual pigment consists of the protein opsin and a particular configuration of an aldehyde of vitamin A. When stimulated by light, the retinaldehyde is converted to an unstable form. The impulse generated as a result of this chemical transformation passes on to the processes of the bipolar cell that connect with the rods and cones on one hand and with the associated processes of the ganglion on the other. It appears that several rods can synapse with one bipolar cell. The long processes of the ganglion cell pass to the brain and relay the impulse generated by the light. These long processes (axons), originating on the inner surface of the retina, converge at a common point penetrate the retina, and continue on to the brain in the walls of the optic stalk. They form the optic nerve.

As we stated earlier, the eye is an extension of the brain itself; the optic nerve is actually a brain tract that links one part of the brain (the eye) with another. Rods and cones are absent at the point in the retina where the fibers of the ganglion cells leave. Consequently, this point is known as the blind spot of the eye. The optic nerve fibers pass to the floor of the diencephalon and cross to the opposite side of the brain at the X-shaped *optic chiasma*. There is complete *decussation* (crossing of fibers) in all classes except mammals. In these animals, *decussation* is incomplete and only the fibers from the medial half of the retina cross to the other side. This leaves about one third to one half of the fibers uncrossed in man. This incomplete decussation is associated with the development of *stereoscopic vision*. Physiologists believe that two duplicate mental pictures form in most animals.

In mammals however, the development of stereoscopic vision enables each brain hemisphere to build up a half-picture of the total image. Interconnections between the two cerebral hemispheres allows the individual to "see" the total picture. In all non-mammalian vertebrates the optic fibers continue in the walls of the thalamus to the roof of the midbrain (mesencephalon). Here they terminate in the optic lobes of the mesencephalon where the primary visual center is located. When the cerebral hemispheres assume their prominence as a co-ordinating and association center in mammals, the mesencephalon loses its importance and no longer acts as a visual center. As a result, most of the optic fibers in mammals are no longer sent to the mesencephalon. Instead, the optic impulses in man are shunted from the thalamus to the new visual centers of the cerebral cortex. It is in the cerebrum that the sensation of sight occurs.

Accessory Structures of Eye

Certain accessory structures protect the eye or aid it in its function. Folds of developing skin, consisting of ectoderm and mesoderm, arise above and below the cornea and give rise to the eyelids of tetrapods. Most fish lack a movable eyelid. The blinking of the eyelids of tetrapods keeps the cornea moist and clean and prevents it from becoming dry and opaque. A transparent third eyelid, the *nictitating membrane*, is present in birds and some reptiles and mammals. It passes from the inner angle of the eye to the outer surface. Rows of hairs, the eyelashes and eyebrows, furnish additional protection to the eye. In land animals, lacrimal glands (tear glands) secrete a watery fluid that keeps the cornea moist and clean and provide nourishment to this avascular structure. Epithelial buds form on the undersurface of the eyelid and during development converge as a group to one area of the eyeball. In most animals they become branched and form an acinous gland that secretes constantly. A lacrimal duct drains the secretions into the nasal passageway.

9

MESODERMAL DERIVATIVES

In vertebrates the mesoderm is differentiated either from embryonic ectoderm or from endoderm. It appears in the form of mesodermal somites. The somites, when first formed, are masses of mesodermal cells with a small cavity in the middle. The cells are arranged radially around the central cavity. With further development, the shape of the somites changes; they become extended in the dorsoventral direction and flattened mediolaterally. In the vertebrates with discoidal cleavage, this leads to the elevation of the dorsal parts of the embryo above the general level of the blastodisc. The flattening of the somite is accompanied by a change in the shape of its central cavity (the *myocoel*); instead of being spherical, the cavity becomes a narrow vertical slit. An inner wall and an outer wall become clearly distinguishable, corresponding to the parietal and visceral layers of the lateral plates.

The inner wall of the somites becomes very much thicker than the outer wall. The fate of the inner and the outer wall of the somites is completely different. The outer wall contributes to the formation of the connective tissue layer of the skin, and is therefore called the dermatome. The inner wall produces skeletogenous tissue and the voluntary striated muscles of the body. The skeletogenous tissue develops from the lower edge of the inner wall of the somite, and this part of the somite is therefore called the sclerotome.

The sclerotome breaks up into a mass of mesenchyme cells. The cells migrate into the spaces surrounding the notochord and the spinal cord, envelop these organs and later differentiate into cartilage, thus forming the bodies and the neural arches of the vertebrae. The hemal arches in the tail region and the ribs are of the same origin. The dorsal part of the inner wall of the somite is the source of somatic muscles, in the vertebrate's body. It is therefore called the myotome.

The cells of the myotome rearrange themselves so that they become elongated in a longitudinal direction. These longitudinally elongated cells differentiate subsequently into the striated muscles fibers. Originally each myotome becomes a muscle segment, separated from the one anterior and the one posterior to it by a connective tissue layer, the vertical myocomma. At one time in its development the whole somatic musculature of the vertebrate consists of such segments arranged in linear order. In the lower vertebrates, the myotome is the largest part of the somite, the sclerotomes being small and rather inconspicuous.

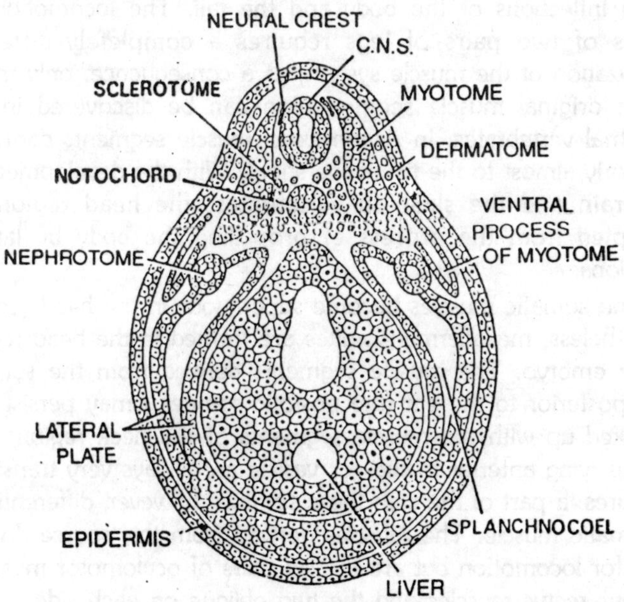

Fig. 9.1. Frog. T.S. of young tadpole through trunk region.

In the Amniota, however, the sclerotomes are much larger. Only the upper edge of the inner wall of the somite adjoining the dermatome becomes the myotome. The size of this part increases rapidly; the myotomes grow downward to assume the same position lateral to the neural tube and the notochord, that they occupy in the amphibians right from the start. Developing as they do from the somites, the myotomes are originally dorsal in position. In the course of the subsequent development, the muscle segments spread downward in the space between the skin on the outside and the somatic layer of the lateral plate on the inside, until the muscle segments on the right and the left side meet ventrally. It is in this stage, with minor alterations that the somatic muscle persist in the fishes and in the aquatic larvae of amphibians.

The segmentation of the lateral and ventral muscles, as well as of the dorsal muscles, is directly derived from the segmentation of the mesodermal mantle in somites. In terrestrial vertebrates, the primitive segmentation of the somatic muscles is more or less obliterated in connection with a change of locomotion. The segmented lateral bands of muscle are adapted to locomotion by lateral inflections of the body and the tail. The locomotion by means of two pairs of legs requires a completely different organization of the muscle system; as a consequence, only traces of the original muscle segmentation can be discovered in the terrestrial vertebrates. In *Amphioxus*, muscle segments *continue* anteriorly almost to the tip of the snout. With the development of the brain and the skull, in vertebrates, the head regions is exempted from the process of propelling the body by lateral inflections.

The somatic muscles become superfluous in the head region. Nevertheless, mesodermal somites are formed in the head region of the embryo. The muscle segments derived from the somites lying posterior to the ear (the postotic somites) may persist and be linked up with the muscle segments of the neck region. The somites lying anterior to the ear vesicle are always very transitory structures: a part of the cells these somites, however, differentiates as somatic muscle. The muscles derived from this source do not serve for locomotion but are the six pairs of oculomotor muscles: the four rectus muscles and the two oblique on each side.

The Axial Skeleton: Vertebral Column and Skull

In most vertebrates, the axial skeleton passes through three phases in its development. In the first phase, the supporting system of the body is represented by the notochord. In the second phase, cartilage develops partly in direct connection with the notochord, partly independently of the latter. This condition is preserved in the adult state of contemporary cyclostomes and elasmobranch fishes (we are not concerned with the question of whether this condition is primitive or secondary). In the remainder of the fishes and in tetrapods, the cartilaginous skeleton is later replaced or supplemented by the bony skeleton. The bony skeleton and its relationship to the cartilaginous skeleton are amply dealt with in courses on comparative anatomy of vertebrates and may, therefore be entirely left out of consideration. We will consider here only the following aspects in the development of the skeleton;

1. The dependence of the formation of the skeletal parts on the adjoining structures.
2. The origin of cells giving rise to the cartilaginous skeleton.
3. The arrangement of the parts of the early cartilaginous skeleton in relation to the other organ rudiments of the embryo.

The material for the cartilaginous axial skeleton in the body and tail of vertebrates is derived from the sclerotomes. The sclerotomes being parts of the somites, are segmental in origin but once they become transformed into mesenchyme the segmental arrangement is largely lost, and the mesenchyme spreads out as a continuous sheath along the notochord, enveloping the spinal cord above and the caudal artery and vein below the notochord in the caudal region. In this continuous mesenchymal sheath, nodules of cartilage, known as the arcualia, appear later in close apposition to the external surface of the notochord.

Typically the arcualia appear in double pairs; one pair is formed dorsolaterally to the notochord and another pair ventrolaterally. The dorsolateral arcualia grow out dorsally alongside the spinal cord and unite above the spinal cord to form the neural arch. The ventrolateral cartilages in the caudal region grow downward and unite underneath the caudal vein to form the hemal arch. In the cervical and thoracic region they give rise to lateral outgrowths-the rudiments of the ribs. At the same time the proximal parts of all four cartilages spread out around the surface of the notochord, contributing in varying degrees to the formation of the body of the vertebra. The cartilaginous neural and hemal arches do not bear a simple relationship to the somites (or sclerotomes); in many groups of vertebrates there are two series of cartilages (two pairs of dorsal elements and two pairs of ventral elements) developed in each mesodermal segment.

In other cases, although there is only one set of cartilages formed, these are situated intersegmentally at the junction of two myotomes (the myotomes retain the original segmentation of the somites). The way in which the original cartilage rudiments co-operate in the formation of the definitive vertebrate will not concern us here. The fact that the cartilaginous (and later bony) vertebral column replaces the notochord functionally suggests that the location and the arrangement of the vertebral cartilages should be dependent on the notochord. In the absence of the notochord the cartilaginous axial skeleton is very irregular. Cartilage, however is not completely absent in those sections of the body which do not have the notochord.

The notochord is thus not indispensable for the formation of axial cartilages. The irregularity of the latter could be a secondary effect of the extirpation of the notochord, seeing that other systems, such as the spinal cord and segmented muscles are greatly distorted as a result of the failure of the embryo to stretch normally. In some experiments in which the spinal cord instead of the notochord was removed from salamander embryos, it was found that the axial cartilages were either completely absent or reduced to insignificant vestiges. This could not be the result of damage to the sclerotomes, since in other experiments very extensive defects of the somites, including the sclerotomal region, were completely restored at the expense of the remaining fragments of the somites. The results of extirpation experiments were corroborated by the transplantation of pieces of spinal cord into an incision on the lateral surface of the somites of another embryo.

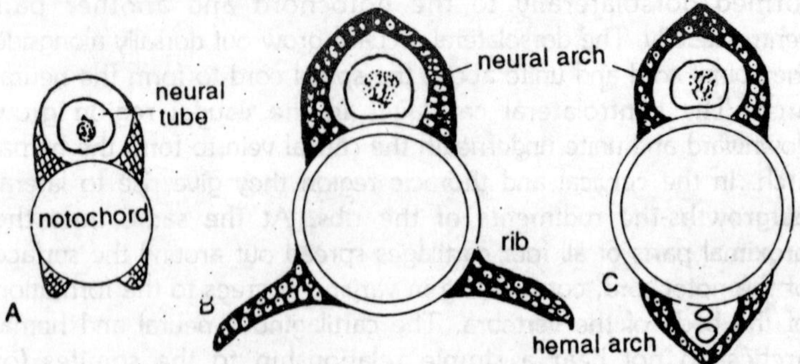

Fig. 9.2. Diagram of the dorsal and ventral pairs of arcualia (a) giving rise to the neural arch and the ribs in the trunk (b) and to the neural arch and the hemal arch in the tail (c).

It was found that complete and well-developed neural arches were formed in association with the grafted spinal cord. The segmentation of the neural arches has not so far been considered. In normal development there is a definite relationship between the segmentation of the longitudinal dorsolateral muscles, the segmentation of the spinal nerves and ganglia, and the segmentation of the vertebral skeletal elements, in particular that of the neural arches. There is a pair of spinal nerves with spinal ganglia corresponding to each muscle segment, the ganglia being situated opposite the median surface of the muscle segment. The neural arches alternate with the spinal ganglia and are thus situated intersegmentally with respect to the myotomes, so that each neural arch (and subsequently each vertebra) is connected to two consecutive muscle segments. The blood vessels also bear a relationship to this common pattern: an intersegmental artery arising from the dorsal aorta along myocomma, close to the vertebral column. Experimental evidence is available to show that this whole system of metamerically arranged parts is originally dependent on the segmentation of the muscle rudiments.

In a salamander embryo at the tailbud stage, the anterior somites are somewhat broader than the posterior ones, so that if a block of several somites in the branchial region is removed and replaced by a block of somite from the posterior trunk region, more somites may be fitted into the wound than had been cut out. Thus somites 7 to 12 could be substituted for somites 3 to 5. As a result, the number of muscle segments on the operated side was increased in comparison with the normal. It was found that the number of spinal nerves and spinal ganglia was increased also, but not necessarily in strict correspondence to the number of muscle segments, one nerve sometimes supplying more than one muscle segment.

On the other hand, the number of neural arches was found to be strictly in accord with the number of spinal ganglia, a bar of cartilage always appearing between two adjacent ganglia. Since the ganglia are formed earlier than the cartilages, there can be no doubt that the neural arches are dependent on the ganglia and not the other way around. The whole chain of reactions would then appear to be as follows.

The neural crest cells which are produced along the whole length of the neural tube become aggregated opposite the median surfaces of the somites and these aggregations become the rudiments of the spinal ganglia. Next, the cells of the skeletogenic mesenchyme, produced by the sclerotomes, spread out over the notochord and neural tube but they are apparently repulsed by the spinal ganglia and thus instead of forming a continuous sheet of cartilage enclosing the spinal cord, they give rise to a series of disconnected element (the dorsal arcualia) alternating in position with spinal ganglia. This is obviously not the complete picture, as it does not account for the cases in which two pairs of cartilages appear between each consecutive pair of spinal ganglia.

It is thus likely that some structures other than the spinal ganglia, the neural tube and the notochord take part in determining the position of early cartilaginous rudiments. The neural portion of the cranium in vertebrates in part bears the same relationship to the notochord and to the neural tube as does the axial skeleton in the posterior parts of the body. But there is little if any trace of a segmentation in the development of the cranium and important parts of the cranium are quite peculiar in this respect. In the lower vertebrates at the earliest stages of development, the cartilaginous cranium consists of several independent rudiments. These are: (1) The trabeculae; (2) the parachordals; (3) the capsules of the sense organs-the nose, the eye and the ear. The trabeculae (or trabeculae cranii) are a pair of elongated cartilages which appear in the most anterior part of the head, in front of the hypophysis.

The trabeculae lie ventrally and ventrolaterally to the diencephalon and telencephalon, and their upper edges are wedged in between the brain on the inside and the rudiments of the nose and the eye on the outside. It has been established both by observation and experiment that the mesenchyme from which the trabeculae are developed comes from two different sources. The anterior part of the trabeculae is formed of neural crest cells migrating forward and downward anterior to the eye cup. The posterior part of the trabecula is mesodermal origin and is derived from the prechordal plate mesenchyme. The parachordals, or parachordal cartilages are derived from the mesenchyme produced by the sclerotomes of the somites in the head region. This mesenchyme spreads out on both sides of the notochord and eventually chondrifies in the form of two longitudinal rods, situated alongside the notochord and ending anteriorly at the same level as the notochord, that is, just posteriorly to the infundibulum and the rudiment of the hypophysis.

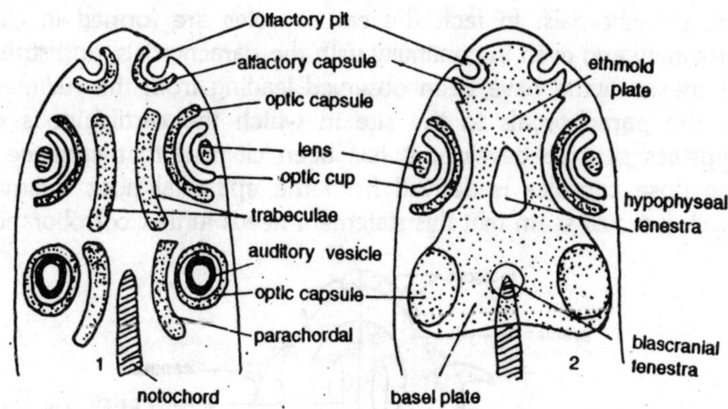

Fig. 9.3. Development of chondrocranium of skull.

The parachordals are very similar in origin and position to the rudiments of the cartilaginous vertebral column, but they lack the segmentation of the latter (possibly because the cranial ganglia, owing to the greater breadth of the neural tube in the head region, lie much farther laterally and away from the region in which development of the parachordals is taking place). The cartilaginous capsules of the sense organs—the nose, the eye and the ear—develop from skeletogenic mesenchyme accumulating around the surface of the epithelial parts of these organs.

The source of the mesenchyme may not be the same in all three cases. It is fairly certain that the ear capsule is formed by mesenchyme derived from the same sclerotomes as give rise to

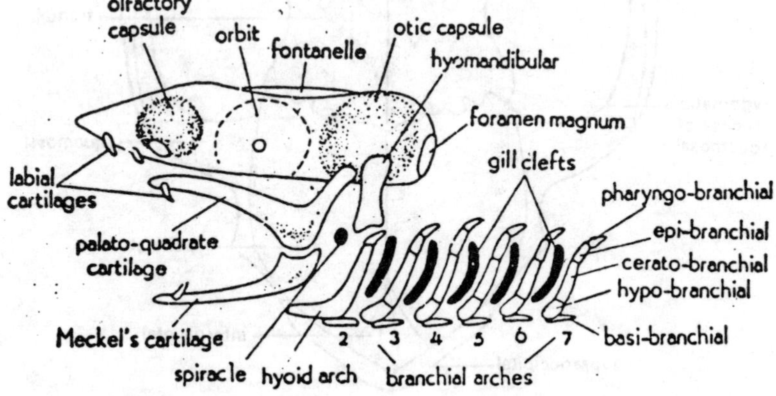

Fig. 9.4. Chondrocranium and splanchnocranium of elasmobranch.

the parachordals. In fact, the ear capsules are formed in close proximity and even in continuity with the parachordals, and strands of mesenchyme have been observed leading from the rudiments of the parachordals to the site in which the cartilaginous ear capsules start to develop. It has been claimed that cartilage of the nose capsules is derived from the epithelial nose rudiment itself. It would seem that this statement needs further corroboration.

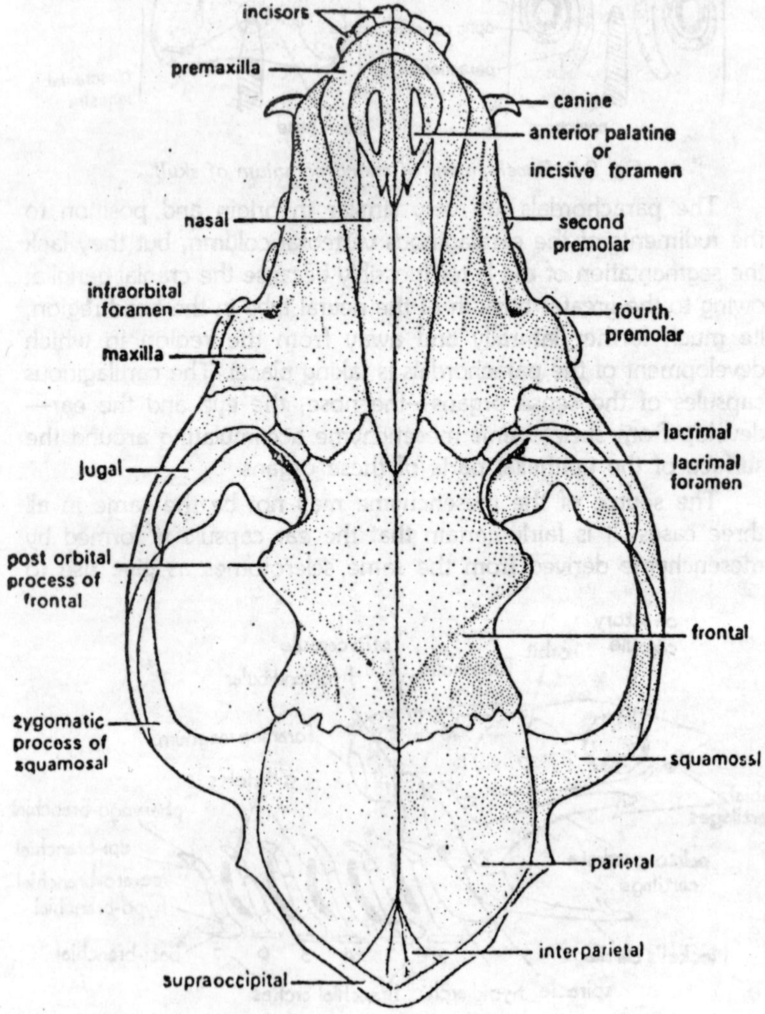

Fig. 9.5. Skull (Dorsal view).

On the other hand, we have seen that the nasal placodes develop in very close proximity to the anterior transverse neural fold.

The neural fold being the source of neural crest cells, it would not be very astonishing if the neural folds could produce the nasal cartilages as well as the epithelial parts of the olfactory organ. In most vertebrates the eyebrow is surrounded first by a connective tissue capsule, the sclera, which becomes cartilaginous is later stages. The source of the cells forming the sclera has not yet been established. The capsules of all three sense organs are

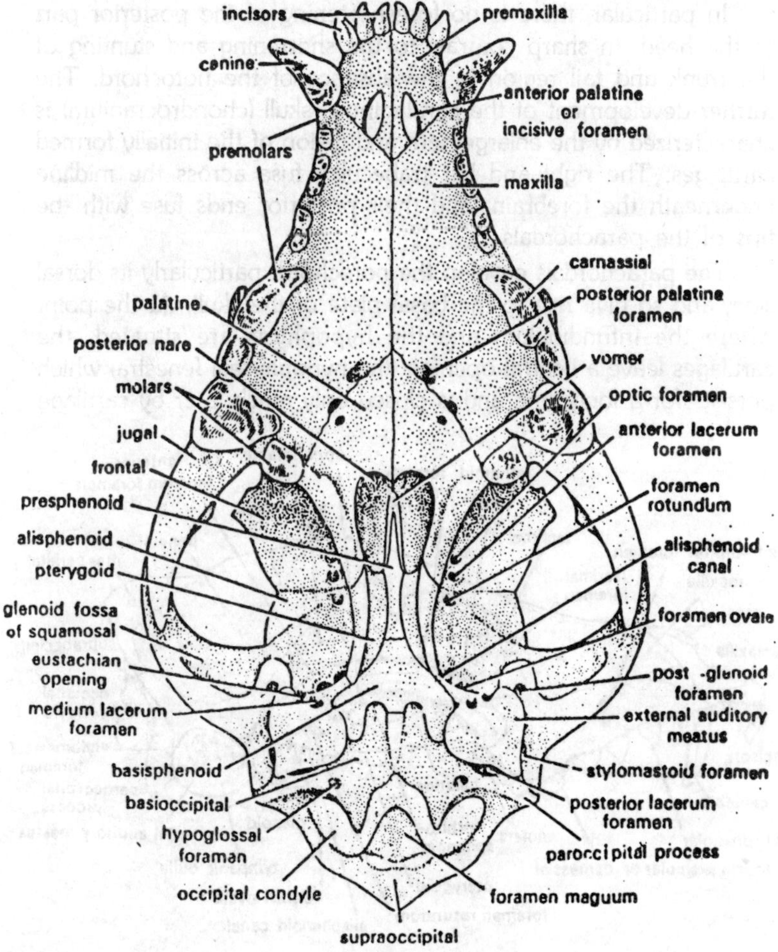

Fig. 9.6. Skull (Ventral view).

dependent in their development on the epithelial parts of the organs. This has been clearly shown in the case of the ear capsule and is very probably true in the case of the olfactory capsule and the sclera of the eye, as these capsules fail to be formed if the epithelial parts of these organs are removed. The parachordals, through spatially intimately associated with the notochord, show a high degree of independence from the latter organ. In experiments in which the notochord rudiment was removed, the development of the base of the skull does not seem to be affected to any extent.

In particular, there is no foreshortening of the posterior part of the head, in sharp contrast to the shortening and stunting of the trunk and tail region in the absence of the notochord. The further development of the cartilaginous skull (chondrocranium) is characterized by the enlargement and fusion of the initially formed cartilages. The right and left trabeculae fuse across the midline underneath the forebrain, and their posterior ends fuse with the tips of the parachordals.

The parachordals envelop the notochord, particularly its dorsal side, and so give rise to the basal plate of the skull. At the point where the infundibulum and the hypophysis are situated, the cartilages leave a ventral opening (the hypophyseal fenestra) which persists for a long time and is closed only much later by cartilage

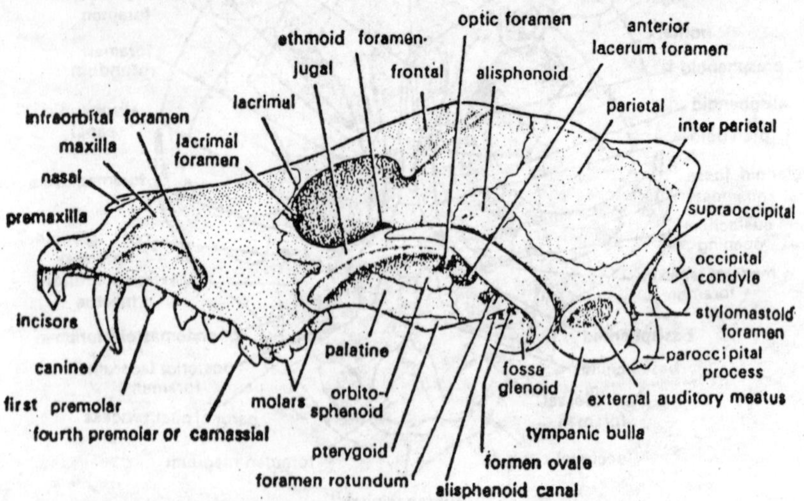

Fig. 9.7. Skull (Lateral view).

or bone on the ventral side. The posterior ends of the parachordals grow upward and eventually fuse above the medulla, thus enclosing the foramen magnum of the skull. The nose capsule and the otic (auditory) capsule become firmly joined to the trabeculae and the parachordals respectively, thus contributing to the formation of the lateral walls of the chrondrocranium.

Gradually the lateral edges of the cartilaginous skull grow upward, and in the more primitive vertebrates (cyclostomes, many fishes and amphibians) form a roof over the dorsal surface of the brain. In teleost fishes and in all amniotes, however, the cartilaginous skull remains incomplete on the dorsal surface, and the cranial roof is formed by bone at a later stage. In higher vertebrates, mammals in particular, the initial stages in the development of the cartilaginous skull may be speeded up in such a way that the trabeculae and parachordals are fused right from the start—a condition which is achieved in lower vertebrates secondarily.

Development of the Paired Limbs

The paired limbs of vertebrates are very complex organs, built up of components derived from several different sources—from the lateral plate mesoderm, the epidermis and the somites, to name only the main components. Nerves and blood vessels are of course also indispensable components of differentiated limbs. The first trace of the development of limbs may be found in the lateral plate mesoderm; the somatic layer of the lateral plate becomes thickened just underneath its upper edge. The cells of this thickening soon lose their epithelial connections and are transformed into a mass of mesenchyme without the somatic layer having lost its continuity. It is therefore a case of migration of mesenchyme cells from an epithelial layer, rather than that of the breaking up of epithelium into mesenchyme.

The mesenchyme accumulates between the remaining lateral plate epithelium and the epidermis and soon becomes firmly attached to the inner surface of the epithelium. The thickening of the lateral plate mesoderm and the subsequent formation of a mass of mesenchyme under the epithelium may coincide rather closely with the position of the two pairs of limbs, that is, they may appear in two disconnected regions-just behind the branchial region and just in front of the anus. This is the case in the amphibians. In other vertebrates, however the thickening and the mesenchyme accumulation may spread far beyond the actual region of limb development. In fishes the early limbs rudiments are more elongated anteroposteriorly in the earlier stages than in the later stages of development.

In the amniotes the thickenings and the mesenchyme gatherings are continuous throughout the whole length of the body in the form of horizontal ridges—the Wolffian ridges. However, the most anterior and the most posterior parts of the ridge are thicker than the intermediate part, and it is only these anterior and posterior parts that develop progressively, giving rise to the forelimbs and hindlimbs. The intermediate part of the Wolffian ridge later disappears. The epidermis over the mesenchyme mass becomes slightly thickened and bulges outward. This happens over the Wolffian ridges as well, but in the intermediate parts of the ridge the epithelial thickening disappears together with the gathering of mesenchyme. In the regions where the fore-and hindimbs are to develop, the protrusion, consisting of a thickened epithelial covering and of an internal mass of densely packed mesenchyme, increases and becomes the limb-bud. Of the two components contributing to the formation of the limb-bud the mesoderm is determined as such at an early stage shortly after the closure of the neural tube. Pieces of lateral plate may be cut out in this stage and transplanted under the epidermis on the flank or on the head.

The local epidermis will then become the epithelial component of the limb-but and a limb will develop heterotopically. The presumptive epidermis of the limb in the same stages, that is, before a limb-bud has been formed, does not possess any special properties and, if transplanted alone, will not give rise to a new limb. Epidermis from any part of the body is able to co-operate with the presumptive limb mesoderm in forming a limb-bud. This can be shown by removing the epidermis in the limb region and then covering the wound with a flap of epidermis taken from any part of the body. The epidermis is, however, by no means a passive components in limb development. This is especially clearly shown by some peculiarities of limb development in higher vertebrates.

In the amniotes, the limb-bud becomes slightly flattened at an early stage, and an epidermal thickening develops along the edge of the flattened bud. The thickening is in the form of a very sharply defined ridge and sometimes (in reptiles) even takes the form of a solid fold of the epidermis. In cross section the ridge looks like a nipple. It is referred to as the ectodermal apical ridge. The cells of the ridge differ from the ordinary epidermal cells not only in their arrangement but also in their physiological properties. It was found that they contain more ribonucleic acid and more glycogen, and they differ very conspicuously from surrounding epidermal cells in their high content of the enzyme alkaline phosphate. All these biochemical properties may be taken as indications of active metabolism. The apical ectodermal ridge is indispensable for the normal outgrowth of the limb rudiment.

If the ectoderm of the apical ridge of a wingbud of a three-day chick embryo is removed without causing damage to the underlying mesoderm, the distal parts of the wing fail to be formed, although the proximal parts develop quite normally. If the ectoderm covering the limb-bud in a chick embryo is removed and replaced by epidermis from another part of the body, the later is found to be incapable of developing an apical ridge. The result is that the development of the whole of the distal part of the limb is suppressed.

The girdle may, however develop normally and a short piece of cartilage representing the proximal part of the humerus or femur may be formed. This is in some contradiction to the conditions found in amphibian embryos, where flank or head ectoderm, as was indicated, may participate in the development of a limb, but then there are no apical ridges on amphibian limb-buds. The investigation of the role of the epidermal apical ridge in the development of limbs sin birds was ascended further after it had been discovered that the mesoderm and ectoderm of a limb-bud may be separated very neatly by chemical instead of mechanical means. Treating a limb-bud with a trypsin solution causes the epidermis to separate from the mesodermal core of the bud.

The mesoderm after this treatment is not fully viable, but the epidermis is quite healthy and may be used to cover the mesodermal part from which the epidermis is removed by immersion in a solution of Versene (the large treatment destroys the epidermis, which comes off in flakes, but leaves the mesoderm in a very good condition.) After the mesoderm of a limb-bud is covered by epidermis of a different origin, the two stick firmly together, and such a composite limb-bud may be implanted onto the flank of a third embryo and there allowed to grow into a limb. In this way it was possible to combine the mesoderm of a legbud with the epidermis of a wing bud and vice versa.

The structure of the developing limb was in every case determined by the origin of the mesodermal component; it was wing if the mesoderm was taken from a wingbud, and a leg if the mesoderm was that of a posterior limb rudiment. The origin of the ectoderm did not affect the nature of the developing limb. This experiment stresses the leading part played by the mesoderm in limb development. The next experiment emphasizes the importance of the epidermis. In a "wingless" mutation in fowls, the forelimb-bud appears approximately at the same time as the normal limb-bud but fail to grow and produce a limb, though parts of the limb girdle may be present. It was noticed that the epidermis covering the wingbud in the affected embryos does not have an apical ridge.

The experiment was therefore undertaken of combining the mesoderm of a normal wingbud with the epidermis of an embryo of the wingless strain. The result was as had been expected; in the absence of an apical ridge the bud did not grow and no distal parts of the wing were formed. The competence for limb development can be shown to be present all along the flank of the embryo between the forelimb and hindlimb region even if it does not manifest itself in normal development.

In amphibians it is possible to induce a supernumerary limb by introducing an inductor into the region between the forelimb and the hindlimb rudiments. Originally the ear vesicle was used as a limb inductor but later it was found that other organ rudiments such as the hypophysis and the olfactory sac can also induce a limb if transplanted into the side of a young embryo.

Under the influence of the induction, the local mesodermal cells accumulate as a compact mass under the epidermis, the epidermis is also made to react, and an additional limb develops. Depending on whether the induced limb lies nearer to the normal forelimb or hindlimb region, it may resemble a forelimb or a hindlimb in its structure. Experiments on limb induction suggest that in normal development, as well, there must be some factor determining which part of the mesoderm competent for limb development actually produces a limb rudiment. Limb-buds are often induced together with other structures when "spino-caudal" inductors are introduced into embryos in the gastrula stage.

By transplanting the presumptive mesoderm together with the notochord into the lateral plate region of an embryo in the neurula stage, it is chord into the lateral plate region of an embryo in the neurula stage, it is possible to induce the lateral plate mesoderm to develop into kidney (normally produced by the stalks of the somite. In these experiments it has often been observed that additional limb-buds developed together with the kidney tubules. It follows that the determination of the limb mesoderm occurs in conjunction with the determination of other parts of the mesodermal mantle. The kidney in itself cannot, however, induce a limb.

Differentiation in the Limbs

After the limb-bud has grown so far that its length exceeds its breadth, the differentiation of the subordinate parts of the limb gets in. We have already noted the slight flattening of the limb-bud. Now the distal portion of the bud becomes flattened even more, and at the same time it becomes distinctly broader than the proximal part of the limb rudiment. The flattened and broadened distal part is the hand (or foot) plate. The edge of the plate is initially circular, but soon it becomes pentagonal, the projecting points indicating the rudiments of the digits. While the tips of the digit rudiments keep growing out farther, extensive necrosis occurs in the intervening sections.

Mesodermal and ectodermal cells (including those of the apical ridge) in these sections die and are consumed by macrophages. As a result the digits become separated by distinct incisions. The five digits appear simultaneously in all amniotes with pentadactyl limbs, but in amphibians, especially in urodeles, the first two digits appear earlier and digits 3, 4, and 5 are formed one after another on the posterior edge of the limb. Where less than five digits are present in adult limb, or more than five in cases of hyperdactyly, this condition is reflected in the structure of the hand (foot) plate. In the early limb rudiments, the future flexor surface is ventral and the future extensor surface is dorsal, but as the limb elongates a rotation takes places so that the flexor surface is turned posteriorly, and eventually it may even face in a posterodorsal direction.

The preaxial edge of the limb, which is originally anterior, is then turned downward. With the elongation of the limb, it becomes bent at the elbow joint (or knee joint). A less pronounced flexion develops at the base of the carpus (or tarsus). The three main sections of the limbs thus become recognizable externally. Concurrently with changes in the external appearance of the limbs, differentiation occurs in the interior of the limb rudiments. The mesenchyme cells which are closely packed in a young limb-bud become segregated into areas in which the mesenchyme is lying more loosely and into other areas in which the mesenchyme cells are crowded.

The latter are the rudiments of the skeletal parts of the limb. The concentrated masses of mesenchyme in due course become converted into procartilage, and then, by further deposition of intercellular matrix into cartilage. Whereas in the initial stage of mesenchyme concentration large sections of the limb skeleton are represented by a common mass of mesenchyme, in the procartilage stage individual elements of the skeleton are laid down as separate units; these may fuse together later. The differentiation of the limb skeleton generally proceeds in a proximodistal direction, though some deviations from this order are fairly general occurrence. In amphibians, the first skeletal part to become recognizable is the stylopodium (humerus or femur). Parts of the zeugopodium (radius and ulna in the forelimb, tibia and fibula in the hindlimb) are laid down next, and the autopodium differentiates considerably later. The girdle rudiments appear after the stylopodium, but earlier than the autopodium.

In higher vertebrates the girdle tends to be developed simultaneously with the proximal elements of the limb. In the autopodium, the proximodistal sequence is upset by the larger skeletal elements, the metacarpals and metatarsals differentiating more rapidly than the smaller elements namely the carpals and the tarsals. In the digits, however, the proximal phalanges are laid down earlier than the distal ones. The blood vessels appear in the limb-bud at an early stage and although the pattern of the arteries, veins and capillaries is too variable to deserve much attention here, one blood vessel may be mentioned. It is situated along the edge of the hand (or foot) plate, just underneath the ectodermal apical ridge, and possibly it is responsible for supplying nutriment to this rapidly growing area of the limb. The pattern according to which the various constituent parts (bones, muscles, blood vessels, nerves) are arranged in each limb is asymmetrical—proximal and distal ends, dorsal and ventral surfaces, anterior and posterior sides of the limb being different from one another.

The first indication of this asymmetrical pattern can be noted when the tip of the limb-bud grows out in a slanting posterior direction instead of growing straight out away from the side of the body. Special experiments have been carried out to find out how the asymmetry of the limb, and thus the basic pattern of its differentiation is determined. Experiments were originally carried out on the embryos of the salamander *Ambystoma punctatum*. Forelimb rudiments were transplanted at different stages after the end of neurulation in such a way that either one limb rudiment axis or two or all three were inverted (disharmonious) with respect to the host's body. The inversion of the proximodistal axis could only be carried out with the mesodermal part of the limb rudiment, but this is not of importance as it is the mesoderm that is the carrier of the limb determination.

The transplantations were done either *orthotopically*, that is in place of a normal limb rudiment, or *heterotopically*, on the flank. The experiments showed that the three axes of the limb rudiment were not determined simultaneously. In the earliest rudiments immediately after neurulation, the anteroposterior axis is already fixed; limb rudiments, transplanted disharmoniously, with this axis inverted, had the tip of the limb-bud growing forward instead of backward and the limbs later had their postaxial (ulnar) side placed anteriorly. At the same time, the inversion of the dorsoventral axis of the transplanted limb rudiment could still right itself; that is, the original upper part of the rudiment developed into the ventral (palmar) surface of the limb and the original lower part of the rudiment developed into the dorsal surface of the limb. This shows that the pattern of the limb differentiation was not yet fixed in respect to its dorsoventral axis and that this pattern could be imposed upon the limb rudiment by the host, that is, by the parts which were in connection with the limb rudiment in its new position.

In a later stage of development when the tail rudiments begins to elongate; the dorsoventral axis of the limb was also found to be determined; if the limbs, were transplanted in an inverted position, with the dorsoventral axis of the limb rudiment the inverse of the dorsoventral axis of the host, the limb developed with an abnormal orientation, its plantar surface facing upward. The surroundings could no longer change the pattern of limb differentiation in respect to the dorsoventral axis, as well as in respect to the anteroposterior axis. At the same time that the dorsoventral axis of the limb rudiment is determined the proximodistal axis may still be inverted without impairing the normal development of the limb.

Only in a later stage, when the limb-bud begins to be visible from the outside, does the proximodistal axis show signs of being determined, and if the limb mesoderm is transplanted with the axis inverted the limb shows abnormalities in its development. (A limb cannot actually grow inward into the body because it would then be no longer in contact with the epidermis, and this contact is indispensable for the differentiation of parts of the limb). In cases when the transplanted limb rudiment grows out in a disharmonious orientation to the host, it is often observed that a kind of regulation occurs by means of the formation of a second limb-bud, whose orientation and differentiation is harmonious with the host's body.

The appearance of the second limb-bud is the result of a sort of splitting of the original rudiment and is due to an influence of the host on the graft. This influence is not strong enough to invert the axial structure of the transplanted rudiment as a whole, but is sufficiently strong to divert a part of the cells of the transplanted rudiment and to cause them to take on the axial structure of the host. If the original disharmonious limb were now to degenerate (as sometimes happens), the host would be in possession of a normal set of limbs. The splitting of a limb rudiment (or any other organ rudiment) to produce two similar rudiments is known as reduplication. That reduplication is possible shows that the cells of the rudiment are not determined each to fulfill a definite part in the developing organ, even though the axial pattern of the rudiment as a whole is determined. From this it is further inferred that the determination of the axial pattern is not a matter of determining what each cell or group of cells has to do in the process of development of the organ, but it is a matter of polarity, of a heteropolar structure of the rudiment as a whole. This immediately recalls the polarity of the egg in the early stages of development, and some similarity between the rudiment of an organ and the early egg as a whole is indeed shown.

Both posses the ability to develop a number of subordinate parts, each within its own scope of action (the whole animal in the case of the egg, one organ in the case of an organ rudiment), and without these subordinate parts being represented by discrete particles in the initial system. Just as the egg can be split mechanically into parts and each part will develop into a miniature whole, so an organ rudiment, in a suitable stage of development, may be split mechanically into halves, and each half will develop into a complete organ. We have already mentioned similar results in the case of the eye rudiment. The same applies to the limb rudiment. The limb rudiment up to the stage of the limb-bud formation may be cut in two and each half transplanted separately, or the two halves left in place and kept apart by inserting a piece of extraneous tissue between them. Each will develop into a whole limb. The two limbs resulting from such a splitting may later grow to the normal size.

Splitting of limbs may sometimes occur accidentally in young amphibian larvae developing in nature, or it may possible be caused by the pressure of folds of the amnion in higher vertebrates, and abnormalities will be the result: limbs that are completely or partially reduplicated. If the splitting of the rudiment is due to a

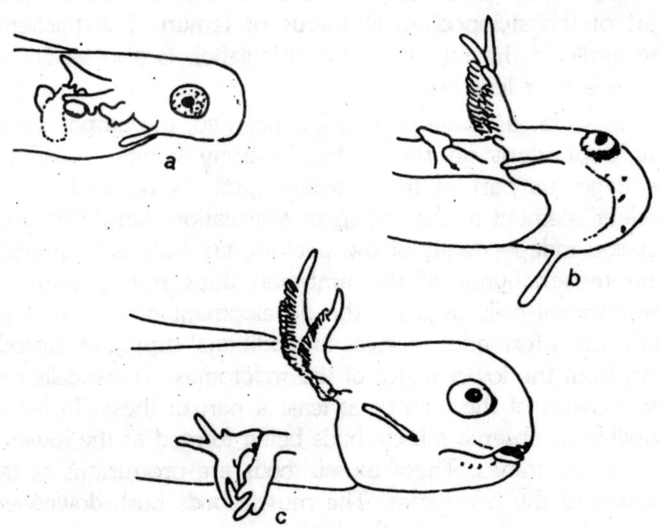

Fig. 9.8. Splitting of the forelimb rudiment, with a strip of extraneous tissue (T) inserted across it (a) to produce two limbs (b, c).

mechanical cause rather than being spontaneous, the two halves retain the same axial structure (polarity), thus being replicas of each other. The limbs girdles normally develop in intimate connection with the limbs themselves. The mesodermal material for the girdle is derived from the peripheral parts of the mesenchyme mass, the central part of which becomes the limb-bud. However, in their determination the girdles are partially independent of the limbs.

The limb girdles may develop independently of the limbs, if the development of the limb itself is suppressed in some way or other, as by the removal of the limb-bud. In limb induction experiments, limb girdles sometimes are found where the limb itself fails to develop. The development of the girdle is in no way dependent on the existence of an interaction with the epidermis of the skin, as is that of the limb itself. In birds, the limb girdle may also develop fairly well even if the distal part of the limb is absent as a result of a failing co-operation between mesoderm and ectoderm, as in case where the epidermis over the limb-bud lacks an apical ridge. If the girdle (shoulder girdle or pelvic girdle) develops in the absence of the limb, it may be normal in its peripheral parts, but the fossa for the articulation with the humerus or femur does not develop in the absence of at least the proximal part of the stylopodium (humerus or femur). The thickening of the girdle in the region of the articulation is also lacking in the absence of a limb.

In so far as these parts are concerned, the amphibian girdles are dependent on the limbs. In bony fishes, in which the cartilaginous part of the shoulder girdle is reduced to a small skeletal element in the region of articulation, bony cleithrum can develop independently of the pectoral fin (Balinsky, unpublished). The mesenchyme of the limb-bud does not contain all the mesodermal cells used for the development of a limb; after the limb-bud is formed it receives an additional supply of mesodermal cells from the lower edges of the myotomes. These cells produce the muscles of the limb or at least a part of these. In fishes it is possible to observe muscle buds being formed at the lower edges of the myotomes. These muscle buds are protrusions of the cell masses of the myotomes. The muscle buds push downward and outward until they enter the limb-buds.

In the limb-bud itself, the muscle buds derived from different myotomes fuse into a common mass of cells (myoblasts) from which the muscles of the limb subsequently develop. In the higher vertebrates, beginning with amphibians, muscle buds are not found, although the migration of individual cells from the myotomes into the limb rudiments may take place. In the fishes, if the lateral plate mesoderm of the limb region is transplanted to near the midventral line on the abdomen, it gives rise to fins which are devoid of muscles, thus proving that the lateral plate mesoderm is not capable of producing the limb muscles. Muscles are developed, however, if a part of the somite is included in the graft.

In amphibians, on the other hand, when the rudiment of the limb, consisting of lateral plate mesoderm with or without epidermis, is transplanted heterotopically, it will develop into a complete limb with muscles. This, however does not definitely exclude the participation of the myotome material, since under experimental conditions some regulation could have taken place, just as half the rudiment may produce a whole limb in cases or reduplication. The nerve supply to the developing limbs has been dealt with in the section on the differentiation of the nervous system.

Development of the Urinary System

The excretory organs in vertebrates are essentially aggregates of uriniferous tubules, connected originally at their proximal ends with the coelomic cavity by ciliated funnels (the nephrostomes) and communicating to the exterior by a system of ducts which, in the lower vertebrates, open into the cloaca. Both the tubules and the ducts are of mesodermal origin, and they develop from the stalks of the somites (the nephrotomes). In the most primitive vertebrates, such as the cyclostomes, and also in the Gymnophiona one uriniferous tubule develops from the nephrotome in each mesodermal segment. The nephrotome, prior to the development of a uriniferous tubule, is a strand of cells connecting the somites to the lateral plate mesoderm.

The cells become separated into the parietal and visceral layer and the cavity, which we may call the *nephrocoele*, is for a while continuous with both the myocoele and the definitive coelom between the two sheets of the lateral plate mesoderm. The connection of the nephrocoele with the cavity of the somite is soon obliterated, but the connection with the cavity of the lateral plate persists in the more primitive type of vertebrate excretory organs and becomes the nephrosome. The dorsolateral wall of the nephrotome becomes drawn out into a hollow tube, the cavity of which is an extension of the nephrocoele. The tube, which is in open connection with the coelomic cavity, becomes the uriniferous or renal tubule. The distal (outward) ends of the most anterior tubules (that is, the tubules formed in the anterior part of the trunk) soon turn backward and then fuse with each other, thus giving rise to the common excretory duct, known as the *pronephric duct*. It is an essential feature in vertebrates that the renal tubules are associated with bunches of fine blood vessels (the *glomeruli*), through the endothelial walls of which the blood plasma containing excretory products is filtered into the uriniferous tubules or into the coelom in the immediate vicinity of the nephrostomes, so that nitrogenous waste products may be carried through the tubes and the excretory ducts and removed from the body.

Either the mass of blood vessels is invaginated into the wall of the renal tubule, which enlarges to contain the glomerulus and becomes the *Bowman's capsule*, or else the blood vessels form a bulge on the wall of the coelom. In this case the structure is referred to as the *external glomerulus*, or, if glomeruli of several segments are joined together, as the *glomus*. It is believed that the segment of the coelomic cavity into which the glomus projects is itself derived from an expansion of the nephrocoele, or several nephrocoeles. The basic pattern of development of the excretory organs as described above becomes modified in various degrees.

As with many other organ systems in vertebrates the development of the excretory tubules progresses in a craniocaudal direction, the tubules differentiating earlier in anterior segments than in posterior ones. The tubules which develop and begin functioning earlier, i.e., those of the anterior part of the trunk, tend to show more primitive conditions in their organization, while those developing later are modified to a greater degree. In the most primitive vertebrates, the cyclostomes and some fishes, the differences between the anterior and posterior parts of the excretory system are only gradual, but in the higher vertebrates we may distinguish more or less clearly three sections of this system: the *pronephros*, the *mesonephros* and the *metanephros*. The most anterior nephrostomes give rise to the pronephros, the mesonephros develops from those of the midtrunk region, and the metanephros is derived from the nephrostomes of the posterior part of the trunk.

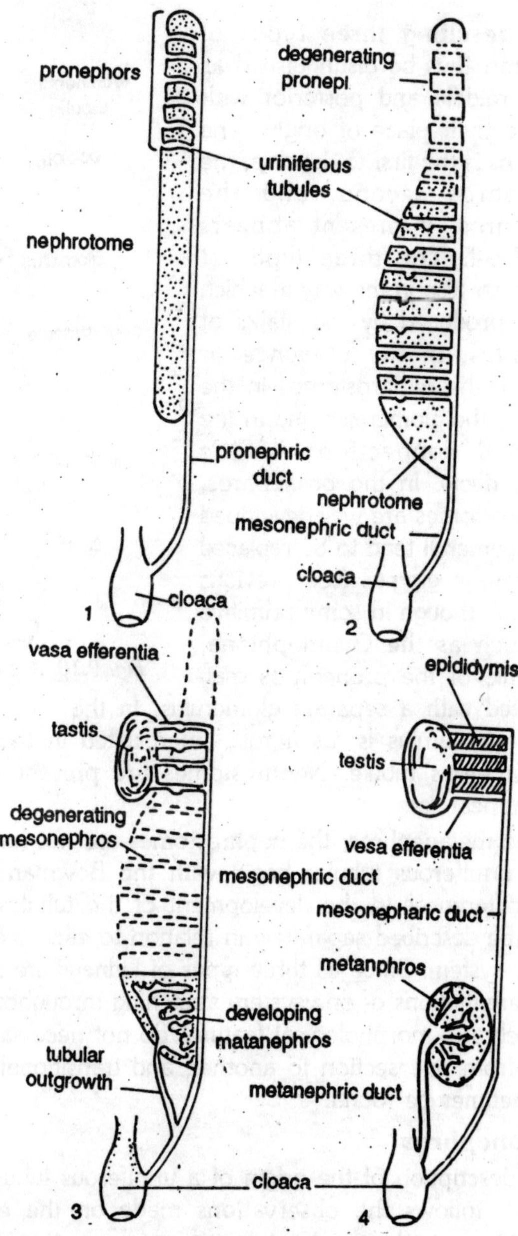

Fig. 9.9. Development of kidneys and their ducts. 1–Pronephros. 2–Mesonephros. 3–Metanephros forming and mesonephros joining testis. 4–Metanephros.

The resulting three types of kidneys can thus be distinguished as anterior, middle and posterior with respect to their place of origin. The pronephros is the first to develop, the mesonephros second, and the metanephros, if present appears latest of all. The three types of kidneys differ also in the way in which they are produced by the stalks of the somites, in the presence or absence of the nephrostomes in the position of the glomerulus and in the origin and connections of the excretory ducts. In the pronephros, the nephrostomes are well-developed and the glomeruli tend to be replaced by a common glomus (see previous description), though in some primitive forms, such as the Gymnophiona, each tubule of the pronephros may be supplied with a separate glomerulus. In the mesonephros, a separate glomerulus is, as a rule, intercalated in the course of each uriniferous tubule. Nephrostomes are present at first but may disappear late.

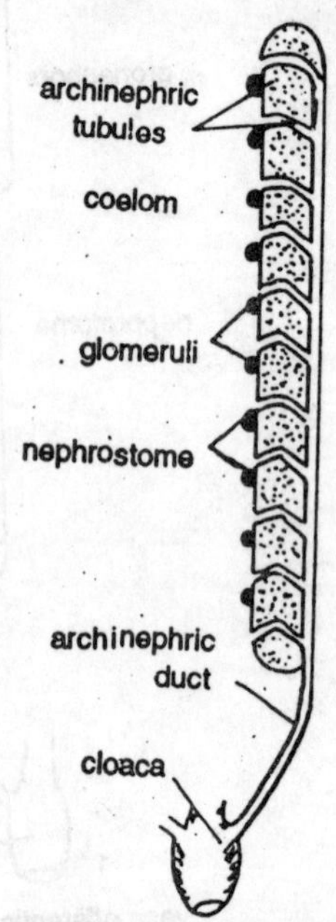

Fig. 9.10. Archinephros.

In the metanephros, the nephrostomes are not formed at all, and the uriniferous tubules begin with the Bowman's capsules. Further differences in the development of the tubules and ducts can best be described separately in relation to each section of the excretory system. Since all three types of kidneys are nothing but local differentiations of one system stretching throughout the body of a vertebrate, morphological features do not necessarily change abruptly from one section to another, and transitional conditions may sometimes be found.

The Pronephros

The description of the origin of a uriniferous tubule given on page 445 follows the observations made on the embryos of Gymnophiona and may be taken as representing the typical development of the pronephric tubules in the most archaic groups of vertebrates. The strictly segmental origin of the tubules is an essential feature of this development. In amphibians, the pronephric tubules are formed from a common mesodermal thickening appearing beneath the third and fourth somites (in frogs). Nevertheless, the number of the pronephric tubules corresponds to the number of segments participating in the development of the pronephros. The rudiment of the pronephros can be traced back in amphibians to the neurula stage.

By means of vital staining, the presumptive material of the pronephros was found to lie in the mesodermal mantle just outside of the edge of the neural plate, posterior to the middle of the embryo. In the neurula stage, mesoderm of this region is capable of self-differentiation when transplanted heterotypically. At the same time, however, other parts of the mesodermal mantle also possess the ability to develop into renal tubules, as for instance when an inductor (the notochord) is transplanted into the lateral plate region of the embryo. The presumptive somites, when isolated from the notochord and cultivated in *vitro* (surrounded by a coat of skin epidermis for protection), may develop into renal tubules even though their normal destiny would be a different one (the development of muscle).

After the end of neurulation (closure of the neural tube), the competence for the development of the pronephric tubules is restricted to the presumptive material of the pronephros. From this stage on ward the pronephric rudiment cannot be replaced, and the removal of the rudiment leads to the absence of the pronephros. The sequence of events during the transformation of the nephrostomes into the pronephric tubules, in frogs, salamanders and higher tetrapods, is not so clear as in the lower vertebrates. It is carried out by means of a rearrangement of cells, that is, by morphogenetic movements. The end result is, however, the same; several pairs of pronephric tubules are formed; they open by means of the nephrostomes into the coelomic cavity and fuse distally to form the pronephric duct. The formation of the pronephric duct, by fusion of the distal ends of the pronephric tubules, is a very important phase in the development of the excretory system, as this duct not only serves the pronephros out is also instrumental in providing pathways for the outflow of urine from the mesonephros and metanephros and, in the males, for the passage of spermatozoa.

In the salamanders, it has been shown that the pronephric duct develops right from the start from a more caudal part of the mesoderm than the pronephric tubules; while the two. Once they have been formed, both the pronephric tubules and the pronephric duct elongate very considerably. The pronephric tubules, as a result of their elongation, are thrown into numerous loops and eventually form a more or less spherical body, consisting of tangled and interwined tubules. The pronephric duct, on the other hand, remains straight and while it elongates, its posterior free end pushes itself backward along the lower ends of the somites. This backward movement of the tip of the pronephric duct ends when the duct reaches the cloaca and fuses with its wall, while the lumen of the duct opens into the cavity of the cloaca. The backward elongation of the pronephric duct may be interrupted in various ways, as by making a deep incision across its path.

If the wound remains gaping, the pronephric duct cannot spread beyond the wound and the duct is not continued into the posterior part of the body. If, however, the wound is too small or is covered to a sufficient extent, the tip of the elongating duct may find its way around the wound and penetrate into the posterior part of the body. Behind the wound, the duct returns to its level under the lower edges of the somites and eventually reaches the cloaca. This experiment shows that the elongation of the pronephric duct is largely independent of the surrounding parts, although the latter do seem to direct the duct by furnishing a suitable path for its movement. The same is borne out with out most clarity by some further experiments in which embryos in the closed neural tube stage bisected transversely, one half reversed, and the two parts caused to heal together with opposite dorsoventral orientation.

When the pronephric ducts reached the level of the operation in their backward elongation, the tip of the duct changed its direction and struck a new path across the lateral body wall, until it reached the somites of the inverted posterior part of the embryo. It then elongated along the edge of the somites and reached the cloaca as usual. The development of the pronephric duct shows a great similarity to the development of the lateral line. In both cases the migration of the posterior end of the organ rudiments along the length of the body is due to intrinsic tendencies of the cells of the rudiment, but the organs and tissues with which the rudiments come in contact influence the path taken by the migrating cells. The kidney of the most primitive living vertebrates, the cyclostomes and elasmobranch fishes, in their development and organization in the adult, show features characteristic of the pronephros, although they may perhaps be more correctly described as representing a stage in which the differentiation into pronephros and metanephros has not yet taken place.

The typical pronephros is a functional kidney of the larval stages of bony fishes and amphibians. In frog tadpoles, the pronephric canals form a bulky convoluted mass on both sides in the anterior part of the trunk at the level of the forelimbs, on the median aspect part of the pectoral girdle. In frog tadpoles approaching metamorphosis, the pronephros gradually degenerates; the glomus shrinks, the nephric tubules are resorbed and the anterior portions of the pronephric ducts are also resorbed. The function of excretion is taken over by the mesonephros. In embryos of amniotes, the pronephros develops in the anterior trunk region, but it is not functional at any stage. In the human embryo, about seven pairs of rudimentary pronephric tubules are formed. These soon degenerate, but not before they give rise to the pronephric ducts, which remain after the pronephric tubules have disappeared.

The Mesonephros

The mesonephric tubules are derived from the nephrostomes, as in the case of the pronephros, but their interrelation is by no means so simple. Only in the more archaic groups of the vertebrates (selachians, Gymnophiona, ganoids) are the renal tubules formed directly from the nephrostomes. In most amphibians and in all higher vertebrates, the mesodermal cells of the nephrotomes dissolve into a mass of mesenchyme stretching on each side of the body along the dorsal edge of the lateral plates from the level where the pronephros ends to the pelvic region. This elongated mass of mesenchyme is known as the nephrogenic cord or nephrogenic tissue.

The mesonephric tubules are developed from the nephrogenic tissue by a secondary aggregation of the mesodermal cells. The aggregating cells began by forming epithelial vesicles which stretch and elongate to become tubes. One end of such a tube becomes connected to the pronephric duct, while the other end becomes invaginated to form the Bowman's capsule; or, in rare cases, it opens into the coelomic cavity by a nephrostome while a Bowman's capsule is formed higher along the course of the tubule. The

Bowman's capsules become supplied by small branches from the dorsal aorta. The number of the mesonephric tubules does not correspond to the number of segments several tubules being developed in the region of each segment.

Even if at first only one tube is formed in each segment, it soon gives rise by budding to secondary and tertiary tubes, and so on. If present at first, nephrostomes may disappear later. In the frogs, the nephrostomes lose their connection with the rest of the nephric tubule but open secondarily into the veins (connecting the veins to the coelom). The nephrogenic tissue is not at once determined for differentiation as mesonephros. If the presumptive material of the mesonephros of a newt embryo in the embryo, the grafted tissue gives rise to excretory tubules of the pronephric type. On the other hand the same material taken from a tailbud stage embryo on grafting will produce mesonephric tubules even if it comes to lie at the site where a pronephros should have developed. The mesonephric tubules do not produce a duct of their own.

As the tubules are formed their distal free ends join up with the pronephric duct, which thus becomes the duct of the mesonephros as well and is then called the *mesonephric* duct (or Wolffian duct). As the mesonephric tubules reach the mesonephric duct, the walls of the latter bulge out, forming collecting ducts. The mesonephric tubules especially the ones formed later, open into these collecting ducts rather than into the mesonephric duct itself. The development of the mesonephros has also been found to be dependent on the pronephric duct in another way. The nephrogenous tissue develops into the mesonephric tubules only if stimulated to this by the pronephric duct.

In the preceding section experiments have been mentioned in which the penetration of the pronephric duct into the posterior half of the body was prevented by placing an obstacle (in the form of a gaping wound) in its path. When the operation is successful—that is, if the duct does not reach the region where the mesonephros normally develops—the nephrogenous tissue fails to form the renal tubules or forms only poorly developed tubules. Apparently some stimulus (induction) from the pronephric duct is necessary for the normal development of the mesonephric tubules from the nephrogenous tissue. The most anterior mesonephric tubules degenerate later in many animals, so that the kidneys becomes shortened at its anterior end and somewhat more compact.

In many fishes and in amphibians, the posterior part of the mesonephros shows some traits found in the metanephros: complete absence of nephrostomes, joining of the nephric tubules to large collecting ducts—out growths of the mesonephric duct. For this reason, some authors prefer to regard the kidney in these animals as a joint mesonephros and metanephros, giving if the special name of *opistonephros*. In fishes and amphibians, the mesonephric (or opistonephric) kidney is the excretory organ of the adult animal. In reptiles and birds, on the other hand, the mesonephros functions only during the embryonic period of development and loses its excretory function at the time of hatching.

In the mammals, the placenta takes on the task of removing excretory products from the blood of the embryo and passing them into the blood of the mother, from where they are removed by the maternal kidneys. In mammals such as the pig, in which the connection between the fetal and maternal tissues is not very close, the mesonephros is still active as an excretory organ, and a certain quantity of urine eventually reaches the cavity of the allantois and the amniotic cavity. In mammals having a very close connection between the fetal and maternal tissues (as in rodents with a hemochorial placenta), there is no sign of excretory activity in the mesonephros, which is thus a rudimentary organ like the pronephros.

The Metanephros

The metanephros develops from the posterior part of the nephrogenic cord, the part adjacent to the cloaca. There is no trace of a relation of the tubules to the nephrostomes, although the nephrogenic tissue is primarily derived from the latter. As in the case of the mesonephros, the metanephric does not develop a duct of its own but uses the mesonephric duct as a means of removing the urine that it excretes. However, the connection between metanephros and *mesonephric* duct is established not directly but by means of a special outgrowth or branch of the duct. Before the metanephros starts differentiating, a bud is formed on the mesonephric duct a small distance in front of the point where the duct joins the cloaca. The bud elongates in the direction of the posterior part of the nephrogenic tissue. The duct formed in this way becomes the *ureter* and the bud from which it develops is the *ureteric bud*. Having reached the nephrogenic tissue, the end of the ureter begins to branch, the branches later becoming the collecting tubules of the kidney. The point at which the branching begins expands to form the renal pelvis.

The nephrogenic mesenchyme accumulates around the tips of the collecting tubules and in due course differentiates into the renal tubules with their glomeruli. No trace of nephrostomes can be discovered in the metanephros at any stage. As in the mesonephros, the conversion of the metanephrogenic tissue into a system of renal tubules is dependent on a stimulus from the excretory duct-in this case from the ureter and its branches, which develop from the ureteric bud. In experiments on chick embryos, in which the backward growth of the pronephric duct is prevented by destroying its growing tip, no mesonephric duct develops in the posterior part of the body and naturally no ureteric bud is formed. The result is that no metanephric kidney is formed on the operated side of the embryo. A confirmation of results gained by operation may be found in certain defects caused by mutations. The metanephros is the functional kidney in the postembryonic life of reptiles, birds and mammals.

In lizards, the fully differentiated metanephros largely retains the position of the metanephrogenic part of the nephrogenic cord: the functional kidney lies at the level of the hindlimb and even projects into the base of the tail. In birds and mammals, however, at an early stage the metanephric kidneys become displaced in an anterior direction. The metanephros shifts past the posterior end of the mesonephros and eventually comes to lie at about the same level as the latter: in the lumbar region, in mammals, and filling the space underneath the synsacrum and the pelvic girdle in birds. Concurrently with the displacement of the kidneys, the ureters elongate to a very great extent as compared with their length at the time when the ureteric bud first contacts the metanephrogenic tissue. The excretory ducts of the kidney in most vertebrates acquire close relations with the reproductive organs. These relations and the modifications of the excretory ducts in this connection will be dealt with together with the development of sexual organs.

Development of the Heart

The heart in vertebrates develops from the mesoderm forming the ventral edges of the lateral plates in the pharyngeal region of the body. It will be most convenient to describe first the development of the heart in animals such as the amphibians and to consider later the development of the heart in higher vertebrates.

The Heart in Lower Vertebrates

During gastrulation in the amphibians the sheet of mesoderm advances forward from the blastopore, penetrating between the ectoderm and the endoderm. The rate of movement of the mesoderm is greatest in the dorsal region of the embryo, intermediate laterally and least ventrally. In the neurula stage, the dorsal and dorsolateral parts of the mesodermal mantle have reached the head region of the embryo; ventrally there remains an approximately triangular area in which there is no mesoderm intervening between ectoderm and endoderm. This triangle, roughly corresponding to the oral and pharyngeal region of the embryo, has its apex posteriorly and the broad base anteriorly.

The posterior part of the area is the site of the development of the heart, and it is later filled in by the mesoderm participating in the formation of the heart. The most anterior part of the area remains free of mesoderm, and here the mouth breaks through after a fusion between the endoderm and the ectoderm has taken place. The presumptive material of the heart is found in the edges of the mesodermal mantle bordering the mesoderm free area in the right and on the left. By vital staining in the neurula stage, it has been shown that the presumptive material for the heart is located in the edge of the mesodermal mantle, rather high on the flank on each side, where it adjoins the parts of the neural folds which become the hindbrain portion of the neural plate. If the mesoderm of this area is stained with nile blue sulphate or neutral red (by lifting a flap of ectoderm and mesoderm and applying a piece of cellophane soaked in stain to the *inner* surface of the mesodermal layer), it can be observed that the mesoderm shifts in later stages in a ventral direction and eventually comes to lie midventrally in the area where the heart can be seen to differentiate.

The particles of vital stain remain in the cells long enough for it to be ascertained that the stained material on each side of the embryo gives rise to exactly one half of the heart rudiment, the other half being derived from a similar area of mesoderm on the other side of the embryo. Parts of the free edge of the mesodermal mantle may be excised in the neurula stage and cultivated in a saline solution. To prevent the mesoderm from disintegrating, it is sometimes isolated together with a flap of ectoderm, the ectoderm then closes into a vesicle with the mesoderm inside.

The mesodermal cells under these conditions differentiate into muscle tissue, and this begins to pulsate rhythmically—an unequivocal indication that the developed muscle tissue is cardiac muscle, as only this type of muscle is capable of autonomous rhythmical contraction. It has been claimed that in some experiments the heart rudiments explanted in neurula and even late gastrula stages can even produce pulsating tubes having an obvious resemblance to normal hearts. In is rather remarkable that there is a definite discrepancy between experiments using the vital staining technique and explantation experiments. While vital staining indicates that the material going into the formation of the heart lies in the upper part of the mesodermal mantle, heart differentiation has been observed in explanted pieces taken from different parts of the free edge of the mesodermal mantle: from the upper part from the middle part and from the ventral part.

It would seem that mesoderm in a wide region of the early embryo has a potentiality for heart development. From this wider area the potentiality for heart development is later restricted to the actual material participating in the formation of the heart in normal development. In these early stages, the development of the heart is in some way dependent on the endoderm. By cutting through the ectoderm and the mesodermal mantle and removing the whole of the endoderm, it is possible to produce newt embryos which consist of ectoderm and mesoderm only. The embryos survive the operation quite satisfactorily, but their term of life is limited because they do not possess the food supply normally contained in the yolky cells of the endoderm.

The endodermless embryos show various defects in the ectodermal and mesodermal organs owing to the lack of inductive influences emanating from the endoderm. One of such defects is the complete absence of the heart, although the mesodermal layer, from which heart rudiment is derived may remain intact. The dependence of heart development on the endoderm and also on other adjacent tissues has been tested in hanging drop cultures the techniques used being similar to those used in studying the

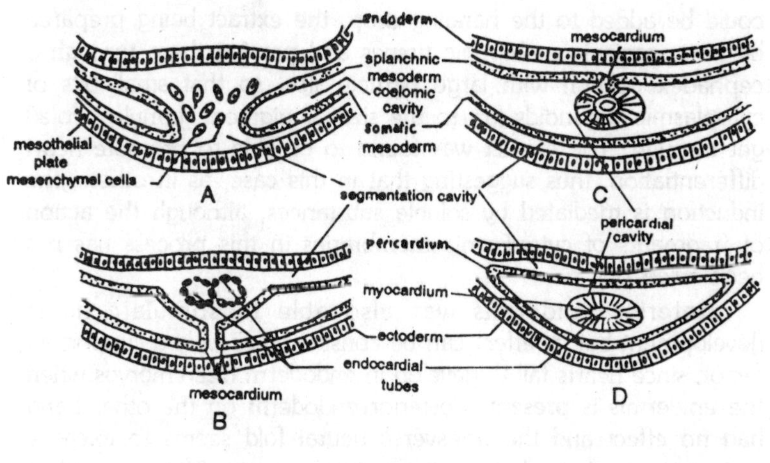

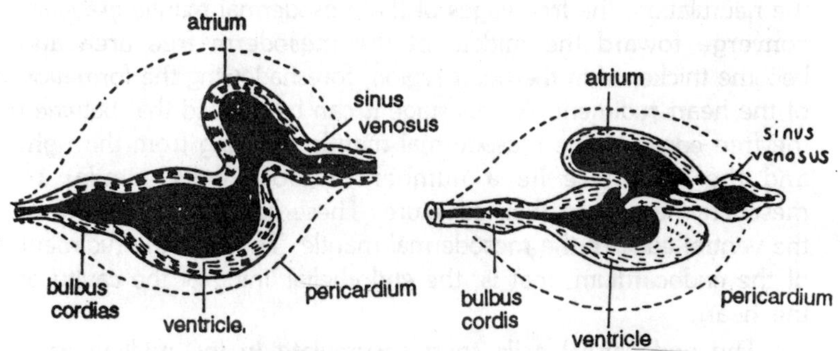

Fig. 9.11. Differentiation of various regions of heart.

induction of nervous tissues. Pieces of mesoderm of newt embryos containing the presumptive heart mesoderm were explanted alone or with other embryonic tissues in "Niu and Twitty solution." If explanted alone, the presumptive heart mesoderm showed signs of progressively increasing determination, depending on the stage at which the material was removed from the embryo.

There was no heart differentiation in material taken from gastrulae, but during neurulation the percentage of positive cases as shown by spontaneous contractions increased rapidly and became maximal (though not quite 100 percent) when late neurulae with closing neural folds were used. Addition of bits of anterior endoderm—the suspected heart inductor—increased the percentage of positive cases and accelerated the beginning of contractions. Instead of bits of intact endodermal tissue, an extract could be added to the hanging drop, the extract being prepared by homogenizing embryonic tissues and passing them through a cephadex column with large particle size, so that small bits of cytoplasmic organoids (up to the size of pigment granules) could get through. The extract was found to be able to promote heart differentiation, thus suggesting that in this case, as in others, the induction is mediated by soluble substances, although the action of fragments of cytoplasmic membranes in this process has not been ruled out.

Anterior epidermis was also able to stimulate heart development but its effect can be considered as only a supporting factor, since hearts fail to develop in endoderm free embryos when the epidermis is present. Posterior endoderm on the other hand had no effect and the transverse neural fold seems to exercise even a depressing effect on heart development. After the end of the neurulation, the free edges of the mesodermal mantle gradually converge toward the middle of the mesoderm free area and become thickened in the heart region, foreshadowing the formation of the heart rudiment. At this stage it can be noticed that between the free edges of the mesodermal mantle covering from the right and the left there lie a number of loose cells, similar to mesenchyme cells in their structure. These cells are derived from the ventral edge of the mesodermal mantle. They are the rudiment of the endocardium, that is, the endothelial lining of the cavity of the heart.

The endocardial cells soon accumulate in the midline as a longitudinal strand and eventually become arranged in the form of a thin-walled tube. The lumen of the tube is the cavity of the heart. The endocardial tube bifurcates at both end; at the anterior end its two prolongations are the ventral aortae, and at the posterior end it receives the two vitelline veins-the first venous blood vessels to reach the heart. All these vessels are at first similar to the endocardial tube in that they consist only of a thin layer of endothelium produced by the mesenchyme cells joined together. While the endocardial tube is being formed, the edges of the mesodermal mantle, which at this stage may also be called the edges of the lateral plate mesoderm, close in along the midline and fuse with each other. The fusion first occurs under the endocardial tube, between it and the ectoderm.

Very soon, however, the visceral layer of mesoderm envelops the endocardial tube on the dorsal side as well. By fusion of the mesodermal layers of the right and left side, epithelial partitions are formed above and below the endocardia tube. In analogy to the dorsal and ventral mesentery, these are called the *dorsal* and *ventral mesocardium*. The ventral mesocardium is of a very ephemeral existence; it is very soon perforated, and the coelomic cavities of the right and left side become continuous underneath the endocardial tube. The dorsal mesocardium persists longer but is also dissolved at a later stage. The coelomic cavities expand in the heart region to form the pericardial cavity.

The pericardial cavity is initially only a part of the general body cavity, the coelom. It becomes completely separated from the remainder of the coelom, largely owing to the failure of the coelom to develop in the region of the branchial pouches which lie immediately dorsal to the heart rudiment. Posteriorly the connection of the pericardial cavity with the rest of the coelom becomes occluded by the developing liver, to which the posterior end of the endocardial tube becomes closely connected. A connective tissue wall developing in this position (at the anterior boundary of the liver) is the *septum transversum*. From the above description, it will be clear that the pericardial cavity is lined by lateral plate mesoderm. The parietal layer of this mesoderm persists as the epithelial wall of the pericardial cavity, or the pericardium proper.

The visceral layer adheres to the endocardial tube. This layer differentiates as muscle tissue and thus gives rise to the myocardium of the heart. The heart is at first an almost straight tube does not show a sub-division into its various chambers. Later the tube becomes inflected in a very characteristic way. Starting from behind, the tube first runs forward, then bends downward and to the right and eventually again to the left, upward and forward. The heart thus becomes coiled in the shape of an S. The degree of the twisting in higher vertebrates is greater than in lower ones, so that in the former the tip of the first inflection. The tubular heart rudiment becomes constricted in some places and dilated at others and is thus subdivided into its four main parts.

The sinus venosus lie posteriorly; the atrium develops at the tip of the first inflection of the heart; the descending part, from the first to the second inflection becomes the ventricle and the part going forward from the second inflection becomes the course arteriosus. Before this subdivision is performed, however, the functioning of the heart starts; it begins to pulsate at a regular rhythm. The pulsations of the heart start very early in the development of the embryo, even before the peripheral blood vessels are ready to receive the blood stream. As the presumptive heart rudiment is being formed, its capacity for performing the further stages of development independently of the norms surrounding increases perceptibly.

If the heart rudiment is excised after the end of neurulation and transplanted into an abnormal position or allowed to develop *in vitro*, enclosed in a vesicle of skin, the differentiation goes much farther than in the experiments referred to above. Not only pulsating muscle tissue develops but a cardiac tube is formed, and the tube becomes inflected just as does the heart in normal development. The rudiment of the heart in this stage, however, is by no means strictly determined in all its parts. The left or right half of the heart rudiment may be excised, and the remaining half then develops into a complete whale. What is more, half of the heart rudiment may be explanted or transplanted, and it still develops into a complete heart.

The ability of a half of the heart rudiment to form a whole heart may be used to produce two hearts in the same embryo. For this an incision should be made lengthwise before the two halves of the heart rudiment unit in the middle. Inserting a piece of extraneous tissue, a somite for instance, or leaving the wound

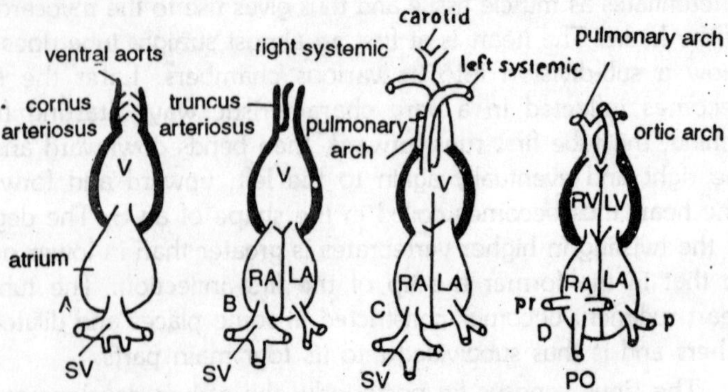

Fig. 9.12. Evolution of vertebrate heart, A—fish, B—frog. C—lizard, D—bird and mammal, LV, left ventricle. RV, right ventricle, LA, left auricle. RA, right auricle. V, ventricle, SV, sinus venosus. P, pulmonary vein, po, postcaval, pr, precaval.

to gape may prevent the halves of the heart rudiment from coming together. Two complete pulsating hearts develop under these conditions the reverse can also be done; a complete heart rudiment may be superimposed on the intact presumptive heart mesoderm of a host embryo. The two rudiments then fuse into one whole and produce a normal heart. The main points emerging from the foregoing will now be summarized:

1. Two heart rudiments may fuse into one organ of normal structure.
2. The functioning of the heart begins at an early stage of development.
3. The heart develops from a paired rudiment, uniting in the midtive secondarily.
4. The heart develops in a very anterior position, in the pharyngeal region. The position of the heart in the thorax (as in an adult tetrapod) is thus secondary and is due to a displacement of the organ in later development.
5. Each half of the heart rudiment is able to differentiate even before the two halves fuse.

The Heart in higher Vertebrates

With all this in view, it is easy to understand the development of the heart in the vertebrates having yolky eggs and partial cleavage. Because of the physiological requirements of the developing embryo (the necessity of establishing circulation), the heart in meroblastic vertebrates develops precociously, before the body of the embryo becomes separated from the yolk sac. The embryo in this stage is to be found toward the surface of the yolk, and its lateral plate mesoderm is to be found toward the outer parts of the blastodisc. The lateral plate mesoderm is prevented from uniting on the ventral side of the embryo by the intervening yolk. As a result, the two halves of the heart rudiment begin differentiating independently of one another. Two endocardial tubes are actually formed; each becomes invested by the myocardium and surrounded laterally by the pericardial cavity.

When the body folds undercut the anterior end of the embryo and the foregut becomes separated anteriorly from the yolk sac, the right and left heart rudiments become able to meet in the midline under the pharynx. The two endocardial tubes come to lie alongside each other and soon fuse into one tube in the cardiac region, while in front and behind the heart the endothelial tubes remain separate, thus leading to a state described earlier for the lower vertebrates. The visceral wall of the right and left pericardial cavities also meet and fuse above and below the endocardial tubes, forming a single pericardial cavity. The single endocardial tube is now completely surrounded by the myocardium.

In view of the experimental results mentioned above, the fusion of two heart rudiments to form one single organ is not at all surprising. Of all the organs of a vertebrates, the heart is the one which starts its definitive function earliest. It is also greatly dependent in its development on function. It is essential for the development of the heart that a blood stream should actually flow through it. It is true that a heart of a typical shape with recognizable parts will develop even in complete isolation as in the explantation experiments, but such an isolated heart soon stops developing further. Also, a heart in its normal position is arrested in its development if in some way or other (interruption of the afferent blood vessels) it is deprived of circulation. The degree of the heart's development and growth appears to be dependent on the volume of blood passing through it or on the size of the animal which the heart supplies with blood.

Heart rudiments have been transplanted reciprocally between large and small species of salamanders and it was found that the hearts of small species grew beyond their usual size in large hosts, and the hearts of large species were undersized in small hosts. In every combination the transplanted heart grew to approximately the same size as the host heart should have grown. In the vertebrates developing a pulmonary circulation, the heart becomes separated to a greater or lesser degree into a right half, carrying blood to the lungs and a left half, receiving the blood from the lungs by way of pulmonary veins and sending it to the rest of the body. The first indication of this separation is found in the lungfishes (Dipnoi). In the amphibians, the atrium becomes subdivided by a partition wall arising between the points of entry into the atrium of the sinus venosus and pulmonary vein. After the separation is completed, the right atrium receives blood from the sinus venosus and the left atrium from the lungs by way of the pulmonary vein. Both atria, however, pour out their blood into the ventricle, which remains undivided.

The Mammalian Heart

The most elaborate transformations are found to occur in the development of the heart in mammals. The task of directing the blood into its various channels is complicated not only by the presence in the adult state of a double circulation the systemic and the pulmonary circulation) but also by the necessity of providing for a rapid change from fetal to postnatal circulation. The subdivision of the heart into right and left halves is dependent on the co-operation of several structures and carried out in stages. The atrium separated into right and left halves first by a primary partition wall (septum primum), which is similar to that found in amphibians.

As soon as this partition wall subdivides the cavity of the atrium, however, it is perforated by an opening allowing for intercommunication between the right and left atrial cavities. Consequently, the left atrium is not deprived of a blood flow (on which, as we have seen the development of the heart is dependent in a high degree) during the intra-uterine life, when the amount of blood passing through the lungs is very small. At a later stage, a second partitions wall (septum secundum) develops in the atrium just to the right of the first, but this remains incomplete, a large foramen ovale remaining open in its lower part.

The foramen ovale does not quite coincide in position with the opening in the primary partition wall; nevertheless, blood may pass through both openings so long as the blood pressure in the right atrium is greater than that in the left atrium, as it actually is during fetal life. After birth, however, when the pulmonary circulation is fully established the pressure in the left atrium becomes higher than that in the right atrium. The primary partition wall becomes pressed against the secondary wall, and as a result the previously existing passage becomes closed. Thus the atria become finally separated.

Eventually the primary and secondary partition walls of the atrium fuse together. The atrium originally communicates with the ventricle by way of the atrio-ventricular opening, which is a narrowed part of the heart tube. The separation of this single opening into two is performed by means of connective tissue outgrowths developing on the dorsal and ventral edges of the opening. The outgrowths, known as the *endocardial cushions*, fuse in the middle and are also joined by the lower edge of the primary interatrial partition wall; as a result two canals are left, one leading from the right atrium into the right ventricle and the other leading from the left atrium into the left ventricle. The

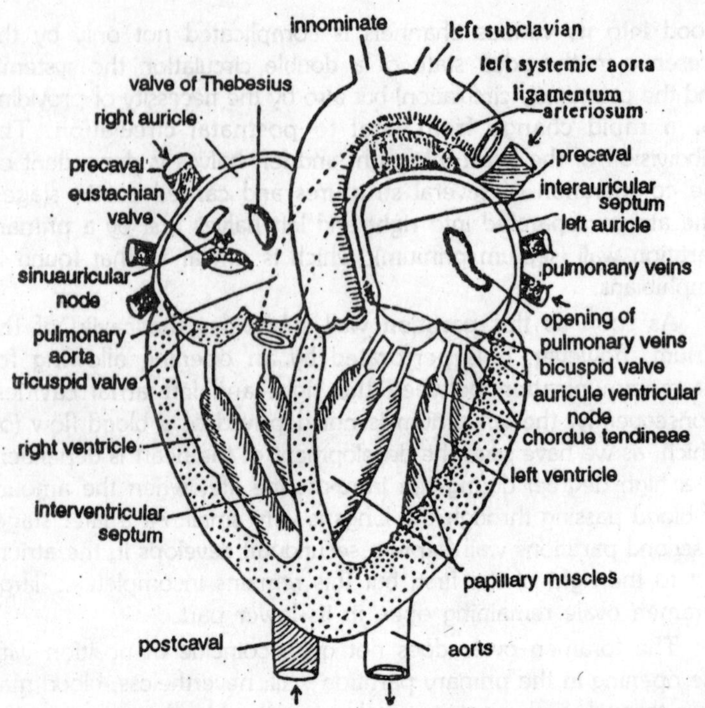

Fig. 9.13. Heart of mammal (ventral dissection).

endocardial cushions also participate in the formation of the atrio-ventricular valves. The ventricle becomes partially separated into left and right halves by an indentation of the myocardium at its tip (at the apex of the posteroventral inflection of the heart tube).

The developing partition wall is therefore largely muscular but is incomplete. It supplemented by a membranous (connective tissue) part arising in conjunction with a wall subdividing the conus arteriosus into two channels: one to serve for systemic circulation and the other for pulmonary circulation. The latter partition wall cuts in backward between the origin of the fourth and sixth pairs of aortic in such a way that ventral channel leads into the arches of the third (carotid) and fourth (systemic) pairs, while the dorsal channel leads into the sixth (pulmonary) pair of arches. The partition wall in the conus arteriosus is spirally twisted so that the ventral channel becomes connected to the left ventricle and the dorsal channel to the right ventricle. It will be seen that the blood entering the right atrium from the sinus venosus, including the blood from the placental circulation is directed into the pulmonary arch (expect for the volume of blood passing from the right into the left atrium via the foramen ovale).

The pulmonary blood vessels, during the fetal life, cannot take all this blood, neither is there any necessity to have the blood pass through the lungs as they are as yet not serving as respiratory organs. The greater part of the blood passing through the right half of the heart is therefore returned to the systemic circulation by way of the arterial duct (ducts arteriosus) which connects the pulmonary arch to the dorsal aorta. After parturition, the arterial duct becomes closed by a violent contraction of its muscular wall, and all the blood from the right part of the heart is forced into the blood vessels of the lungs. The double circulation thus becomes finally established. It is a further peculiarity of the mammalian heart that the opening between the sinus venosus and the right atrium broadens out to such an extent that the sinus venosus becomes incorporated into the wall of the right atrium. As a result the main veins (anterior and posterior venae cavae the coronary vein), which originally entered the sinus venosus, now have separate openings into the right atrium. Similarly, the proximal part of the pulmonary vein is incorporated into the wall of the left atrium, so that there are two, or sometimes even four, pulmonary veins entering the left atrium independently of each other.

Development of the Blood Vessels

In the adult vertebrate blood vessels large and small permeate almost all parts of the body. Supernumerary parts of the body, whether they are grafts or results of induction, become vascularized, and the blood vessels, which in this case are additional to the ones produced in normal development, link up with host arteries and veins and are included in the host's blood circulation. It would appear that blood vessels are attracted to penetrate any parts which are in need of a blood vessel supply. Furthermore, it appears that a suitable situation for blood vessels is duly reserved in the pattern of organization of any organ. It must be noted, however, that in vertebrates the blood vessels are invariably situated in spaces occupied by connective tissue or its derivatives. Where capillaries seemingly penetrate into epithelium, this is actually tantamount to the invasion of channels or other spaces between epithelial cells by connective tissue from which the endothelial cells of the blood vessels are derived. The rudiments of the blood vessels are laid down as aggregations mesenchyme cells.

The cells participating in the formation of blood vessels are called angioblasts. They are probably always of mesodermal origin. We have seen that in the vascular area of amniotes the walls of the blood vessels are developed from the blood islands in conjunction with the development of the first blood cells. Similar, though more concentrated, blood islands occur in vertebrates with holoblastic cleavage (in particular in amphibians) on the ventral side of the abdomen. In the amniote body proper and in all areas except for the ventral body wall in the anamniotes, the blood vessels develop independently of blood corpuscles. The aggregations of angioblasts become arranged in the form of a flat epithelium surrounding a cavity. The epithelium is the endothelium of the blood vessel; the outer layers of the blood vessel walls are differentiated much later. A student of the anatomy of the adult vertebrates is accustomed to find the blood vessels in the form of tubes of a constant diameter over relatively long stretches. The first blood vessels laid down in the embryo are only rarely in the form of straight tubes. Over large areas the blood vessels are initially laid down in the form of network.

The further development of individual canals in the network depends on the amount and the direction of blood flow. Those channels which happen to come in the line of the greatest blood flow become increased in diameter, develop the connective tissue and the muscular layers and become arteries or veins. The channels that receive less blood remain in the form of a capillaries or degenerate and disappear completely. Another form in which early blood vessels appear is as sinuses—extensive irregular spaces surrounded by endothelium. Large sections of some of the larger veins appear in this form. Some of these sinuses later acquire a more regular tubular form: other remain as such in the adult animal—for instance the postcardinal "vein" becomes a tubular blood vessel in urodeles but remains essentially in the form of a sinus in the dogfish. Once the network of endothelial tubes—the rudiments of the blood vessels—has been established, new blood vessels continue to be formed by sprouting and outgrowth of those already present.

Lateral branches may form on already existing capillaries, or a capillary may become interrupted and each free end grow out in a new direction. Two outgrowth from different blood vessels may contact each other and fuse, thus establishing a new channel for circulation. As a result of the extreme plasticity of the blood vessel system, the eventual arrangement of arteries, veins and capillaries in any part of the body is largely dependent on the amount and direction of blood flow in the part of question. In their first formation the embryonic blood vessels also appear in conjunction with other rudiments, suggesting a dependent mode of differentiation, though the dependence is not of a functional nature since the blood vessels develop prior to the establishment of circulation. If the heart rudiment in an amphibian embryo is removed before circulation starts the main blood vessels continue to develop for sometimes, until the embryo dies. The exact position of the blood vessel rudiments is probably determined by a process similar to an induction, though this has not been shown conclusively. Certain locations in the embryo offer preferential conditions for the development of blood vessels. These locations are : (1) between the visceral mesoderm and the endoderm (2) around the kidneys, especially the pronephros and the mesonephros.

Close networks of capillaries are always found in these two locations and they are the site of development of some of the major blood vessels. The heart obviously belongs to the first group. We have seen that the determination of the heart occurs at an early stage of development and that as a consequence it shows a high degree of autonomy in its further differentiation. It is noteworthy, nevertheless that the heart does not develop in embryos from which the endoderm has been removed. The blood vessels which form the (paired) continuation of the heart tube anteriorly and posteriorly conjunction with the endodermal parts. The two posterior vessels are the vitelline veins, which collect the blood from the network on the surface of the gut and, in amniotes, from the network on the yolk sac.

The anterior prolongations of the heart tube are the ventral aortae, which become connected along the partition walls between the endodermal pharyngeal pouches with the dorsal aortae, a pair of blood vessels developing on the dorsal side of the endodermal gut. The vessels conveying blood from the ventral aortae to the dorsal aortae are the aortic arches. The aortic arches develop in a cranio-caudal sequence, and the first to appear is the mandibular arch which ascends between the edge of the mouth and the first (spiracular) pharyngeal pouch. The second aortic arch develops between the first spiracular and the second (first true branchial) pouch, and subsequent aortic arches develop between the more posterior branchial pouches. Six aortic arches are laid down in the embryos of all vertebrates except for some very archaic forms (cyclostomes and some selachians), whose number of gill clefts is larger than six. The dorsal aortae become extended forward into the head and backward throughout the length of the trunk and to the base of the tail.

In the trunk region the two dorsal aortae later fuse into one unpaired dorsal aorta, while in the branchial and head region the two aortae remain separate. In the trunk the dorsal aorta gives off at the level of each myocomma an intersegmental artery. Larger arteries convey blood from the dorsal aorta to the viscera, to the limbs and, in embryonic stages to the yolk sac (vitelline arteries) and to the allantois (umbilical arteries). The blood vessels developing in conjunction with the excretory organs are the *cardinal veins*. The first rudiments of these appear, in Amphibia, in the form of a venous sinus around the pronephros. Very soon prolongations of this sinus are found anteriorly and posteriorly.

The posterior prolongations is the postcardinal vein, which develops along the groove between the somites and lateral plates dorsolaterally to the nephrostomes and later becomes closely associated with the mesonephric kidney. The anterior prolongation is the anterior cardinal vein. This runs forward on the same level as the postcardinal vein, just above the pharyngeal pouches into the head. The anterior cardinal vein and postcardinal vein join at the level of the anterior edge of the pronephros forming the common cardinal vein (*duct of Cuvier*), which runs inward to join the vitelline veins where they enter the heart. At a later stage, a second pair of veins develops in tetrapods on the median side of the mesonephric kidney—the *subcardinal veins*. This basic pattern of the main blood vessels is laid down with great regularity in embryos of all classes of vertebrates, but it becomes greatly modified as development goes on, especially in the higher vertebrates.

In the fishes and in the amphibian larvae, a number of aortic arches from the third onward supply blood to the gills. The blood flow from the ventral portion of each aortic arch involved is directed intc a network of capillaries which develop in the gill lamellae attached to the gill septum or in the filaments of the external gills. From these, the blood then returns by way of collecting vessels into the upper part of the arch and through it to the dorsal aorta. The middle part of the aortic arch becomes greatly reduced in diameter. It becomes lost in the capillary network in the branchial septum but may not be completely interrupted, though the amount of blood which passes through it is small compared with the volume of blood going through the gills.

In amphibians which lose their gills on metamorphosing into the adult stage, the connection between the ventral and dorsal portions of the aortic arches is reinstated by the enlargement of the narrow blood vessels connecting them during the larval stages. The capillary system in the gills is reduced and disappears, and the aortic arches again assume the shape of simple tubes leading from the ventral aorta to the dorsal aorta. Of the six pairs of aortic arches, the anterior two pairs becomes degenerate to a greater or lesser extent, but even in the higher vertebrates remnants of the ventral portions of these arches take part in the formation of the branches of the external carotid in the ventral region of the head.

The anterior portions of the ventral aortae, which gave rise to the first and second aortic arch, become the external carotids. In the terrestrial vertebrates (including amphibians after metamorphosis), the third pair of aortic arches become the carotid arches. The anterior parts of the dorsal aortae now serve to forward the blood from the carotid arch to the dorsal part of the head and to the brain. This blood vessel becomes the *internal carotid artery*. The sections of the ventral aortae between the points of origin of the fourth and the third (carotid) aortic arch carry blood both to the carotid arches (now the internal carotids) and to the external carotids. They thus become the common carotids. The fourth pair of aortic arches becomes the main channel for the blood flow from the heart to the dorsal aorta and hence to the body. These are the systemic pair of arches (in amphibians and reptiles). The fifth pair of aortic arches degenerate in all tetrapods expect in urodeles, where they carry part of the blood to the dorsal aorta.

The general direction of blood flow in the dorsal aorta is from the anterior to the posterior end. This direction of blood flow is patent from the level at which the fourth (systemic) arches join the dorsal aortae. However, the anterior prolongations of the paired dorsal aortae serve for supplying blood to the head. This blood enters the dorsal aortae mainly through the carotid arches; the anteriorly directed flow of blood starts therefore from the point of entry of these arches. Between the level where the carotid arches join the dorsal aorta to the level where it is joined by the systemic arches, there is an area of blood stagnation.

As in other cases, where a blood vessel is devoid of blood flow, the sections of the dorsal aortae between the carotid and the systemic arches tend to become atrophied; the vessels become attenuated or even completely interrupted. In some urodeles and

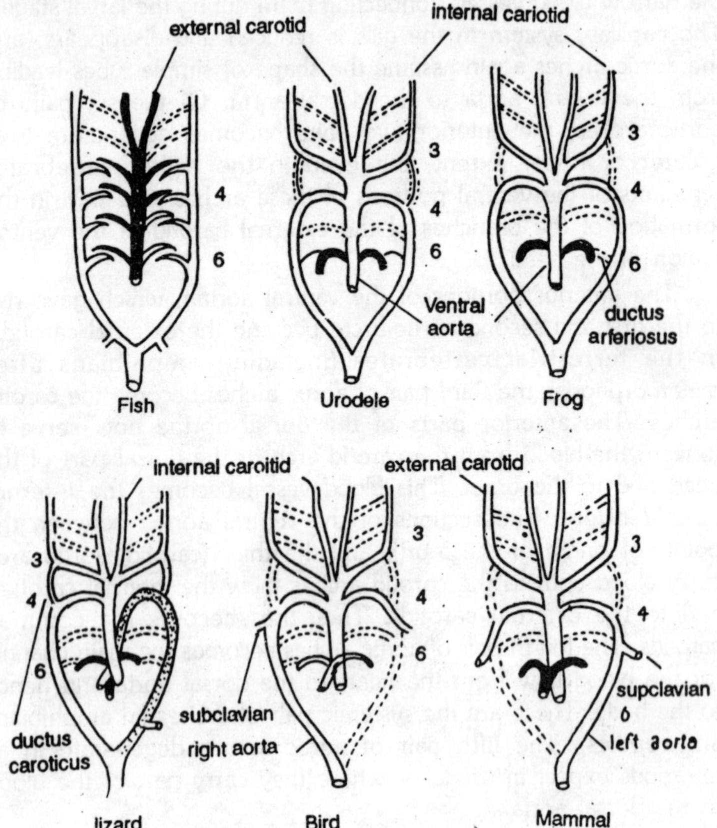

Fig. 9.14. Modifications of aortic arches in vertebrates. 3–carotid arch. 4–systemic arch. 5–abortive. 6–pulmonary arch.

reptiles this section of the dorsal aorta, known as the *carotid duct*, persists as a narrow vessel in the adult stage, but in most terrestrial vertebrates the duct, though existing in the embryo becomes completely reduced and the anterior sections of the dorsal aortae, now becoming a part of the internal carotid artery, are completely separated from the remainder of the dorsal aorta. The sixth pair of aortic arches gives off blood vessels to the lungs (and to the skin in amphibian). It is thus the *pulmonary arch*.

The upper part of the pulmonary arch, connecting the base of the pulmonary artery to the dorsal aorta, is known as the *arterial duct* (ductus arteriosus or ductus Botalli). In the larvae of amphibians and in the embryonic stages of amniotes, the whole of the pulmonary arch is a fairly large vessel carrying blood from the ventral aorta to the dorsal aorta. This condition changes drastically, however when the lungs start functioning as respiratory organs. The pulmonary arteries, which had been insignificant blood vessels till that stage, greatly increase in size and take over the blood flow entering the pulmonary arch, while the arterial ducts become reduced. They remain as narrow blood vessels in adult urodeles and some reptiles, but in adult frogs and the majority of amniotes the ducts are closed and reduced to a fibrous strand connecting the pulmonary and systemic arteries. In amniotes—and in mammals in particular—the pattern of aortic arches is modified still further.

The fifth aortic arch is already greatly reduced at the time of its formation. It is still recognizable as such in embryos of reptiles and birds but is completely lacking in mammals. A characteristic feature of amniotes is the asymmetry in the development of their systemic arches. Although the aortic arches are laid down as symmetrical pairs of blood vessels, the left systemic arch degenerate in birds and the right systemic arch degenerates in mammals. In mammals originally both systemic arches reach the paired dorsal aortae, and the latter fuse together posteriorly into the single dorsal aorta. Before their fusion the paired aortae give off the subclavian arteries laterally. When the right aorta becomes reduced, it is interrupted at its posterior end after the origin of the subclavian artery and just before its connection to the unpaired aorta. In this way, the right systemic arch becomes the channel carrying blood to the right subclavian artery.

In later stages, the distinction between the aortic arch and the right subclavian artery subclavian artery is lost, and the whole vessel is then known as the right subclavian artery. It is therefore, not entirely homologous to the left subclavian artery, which starts, as in lower vertebrates, from the left paired dorsal aorta. The section of the paired ventral aorta between the branching of the common aorta and the origin of the right common carotid artery becomes what is known as the *innominate artery*. The reduction of the aortic arches is obviously dependent on the blood flow through these vessels, in the same way as the blood flow has been found to mold the development of vessels from the original capillary networks. The following experiments prove this.

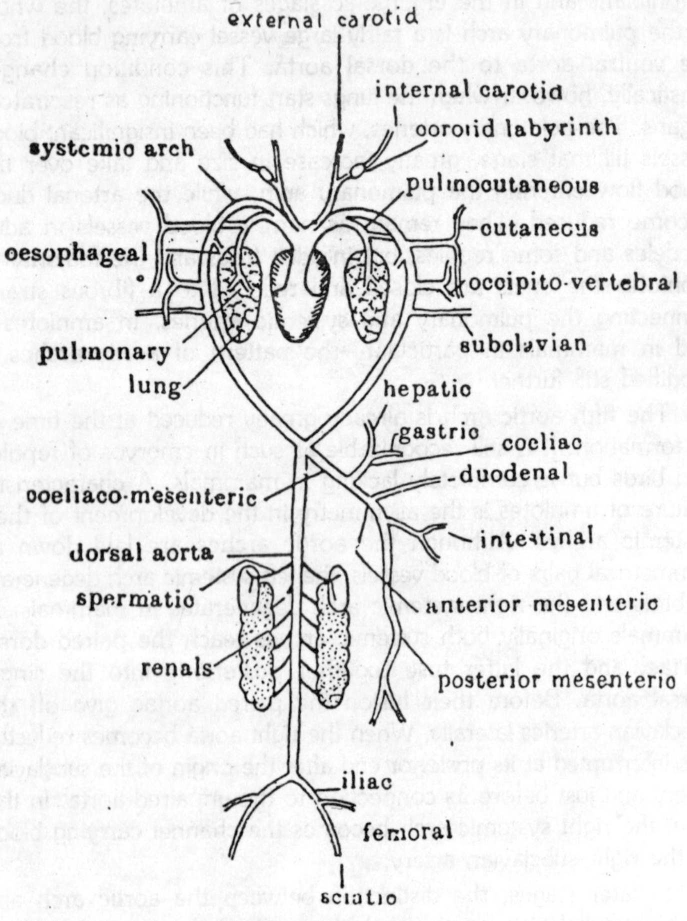

Fig. 9.15. Arteries of frog.

In frog tadpoles, all four (third to sixth) aortic arches persist up to metamorphosis, and only then the transformations occur which lead to the adult condition. In the tadpole of the clawed toad, *Xenopus laevis*, the systemic arch on one side was destroyed shortly before metamorphosis. The result was that the carotid arch of the same side took over the work of the systemic arch: the connection of the carotid arch to the dorsal aorta, which is interrupted in normal development persisted and the blood from the heart flowed to the dorsal aorta by way of the carotid arch.

Even more interesting, perhaps, is an experiment performed on the chick embryo: the normally persisting right systemic arch was ligatured. The effect was that the left systemic arch now took over the blood flow from the heart to the dorsal aorta and became permanent instead of degenerating. At the time of its origin, the ventral aorta is a paired blood vessel diverging anteriorly from the heart. The two ventral aortae may be quite independent, each giving off a number of aortic arches. Subsequently they fuse in various degrees, forming a single unpaired ventral aorta, especially in fishes. In terrestrial vertebrates, however, the two aortae usually remain separate, except for the most posterior section, where the aortae are connected to the heart. In diagrams, it is convenient to show the ventral aorta as being a long tube extending forward from the heart. Actually this representation is not very accurate.

In adult fishes, the ventral aorta is a fairly long blood vessel extending in an anteroposterior direction, but in embryos of all vertebrates the heart is situated underneath (ventral to) the pharynx; the ventral aorta lies more or less in a vertical position and very soon splits to give rise to the roots of the aortic arches, of which only the first two or three pairs lie anterior to the heart, while the rest are developed on a posterior extension of the aorta. The pulmonary arteries, when these develop, may be described as being on the extreme backward prolongation of the ventral aortae; when the last aortic arches area formed, the pulmonary arteries shift upward and can then be seen originating from these last (sixth) aortic arches. A further important change in the arterial system in higher vertebrates accompanies the separation of the heart into right and left halves.

The pulmonary arteries acquire a separate connection to the right ventricle, while the systemic and carotid arteries are linked up to the left ventricle. The separation occurs from the branching points of these arteries backward toward the heart. The ventral aorta becomes split lengthwise by a partition cutting in at the point where the pulmonary arches branch off from the systemic arches, the latter carrying the carotid arteries with them. In fishes, the heart remains in the pharyngeal region, but in terrestrial vertebrates the heart is shifted into a more posterior position in the thoracic cavity of the animal. The shift is, in part, owing to the emancipation of the heart from its connection with the gills and, in part, is a result of the development of the neck as a distinct region of the body. The posterior position of the heart causes a considerable elongation of some of the blood vessels conveying blood to the head. The carotid arteries are the ones affected most. There is, however some variety in the way that the elongation is achieved.

In the amphibians, mammals, birds and most reptiles, the carotid arch is carried backward together with heart (and the other aortic arches). It is then the external and internal carotids which become elongated. In snakes and some lizards, on the other hand, the carotid arches remain in the original position, just posterior to the head, and it is the common carotids which elongate. The systemic and pulmonary arches in all terrestrial vertebrates move into the thoracic cavity together with the heart. The disappearance of the brachial respiratory apparatus and the development of the neck, separate parts of a previously closely knit system and bring them in widely different positions in the body. The ganglia of the cranial nerves originally connected to the branchial apparatus and developing in part from epibranchial placodes in the head, inside the brain case. The remnants of the visceral skeleton (hyoid and branchial arches) are to be found in the neck as the hyoid bone and the cartilages of the larynx and trachea.

The aortic arches lie in the thoracic cavity and close to them also lies the thymus, another derivative of the branchial pouches. The vitelline veins on their way to the heart pass through the region in which the liver is developed at a later stage. When the rudiment of the liver is formed it envelops the vitelline veins. These break up into a system of hepatic sinusoids permeating the liver lobules and thus give rise to the hepatic portal system. The veins in front of the liver and leading to the heart become hepatic vein, and the parts of the veins caudal to the liver give rise to the *hepatic portal vein*.

In amniotes, the development of the vitelline veins is asymmetric, the right vein tending to degenerate posteriorly and the left vein disappearing, in its most anterior part, in front of the liver. Through a system of anastomoses, however, the continuity of the blood flow is preserved, and a single blood vessel may eventually be built from parts of the two vitelline veins. The parts of the vitelline veins collecting blood from the yolk sac of the embryo disappear at the time of birth, but a branch from the vitelline veins—the mesenteric vein—takes over the drainage of blood from the intestinal tract and becomes the main component of the hepatic portal vein of the adult animal. In embryos of reptiles and birds, the vitelline veins are not completely broken up into hepatic sinusoids as they pass the liver but preserve a large unpaired channel, the *ductus venosus*, by means of which blood passes through the liver and reaches the sinus venosus. After hatching, however, the ducts venosus is occluded, and all the blood of the portal vein is forced into the system of hepatic sinusoids. The blood vessels of the cardinal undergo extremely complicated transformation in the course of ontogenetic development and of evolution in vertebrates.

The anterior cardinal veins are least changed; they become the internal jugular vein, and the common cardinal veins become the anterior venae cavae. In some mammal, including man, a transverse anastomosis is formed between the anterior the anterior cardinal veins of the right and left side, whereupon the connection of the left anterior cardinal to the heart disappears, and all the blood from the head, neck and forelimbs is directed into the right common cardinal vein which becomes the single anterior vena cava. The transformations of the veins in the posterior part of the body are very much drastic: the postcardinal veins, which are the main veins of the posterior trunk region in the fishes, and in the early embryos of all vertebrates, are gradually reduced in the embryos of tetrapod. The first stage in this reduction as it occurs during the embryonic development in amniotes, is the formation of a venous network on the ventromesial side of the mesonephric kidneys. This plexus gives rise to veins lying in the same position—the *subcardinal veins*. The subcardinal veins become linked to one another by anastomoses in the midline. Initially, the subcardinal veins join the postcardinal veins anteriorly and use the latter as a channel for blood outflow.

At a later stage, however a connection become established between the network of the subcardinal veins with their associated plexuses, and the system of veins in the liver. This opens a more direct route to the heart for the blood collected from the kidneys into the subcardinal vein. At first, the blood from the subcardinal veins passes through the hepatic sinusoids in the tissue of the liver, but as more and more blood goes through, the sinusoids enlarge and develop into one straight channel passing through the liver directly to the sinus venosus. This channel is the anterior portion of the posterior vena cava. (As result of the reduction of the part of the hepatic tissue, the posterior vena cava later comes to lie on the edge of the liver). The middle portion of the posterior vena cava is formed by the joining together of smaller vessels belonging to the subcardinal vein system.

Eventually, the posterior vena cava, through a system of intermediate smaller vessels, establishes a connection with the iliac veins which collect the blood from the hindlimbs. The posterior vena cava is then established in its final form as it is found in adult amniotes. The postcardinal veins, in the meantime, lose their significance channels conveying blood from the posterior part of the body, and their anterior halves leading to the ducts of Cuvier degenerate, leaving behind only insignificant remnants. The posterior parts of the postcardinal veins remain intact a renal portal vein, in amphibian throughout life, and in embryos of reptiles and birds so long as the mesonephric kidneys remain functional. By means of these veins, the mesonephric kidneys are supplied with blood returning from the hindlimbs through the iliac veins and from the caudal vein (where it is present).

With the replacement of the mesonephric kidney by the metanephros the renal portal veins degenerate and disappear, and the iliac veins gain direct connection to the posterior vena cava, as has been explained. The development of the posterior vena cava is a good illustration of the formative influence of the blood flow on the differentiation of blood vessels: we can observe how the blood flow is not restricted by the arrangement of existing blood vessels but causes the blood vessels to be fashioned in correspondence with the direction and amount of blood flow. Thus, the largest vein of the vertebrate body develops from blood vessels of capillary dimensions where they lie in the line of movement of the blood. It may be noted that in adult amphibians we may find conditions representing consecutive stages in the development of the posterior vena cava.

Salamanders still posses a pair of postcardinal veins which carry part of the blood from the kidneys to the heart via the common cardinal veins. At the same time, they have also a posterior vena cava which is connected at its hind end to the system of veins between the kidneys. The posterior vena cava is much larger that the kidney direct to the sinus venosus. In frogs, the postcardinal veins are reduced in the adult, and all the blood from the kidneys is collected on their mesial aspect into the posterior vena cava. The latter, however, is not yet linked up with the hind legs. This last stage in the development of the posterior vena cava is completed only in the amniotes. In addition to the hepatic portal vein and the inferior vena cava, a third system of veins becomes associated with the liver in amniotes.

The allantoic (or umbilical) veins go toward the heart in the body wall and enter the common cardinal veins. Before joining the latter, the allantoic veins pass very close to the liver. As the liver grows, it makes contact with the allantoic veins, whereupon a connection is established between these and the system of hepatic sinusoids in the liver. More and more blood from the allantoic veins enters the liver, until the main channel into the common cardinal veins becomes reduced and obliterated, so that all the blood from the allantois (from the placenta in mammals) goes to the heart by way of the liver. The right allantoic vein later degenerates in all amniotes, the left remaining as the only functional one. The large amount of blood entering the liver from the left allantoic vein cannot be accommodated in the hepatic sinusoids and eventually makes for itself a broad direct channel which joins the posterior vena cava near the point where it leaves the liver at its anterior end. This direct channel is referred to by students of mammalian (especially human) development as the ductus venosus of reptiles and birds, where it is a section on the pathway of the vitelline veins. After birth and the cessation of placental circulation, the ductus venosus degenerates and becomes transformed into a strand of connective tissue (the *ligamentum venosum*), as often happens with large blood vessels when they cease to be functional.

Development of the Reproductive Organs

The gonads in vertebrates develop from the upper edge of the visceral layer of lateral plate mesoderm in the posterior half of the body. The first rudiment of the gonad appears as a thickened longitudinal strip of mesodermal epithelium lining the body cavity immediately lateral to the dorsal mesentery. This thickening is called the germinal ridge. Initially the ridge is produced by the mesodermal cells, assuming a high columnar shape, but soon the cells become arranged into a compact mass several cells thick, and the ridge protrudes into the coelomic cavity. Still later, the connection between this cell mass and the peritoneal wall becomes constricted laterally, and the gonad remains suspended from the peritoneal wall only by a double layer of peritoneum—the *mesorchium* or *mesovarium*. The genital ridges are situated in the same region of the body as the rudiment of the mesonephros and in close spatial association with the latter, the genital ridge being medial to the mesonephros. Although in the later development of vertebrates the genital apparatus becomes associated with the excretory system, the initial stages of the two organ systems are completely independent of each other. What has been said so far about the formation of the gonad rudiment refers only to its gross morphology.

A more detailed microscopic examination of the germinal ridge epithelium discloses that it consists of two types of cells. The majority of cells are very similar to the other cells of the peritoneal epithelium, even though they may become columnar in the initial stages of the development of the ridge. The cells of the second type have a very different appearance. They are much larger than the ordinary mesodermal cells, they contain large, more or less vesicular nuclei and their cytoplasm differs from that of the surrounding cells in its staining properties amphophil instead of acidophil in birds and, in amphibians, in its higher yolk content. The shape of the cells is nearly spherical, and they do not participate in the epithelial arrangement of the typical mesodermal cells but appear to be interspersed between the cells of the mesodermal epithelium cells of this peculiar type are known as the *primordial germ cells*.

The Origin of the Primordial Germ Cells

According to views held by many embryologists, only the primordial germ cells are destined to give rise to the gametes (eggs and spermatozoa), while the ordinary mesodermal cells differentiate into the somatic cells of the adult gonad: the Sertoli cells in the testes, the superficial epithelium and the follicle cells in the ovaries. Whatever their eventual fate, it appears certain that the primordial germ cells have a different origin from the rest of the cells in the germinal ridges, that they first appear in parts of the embryo other than those in which the germinal ridges lie, and that the primordial germ cells reach the germinal ridges after a more or less extensive migration. In amphibians, the primordial germ cells may be traced back in microscopic sections to a position just above the dorsal mesentery. Here the primordial germ cells lie surrounded by mesenchyme before they migrate into the mesodermal epithelium of the germinal ridges. The median position of the primordial germ cells is probably already a secondary one.

In the birds and reptiles cells identical in their cytological properties with the primordial germ cells have been found in early developmental stages—before the beginning of segmentation of the mesoderm—in the endodermal layer of the extraembryonic part of the blastoderm. Similarity of appearance cannot however, be considered as proof that the cell found in the extra-embryonic endoderm are the same cells appear later in the germinal ridges. To test whether the primordial germ cells are derived from the cells in the extraembryonic endoderm, the latter may be destroyed prior to their migration.

In the chick, the cells in question occupy a crescentic area in front of the head end of the embryo. This area can be destroyed in various ways, as for instance by cauterizing it with a hot needle or irradiating it with radium emanation. As the crescentic area lies well beyond the embryo proper, the development of the latter is unimpaired. At the time when the germinal ridges are formed, no primordial germ cells are to be found in and around the germinal ridges, although the area of the germinal ridges had not been tampered with. The experiments shows that the peculiar cells of the extraembryonic endoderm are actually the primordial germ cells, which are later incorporated into the gonad. This conclusion is further supported by the following experiments: if the primordial germ cells are exposed to very weak irradiation with x-rays, while still in the extraembryonic endoderm, they are damaged to a greater or lesser extent but not killed at once. Some of the damaged cells succeed in reaching the site of gonad formation and are to be found there for a short time. Then all the primordial germ cell perish and there remains a gonad consisting of mesodermal cells only.

The gonad may grow and develop further without showing any signs of producing generative cells. Such a gonad, devoid of generative elements, may be called a sterile gonad. As to the means by which the primordial germ cell travel from the extraembryonic endoderm to their later position in the germinal ridges, it is supposed that in bird they do so largely through the blood vessel system. The cells have been observed to leave the endodermal epithelium, to move slowly into the space between endoderm and mesoderm, and then to penetrate into the blood vessels of the area vasculosa. The primordial germ cells may actually be seen in the blood stream. Although most of them eventually reach the germinal ridges, some go astray and are found in various parts of the body where they eventually degenerate. Further experiments work on the origin of primordial germ cells has been carried out in amphibians.

According to the view held by the majority of embryologists the cytoplasmic material included in the primordial germ cells in amphibians can be traced back as far as the uncleaved egg. Before cleavage begins, cytoplasmic inclusions of a particular kind can be found in the subcortical layer at the vegetal pole of the egg. The inclusions are very rich in RNA. During cleavage these inclusions move upward along the cleavage furrows and surround the nuclei of a group of endodermal cells lying right in the middle of the mass of yolky endoderm. The cells receiving this particular cytoplasmic substance are the primordial germ cells and the substance of the cytoplasmic inclusions has been called the germinal plasm. By irradiating the vegetal pole of newly fertilized frog's eggs with ultraviolet light, it is possible to destroy the "germinal plasm" and in this way to greatly reduce the number of primordial germ cells formed in later stages.

By removing the endoderm in the neurula stage, the embryos are completely sterilized. After the completion of neurulation, the primordial germ cells migrate first to the dorsal region of endoderm, to the dorsal crest part of the endodermal wall lying above the cavity of the gut. From here the primordial germ cells move into the mesenchyme of the dorsal mesentery and eventually into the germinal ridges. The area containing primordial germ cell can be transplanted in the late neurula stage from one embryo to another and if the tissues of the donor embryo can be distinguished from the tissue of the hot it can be shown that the eggs and spermatozoa are actually derived from the grafted material.

In one experiment, the mensoplantation was made between two subspecies of the clawed that *Xenopus laevis* differing, among other things, in the properties (size, colour) of their eggs. Operated female embryos were reared to the adult stage and laid eggs which in colour and size clearly showed the properties of the donor species. Thus the primordial germ cells transplanted with a section of endoderm at the end of neurulation not only found their way into the gonads but also preserved intact their genetic characters, in spite of the fact that during their development and maturation in the gonads they were surrounded by somatic cells of a different subspecies. Surprisingly, it appears that the primordial germ cells in urodele amphibians have an entirely different origin from those in frogs and also from those in amniotes. They develop not from the endodermal cells but from cells originally located in the posteroventral parts of the lateral plate mesoderm. It has been stated earlier that in salamander embryos it is possible to remove the whole of the endoderm in the neurula state.

The endoderm from another embryo may then be inserted into the empty ecto-mesodermal shell. If the operations is carried out heteroplastically, and if the two species are chosen in such a way as to allow the cells of each to be distinguished by some peculiarity then the derivation of the primordial germ cells or any other cells, from the endoderm or the other mesoderm respectively can be determined with certainly. The experiments were carried out with the purpose of investigating the origin of the primordial germ cells using *Triturus cristatus* as one of the components and *Triturus alpestris* or *Ambystoma mexicanum* as the other. The eggs of the latter two species contain in their cytoplasm considerable amounts of pigment granules while in the eggs of *Triturus cristatus* pigment granules are either completely absent or very few in number of operated embryos survived.

The result was very clear-cut one all the primordial germ cells found in the operated embryos had the specific properties (pigmentation) of the species to which the ectoderm and mesoderm belonged and not of the species to which the endoderm belonged. The primordial germ cells were in this experiment, embedded in germinal ridges consisting of cells of the same species. In another series of experiments, the lateral plate mesoderm was transplanted in the same heteroplastic combination and it was found that after such operations numbers of primordial germ cells develop from the graft.

The primordial germ cells, although mesodermal, do not arise from cells in the germ ridges, but as in the case of endodermal primordial germ cells, they migrate from their site of origin (ventrolaterally in the posterior part of the body) to the area above the dorsal mesentery and then into the germ ridges. The conclusion seems inevitable that the primordial germ cells are derived from different germinal layers in the urodele amphibians and in the frogs. In many invertebrates however (*Ascaris*, *Sagitta*, some insects), the primordial germ cells are known to be distinguishable in very early cleavage stages even before the germinal layers become segregated. The primordial germ cells can thus be considered as not belonging to any of the germ layers. This is in contrast to the somatic parts of the gonads and the auxiliary parts of the sexual organs (ducts, etc.) which are mesodermal. As to the mechanism by which the primordial germ cells in the amphibians reach their destination, it is ameboid movement of the cells themselves. There is no reason to suppose that blood vessels play any part in their transport. No experimental investigations have been carried out on the origin of the primordial germ cells in human embryos, but careful observations show that these are first found in the endodermal epithelium of the yolk sac in the vicinity of the allantoic stalk, and that from there the germ

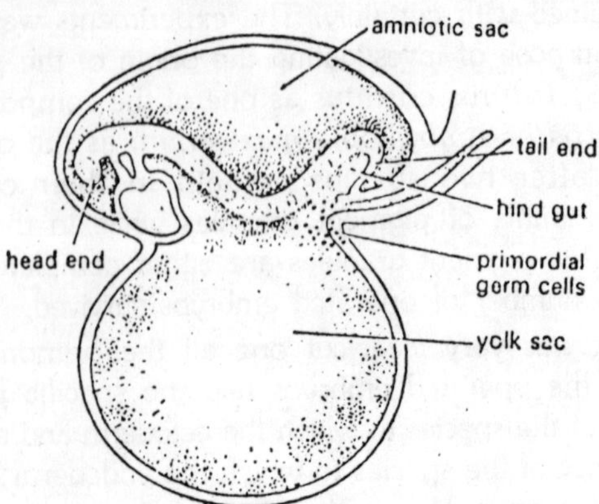

Fig. 9.16. A 4-week old human embryo showing the original position of the primordial germ cells.

cells migrate into the adjoining mesenchyme and eventually take up their position in the germinal ridges. In mice, on the other hand, investigations comparable to those done on birds and amphibians have been carried out.

As in man, the primary germ cells first occur in the endoderm and later migrate into germ ridges. These cells contain large quantities of alkaline phosphatase in early embryonic stages when this enzyme is absent from other cells of the embryo. By using a special histochemical stain for alkaline phosphatase on sections, therefore, it is possible to distinguish the primary germ cells and to follow their migration from the gut into the dorsal mesentery and then into the germinal ridges. Originally there are less than a hundred primary germ cells in a mouse embryo, but during their migration the cells increase in number by repeated mitosis of that eventually 5,000 or more cells are present. The primary germ cell are extremely sensitive to x-rays, and if the female mice are irradiated between the eighth and twelfth day of gestation with a dose of 400r, all or almost all the germ cells are killed off before they reach the germinal ridges. The germinal ridge develop nevertheless and differentiate into gonadal tissue which is, however, completely sterile—devoid of sex cells or their predecessors.

Differentiation of the Gonads

After arriving in the germinal ridges, the primordial germ cells become embedded in the epithelium of the germinal ridge. At the same time, the germinal ridge epithelium becomes convert toward the coelomic cavity and, on its dorsal side, forms a hollow which is filled initially with loose undifferentiated mesenchyme. Subsequently, this loose mesenchyme is partly replaced by compact strands of cells migrating into the gonad from the mesonephrogenic cord—the *primitive sex cords*. The epithelium of the germinal ridge (originally a thickening of the coelomic epithelium) which forms the surface of the gonad is known as the cortex of the gonad. The primitive sex cords (the strands of compactly arranged cells in the interior of the gonad) constitute its medulla. The remainder of the loose mesenchyme serves as the pathway for the blood vessels supplying the gonad. In the foregoing, no distinction has been made between the development of the ovaries and the testes.

In the early stages of organogenesis with which we are concerned here, there is practically no difference in the development of the gonads of the two sexes, and the gonad in

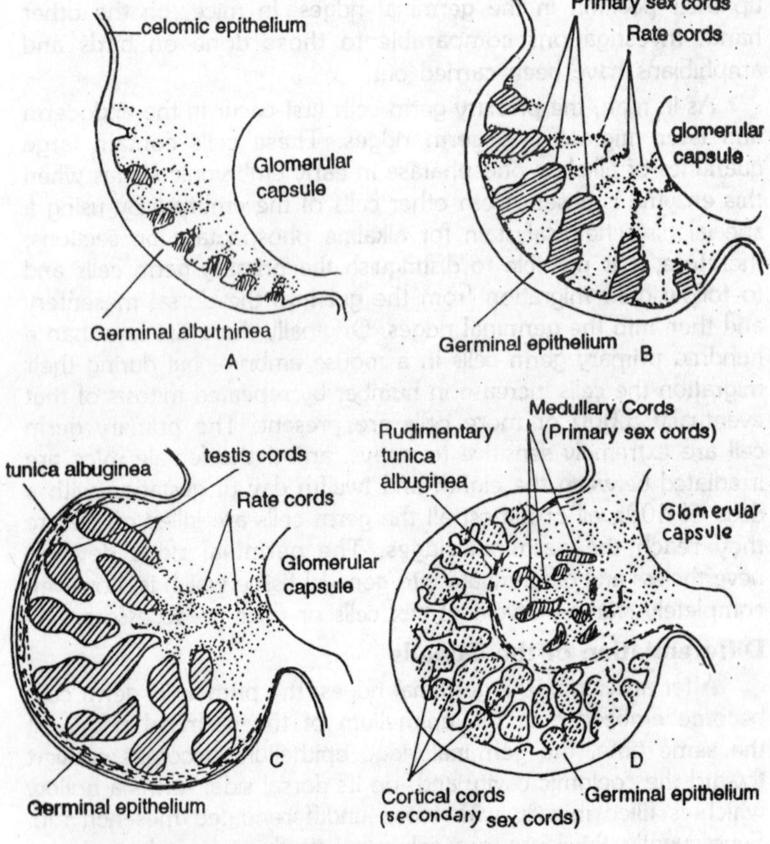

Fig. 9.17. Development of testis and ovary of a mammal.

these stages may be called the indifferent gonad. Only with the onset of histological differentiation do the testes and the ovaries become increasingly different from each other. In the male embryos, the primordial germ cell migrate from the cortex of the gonad, in which they were originally embedded, into the primitive sex cords of the medulla, which forthwith become hollowed out and are thus converted more or less directly into the seminiferous tubules. The primordial germ cell give rise to the spermatogonia and subsequently to the spermatozoa, while the Sertoli cells are derived from the sex cords.

The seminiferous tubules become connected to the rete testis, a system of very thin tubules developing in the dorsal part of the gonad, probably from the same material from which the seminal tubules have been built. The canals of the rete testis, in their turn, form connections to the adjoining tubules of the mesonephros. Thus a pathway is established from the male gonad, via the mesonephros, to the mesonephric duct, which in vertebrates serves as the outlet for the sperm. While the medulla of the testis becomes its functional part, the cortex undergoes reduction and is converted into a thin epithelial layer covering the coelomic surface of the testis. In female embryos, the medulla of the gonad becomes reduced, the primary sex cords are resorbed, and the interior of the mature gonad is filled with loose mesenchyme permeated by blood vessels. The primordial germ cells remain embedded in the cortex, which greatly increases in thickness.

During the growth of the gonad, the masses of cortical cells becomes split up on the inner surface of the cortex into groups and strands of cells surrounding one or several primordial germ cells, these then become the primary oocytes which enter in the initial stages of oogenesis. The primordial germ cells lying nearer the surface of the cortex retain their undifferentiated state longer and serve for the production of eggs at a later stage in the female's

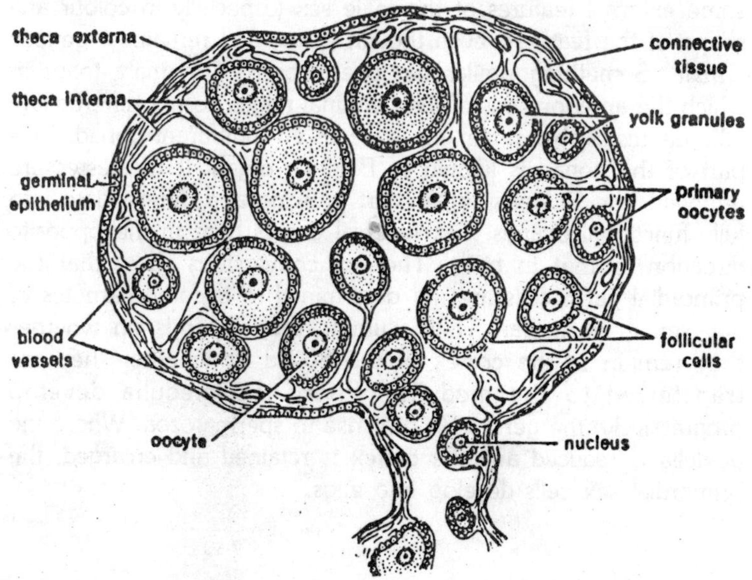

Fig. 9.18. T.S. of ovary of frog.

life. Special features in the development of gonads in some groups of animals will now be noted. In the ovaries of amphibians the degenerating medulla becomes cavitated, so that the mature gonad acquires the form of a sack with a hollow space (or rather with several hollow spaces) in the interior.

The mature eggs bulge into these internal cavities. When ovulation occurs, however, the external surface of the gonad becomes ruptured and the eggs are shed not into the internal cavity but into the coelom. In birds, with some rare exceptions only the left ovary becomes fully developed. From the start more primordial germ cells enter the left germinal ridge, and only here the ovarian cortex acquires a typical structure. The gonad on the right side remains in the state of an indifferent gonad, with the primary sex cords not reduced. The primordial germ cells in the right gonad eventually degenerate. However, if the left gonad is removed by operation, the right gonad enters into a stage of compensatory growth and differentiates progressively. The right gonad then acquires the general structure not of an ovary but of a testis and even starts to secrete male hormones.

Under the influence of the male hormones, the bird shows some external features of the male sex (especially in colour and shape of the feathers) even though it was and remains a genetic female. Something similar has been observed in male toads in which the anterior ends of the germinal ridges do not differentiate fully (as testes) but retain the state of an indifferent gonad. This part of the gonad is known as Bidder's organ. If the testes are removed, the Bidder's organ start growing and differentiate into fully functional ovaries. Sex reversal occurs here in the opposite direction to that in birds. The sex conversions show that the primordial germ cells are not determined to become gametes of any sex in particular. Their ultimate fate depends on whether they remain in the cortex of the gonad or whether they are transferred to the medulla. Where the medulla develop progressively, the germ cells give rise to spermatozoa. Where the medulla is reduced and the cortex is retained and enlarged, the primordial sex cells develop into eggs.

10

GENES AND MORPHOGENESIS

Pleiotropism

The earlier literature of development genetics is full of examples of genes affecting morphogenesis. Mutants have been found that induce the formation of an abnormal tail in the mouse, that lead to cyclopia or anencephaly in guinea pigs, to the lack of limbs in calves, and to deformed beaks or the absence of kidneys in chickens, to mention but a few examples. The study of the sequential developmental steps that lead to such gross abnormalities is very complicated and seldom leads to a clear explanation. Nevertheless, such research is or great importance in view of the frightening number of genetically controlled congenital malformations in man. These range from very harmless malformations, such as webbing of toes to more serious malformations, such as phocomelia.

Congenital malformations affecting almost every developmental event have been observed in humans, and we shall not understand these abnormalities completely until we understand normal development completely. So little is presently known about the genetic control of mammalian development that we cannot in the near future hope to understand the paths that must lead from the altered gene to the complex abnormal phenotypes. Nevertheless, many of the complex syndromes have been traced to simpler events closer to the primary function of the gene. An informative analysis is that of the mouse mutation, congenital hydrocephalus (ch). This syndrome consists in an abnormal number and arrangement of whiskers, failure of the eyelids to be closed in the embryo, abnormal shape of pituitary gland, absence of dermal flat bones of the skull, and cerebral hemorrhage, which leads to the death of these mice immediately after birth. The bulging foreheads of these newborn mice are filled with hemorrhagic fluid, and the animals have blunt snouts. From the genetic information available, a single Mendelian unit is responsible for this complex syndrome. This is not an exceptional case. As a rule, genetics defects that manifest themselves at a morphological level are pleiotropic in nature–that is they cause a multitude of effects. And yet many of these multiple effects can be traced back through development and reduced to a few, and occasionally to one, primary genetic effect. For the case of the *ch* mouse, such a "pedigree of causes" has been elaborated, which makes hydrocephalus directly responsible for the entire group of symptoms. Hydrocephalus itself is brought about by a growth anomaly in the cartilaginous skull so that the skull is shortened.

The brain is forced upward, thereby impeding the proper drainage of cerebrospinal fluid. When the cartilage cells forming the cartilaginous skull were examined histologically they proved to be abnormal, with extensive vacuolization and a highly abnormal matrix. This cartilage abnormally is not restricted to the skull, but is also present in the vertebrae, the sternum, and other parts of the skeleton. The common denominator of all the symptoms is the cartilage anomaly. It leads to a fundamental disturbance in morphogenesis, which, in turn, is responsible for the appearance of many of the secondary and tertiary phenotypic effects. The causal analysis of pleiotropism has been carried somewhat farther in a few mutants, such as the pituitary dwarf mouse (dw).

The two major symptoms of this syndrome are dwarfism and sterility. At the histological level, the thyroid and adrenal cortex are found to be hypoplastic, and a careful investigation of the pituitary glands shows that its anterior lobe is retarded in its growth. The pituitary of the mutant mouse contains only small cells with pyonotic nuclei, but lacks the large acidophils that, in normal animals, produce growth hormone. From the viewpoint of basic research, this observation is not very satisfying, because we have no idea just how the *dw* allele affects the proper differentiation of acidophils. However, the discovery has proved to be very important for therapy, for it gives the investigation the knowledge needed to correct the developmental abnormality. If the primary

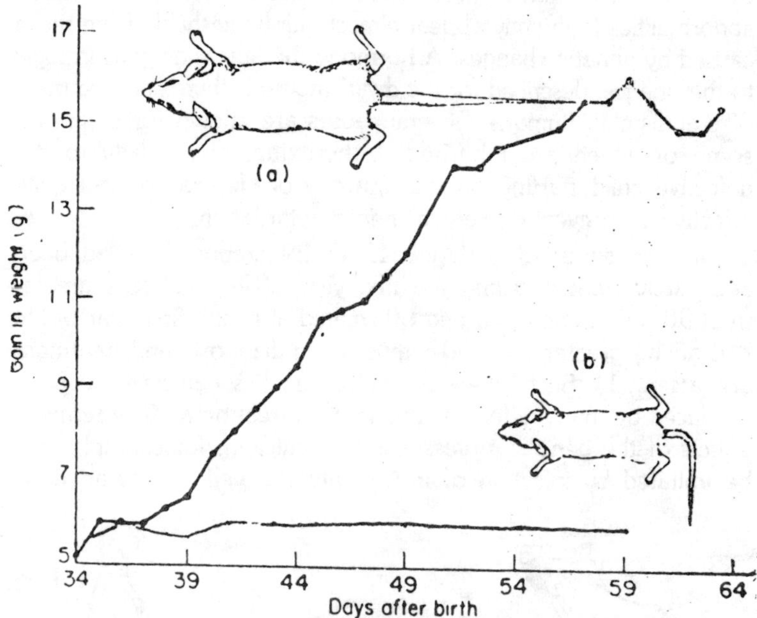

Fig. 10.1. Effect of implantation of normal pituitaries into dwarf mice on their body weight. Normal pituitaries were implanted into dw/dw mice daily, starting 34 days after birth (a). in the absence of this treatment, the mice remained very small.

defect is, indeed, an effect on hormone production by acidophils of the pituitary, then administering normal pituitary hormone should correct some or all of the secondary and tertiary symptoms. This therapy was attempted on dwarf mice. The mice received daily implants of wild-type pituitaries from the day of birth. They reached normal size and, in fact, grew to be sexually mature individuals. The thyroid gland and adrenal cortex also became normal in size during this treatment. The pituitary itself did not become normal, however. This result supports the assumption that the pituitary cells are those primarily affected by the mutant locus.

Phenocopy

Tracing a developmental abnormality back in ontogeny to its earliest expression offers the hope of identifying the primary genetic effect. From such an identification, one might expect to rectify at least the consequences of the defective gene. A different approach consists in attempting to imitate a genetically caused abnormality by interfering with normal development through the use of physical or chemical treatments. Such experimentally produced abnormalities frequently appear almost indistinguishable from those caused by genetic changes. A hormone therapy analogous in logic to the one just described for the dwarf mouse is thyroxine treatment of cretinism in humans. Several causes are known for cretinism, some of which are inherited. L-thyroxine, given daily to the defective child starting on the first day of life, has proved quite effective in preventing severe mental retardation.

In one series of patients, 12 of 29 cretins that had been adequately treated during the first year of life had IQ's greater than 90. (No patient, among 50 treated after the first year of life had an IQ greater than 90 ; thus, early diagnosis and treatment are essential.). Such phenocopies of genetics defects have been produced by many different experimental treatments. The terminal effects of the gene *rumplessness* in a chicken, for example, can be imitated by injection of insulin into the yolk sac at an early

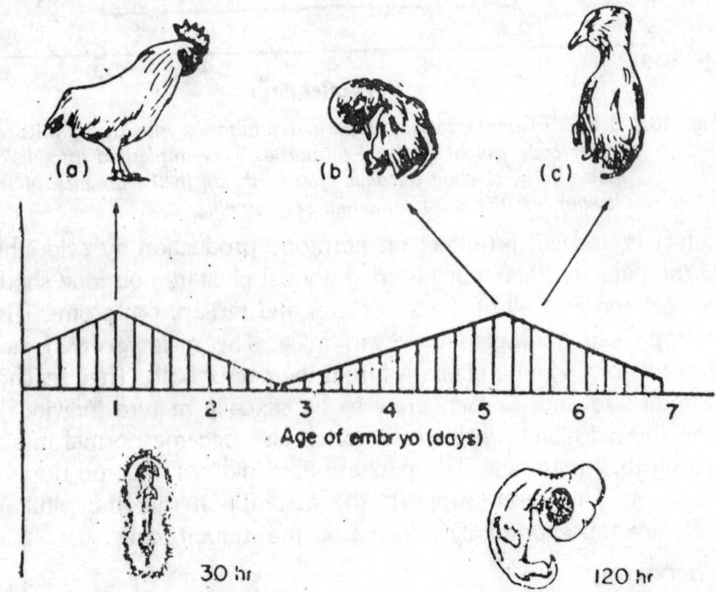

Fig. 10.2. Phenocopies induced by insulin in the chick. The ordinateindictes the relative sensitivity as a function of developmental stage (abscissa). If insulin is given during days 0-3, rumplessness is induced. If it is given around day 5, short upper beak is produced. The lower figures show stages of development representative of these two critical phases.

stage of development. Interestingly enough, the same compound, when injected at a later development stage, leads to the imitation of a different genetic lesion, short upper beak.

The different organ primordia of developing organisms are sensitive to phenocopying agents at particular phases of development. These phases have been called *phenocritical phases*, and similar critical phases have also been postulated for gene action. For example, hybrid arrest occurs at specific developmental stages that are not randomly distributed over the life span of an organism. Time specificity of gene action would be expected to generate critical stages of development. Many investigators have used the phenocopy approach in searching for the primary genetic lesion in congenital abnormalities. They assume that similar effects may have similar causes. Insulin-produced abnormalities are similar to those produced by the mutants short upper beak or rumplessness; therefore, these genes may induce metabolic disturbances similar to those produced by excessive insulin administration.

Initially, much support for this notion came from genetic experiments. Modifier genes were discovered which, when built into the genome of mutants, affected the expression of the mutant phenotype. These same modifier genes also affected the expression of the phenocopying substances. But the phenocopy approach suffered a blow when it was found that very different agents produced the same malformations ; boric acid, for example, was found to produce the same terminal effects as insulin.

The fact that a multitude of different causes can produce the same phenocopy is well documented in a number of experimental systems. In *Drosophila*, for example, both ether and high temperature, applied at an early embryonic stage of development, produce flies that look just like *tetraptera* mutants. (Tetraptera flies have four wing, two on the mesothorax and two in the most posterior segment of the thorax, which in the mutant is not a metathorax, but rather another mesothorax.)

It is not difficult to understand how a variety of substances can produce the same end effect. Developmental processes are long chains of events, many of which are under genetic control, but many of which are specified by the cellular environment. It does not matter very much at which point the chain of events is disturbed; the end result will be compounded of similar defects,

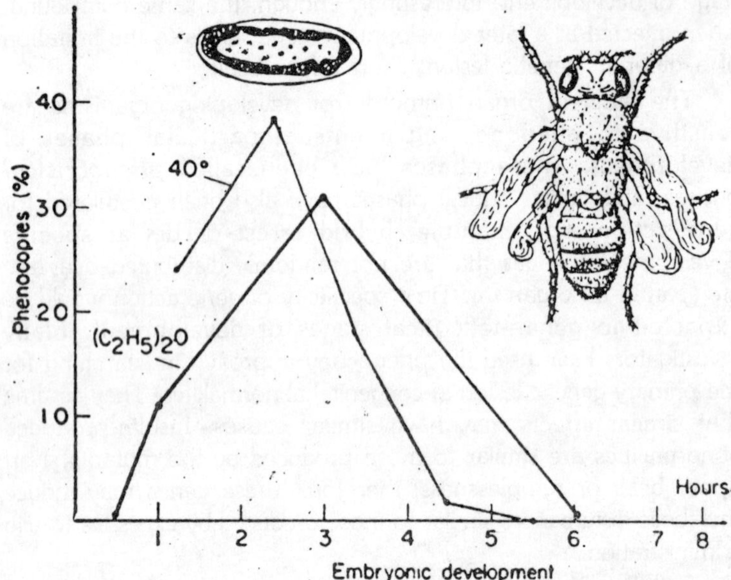

Fig. 10.3. Ether phenocopy of tetraptera in Drosophila. The ordinate indicates the yield of phenocopies resulting from ether $(C_2H_5O)_2O$ (black curve) or heat treatment given at various developmental stages (abscissa). The drawing on the upper left shows a section through the blastoderm at which stage the embryo is most susceptible to the induction of the tetraptera phenocopy.

no matter where the initial defect was introduced into the chain of events.

Embryonic Induction and Morphogenesis

Developmental process are often a result of interactions between different cells types in the embryos. We have previously mentioned the role of hormones in such interactions. Two kinds of genetic interference with such interactions are to be expected: first, genes could control the production and release of the hormonal signal needed for a developmental process ; second, genes could control the ability of a cell to respond to the hormonal stimulus. The hormonal circuitry controlling insect development has been intensively studied.

At least two hormones are directly responsible for growth and differentiation in insects. One of these, the growth and differentiation hormones, controls cell division and enables the cell to undergo specialization. The second, the juvenile hormone, determines whether a cell remains in a larval state or enters metamorphosis. In *Drosophila* these two hormones are produced by a small organ called the ring gland. The identification of this gland was made possible because of a mutant gene, lethal giant larva, which prevents metamorphosis of the larva. This failure to metamorphose suggested that the mutant gene might control the appropriate hormonal stimulus. In a search for the organ that normally produces and releases this hormonal stimulus, a variety of wild-type organs was transplanted into lethal giant larvae. Implanted wild-type ring glands, but no other tissues, initiated metamorphosis in the mutant larvae.

Apparently the cells of the hypodermis responsible for the formation of the puparium have the inherent capacity to function even in the abnormal genotype, but the ring gland of the mutant does not elaborate the hormonal signal required to trigger the initiation of puparium formation. Other cells types of lethal giant larvae also maintained the ability to respond to a normal hormonal stimulus, as shown by a reciprocal experiment. The larval primordia

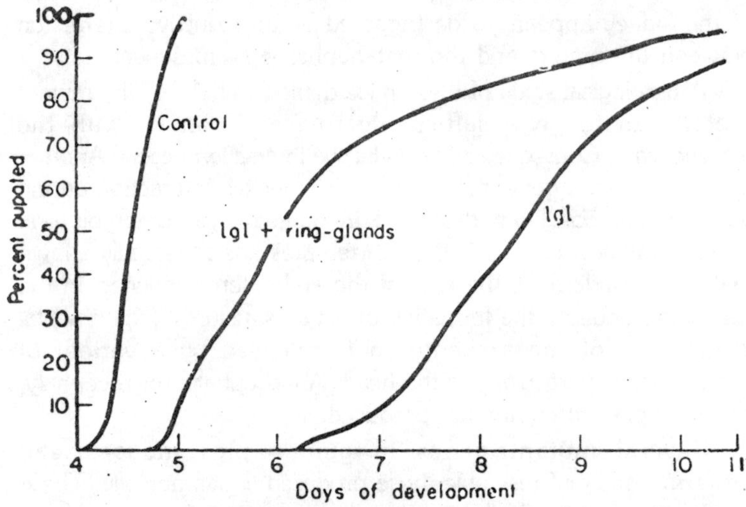

Fig. 10.4. Induction of puparium formation in lethal giant (lgl) Drosophila larvae by implantation of normal ring-glands. The ordinate indicates the percentage of animals that have completed puparium formation at the time indicated on the abscissa. The normal control animals pupate rapidly. The lgl animals with implanted wild-type ring glands develop more slowly but much faster than the lgl larvae with no ring glands implanted.

of ovaries, which in lethal giant larvae never develop, were transplanted to wild-type hosts. In these hosts, the transplanted ovaries differentiated in synchrony with the ovaries of the host; at least the somatic components of the gonads differentiated. The germ cells did not complete differentiation. Thus, two cell types, ring gland cells and germ cells, appear to be directly affected by the mutation. Imaginal disc cells—that is, those primordia giving rise to the adult fly during metamorphosis—also are affected by this mutation; these primordia degenerate during early larval life.

On the other hand, the intestine, nervous system, and the larval muscles are all normal in the mutant. Other tissues, such as the fat body and the salivary glands, are present in the larva, although much reduced in size. The example of lethal giant larva illustrates that release of a hormonal signal and the ability to respond to the signal can both be under genetic control. Many other genes controlling cell and tissue interactions have been extensively investigated. The Danforth short tail mutant of the mouse is particularly instructive. In mice of this genotype, kidneys often fail to develop. During normal development, the formation of the kidney appears to be triggered by an inductive interaction between the ureters and the metanephrogenic mesenchyme.

A histological study of these mice demonstrated that the mutant ureters fail to grow sufficiently to make contact with the mesenchyme. Consequently, no inductive interaction occurs. Another case of a mutant gene interfering with inductive interaction during development has been described in guinea pigs. Embryological experimentation in a variety of vertebrates has repeatedly shown that the mesoderm in the roof of the archenteron induces, in the overlying ectoderm, the formation of neural structures. Experimental disturbance of the archenteron is followed by a variety of abnormalities, particularly in the head. Anencephaly, microcephaly, and cyclopia are commonly produced.

Several mutant genes in guinea pigs induce head malformations that resemble those produced experimentally. These mutations affect solely the head regions of guinea pigs, producing a series of abnormalities ranging from an almost normal appearing skull to extreme anencephaly. It is not known how the mutant genes produce this developmental disturbance, but it is evident that the cells normally giving rise to the skull fail to proliferate as a consequence of the genetic lesion. In other deviant cells, the genetic constitution permits proliferation, but the cells differentiate abnormality. We will discuss examples of this type of genetic interference with normal development in the section on neoplastic transformations.

Imaginal Discs

In our discussion of phenocopies, we have encountered the four-winged mutant, tetraptera, of *Drosophila*. The underlying mutations are called *homoeotic*, meaning that they confer the properties of one body segment on another. Tetraptera is not the only homoeotic mutant. Tetraptera is another, so named because it carries four balancers (halteres) on the thorax instead of the normal two wings and two halteres. Aristapedia is a third homoeotic mutant, characterized by the transformation of antennae into legs. Recently a new homoeotic mutant was discovered and named *Nasobemia*, after a fictitious animal called "Nasobem" in a poem by the German poet Christian Morgenstern. This fictitious animal walks on its nose The mutant *Nasobemia* grows a leg from its

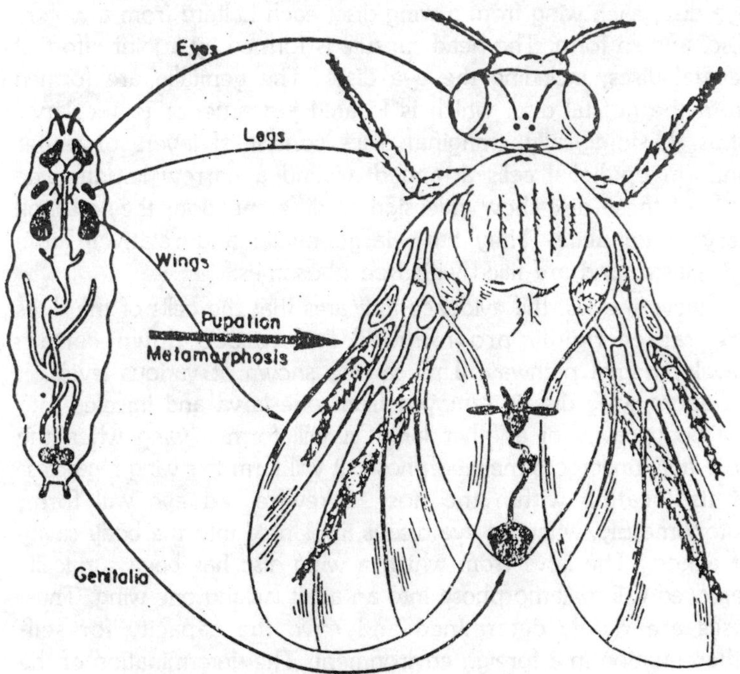

Fig. 10.5. Development of Drosophila imaginal discs.

palpus ! To understand the ontogeny of homoeotic mutants, we must be familiar with insect development. Flies have a complicated life cycle. About one day after fertilization, a larva, the first instar larva, hatches. After another day, this larva, which has grown considerably, sheds its old cuticle and emerges as the second instar larva. The second instar larva grows for about a day molts, and gives rise to a third instar larva. After another two days of feeding and extensive growth, this larva enters pupation. Its cuticle hardens and forms the pupal case or puparium. Inside the puparium a most remarkable process, metamorphosis, starts. In the course of about 4 days, many larva tissues are destroyed and the adult fly, or imago, is built up.

Most structures of the imago are pieced together during metamorphosis form the so-called *imaginal discs*. Imaginal discs are groups of cells, present throughout all the larval stages, but with rise to imaginal or adult structures. For example, each leg is formed from a leg imaginal disc, each compound eye from an eye disc, each wing from a wing disc, each haltere from a haltere disc, and so forth. The head capsule is formed by a joint effort of several discs, including the eye discs. The genitalia are formed from the genital disc, which is located just anterior to the larval anus. Histologically, imaginal discs consist of layers of rather uniform epithelial cells arranged around a narrow lumen. The cells of these disc show little sign of differentiation; they all look very much alike. They have large nuclei and relatively little cytoplasm, and are filled with free ribosomes.

Nevertheless, the evidence indicates that the cells of the discs are rather rigidly programmed to proceed down certain developmental pathway. This can be shown at various levels. If an entire wing disc is removed from one larva and injected into the body cavity of another larva, it will form a wing when the host larva undergoes metamorphosis. It will form this wing regardless of its location within the host. Likewise, an eye will form, autonomously, when an eye disc is implanted into the body cavity of a host. The larva from which a wing disc has been surgically removed will metamorphose into an adult lacking one wing. Thus, discs are rigidly determined and have the capacity for self-differentiation in a foreign environment. The determination of the disc goes even farther. Several parts of the genitalia form from this disc : two vasa deferentia, which become hooked up with the

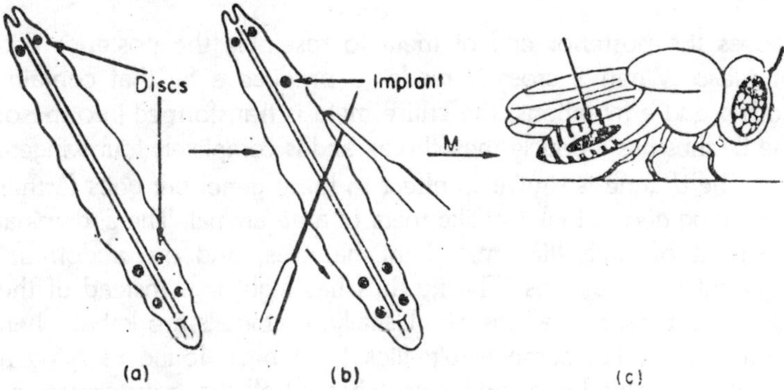

Fig. 10.6. Transplantation of imaginal discs.

gonads; two accessory glands called paragonia; and a contractile ductus that transports sperm into the sperm pump, which, in turn, moves sperm into the external genital apparatus. The latter consists of several chitinous parts, including a penis and several plates carrying bristles and thorns that engage with similar structures on the female fly during mating. The two anal plates surrounding the anus the two claspers are particularly noticeable. Two lines of experimental evidence indicate the each of these organs is already represented within the genital disc at an early developmental stage. First, it pieces cut from a disc are each cultured in a larva then each will give rise, predominantly, to only one structure. Second, exposure of small parts of a disc to an ultraviolet microbeam will prevent the development of individual organs when the disc is reimplanted into a host larva. From the results of such experiments, fate maps have been drawn for several imaginal discs, including the genital disc and the wing disc.

The Bithorax Series

With this information, it is now possible to understand the phenomenon of homoeotic mutations. Five different mutations will be discussed, which alone or in combination affect portions of the mesothorax, metathorax or the first abdominal segment. The mutation *a* causes the anterior portion of meta to develop into a structure that looks like the anterior portion of meso. These animals have two normal wings and two abnormal wings ; the latter appear to be formed by a transformation of the anterior portion of the haltere disc into anterior wing parts. This mutant is called *bithorax*. At the other end of the spectrum, the mutation *e* (*postbithorax*) causes the posterior end of meta to resemble the posterior end of meso. When a cross is made to produce a fly that contains both *a* and *e* mutations, the entire meta is transformed into meso; the fly thus has a double mesothorax and is completely four winged.

The *d* gene is similar in effect to the *e* gene, but goes farther in causing *abd* to look just like meta of an *e* animal. The abdominal segment of such flies may bear halteres, and the abdominal segment now has legs. The fly thus has eight legs instead of the six characteristic of all insects. Usually, *c* animals are lethal when homozygous, but some viable flies have been found carrying a mutation at this locus, and these showed all the transformations that we have listed thus far. Finally, in *b* animals, the posterior portion of meso is altered to resemble the posterior portion of meso is altered to resemble the posterior portion of meta. The diagram shown in figure summarizes the various effects attributable to these five mutations.

This phenomenon is significant for understanding both genetics and development. From the viewpoint of genetics, note that these genes are very closely linked. They are clustered at locus 58.8 on the third chromosome. Genetic recombination analyses demonstrated that these five genes formed a cluster with very rare crossing-over, namely a recombination frequency of approximately 1 :10,000 gametes. We have encountered such cases of clusters of related *loci* before, in the case of the bacterial operon. In *Drosophila*, such clusters are called complex loci, or a pseudoallelic series of genes. Typically, members of such pseudoallelic series show the cis-trans effect—that is, the effect of two mutant pseudoalleles in a single individual depends upon their arrangement in the two homologous chromosomes. The fly of the genotype *a+/+c*, for example, has a mutant appearance, where a fly of a genotype *ac/++* resembles the wild type. This cis-trans effect is a from of possible effect ; clearly, the same genes behave differently when on separate chromosomes. [Since mottling is not observed in the cis-trans position effect, this position effect is called an S-type (stable) effect as opposed to the V-type (variegated) position effect.

When genetic combination of the various mutants of this pseudoallelic series were made, a conspicuous polarity effect became apparent. For example, a trans heterozygous d+/+e does not show a transformation a typical of *d* but, a transformation typical for e (de/++ is wild type). Flies of the genetic constitution a+/+e, on the other hand, show a transformation similar to e, but none of the a type. The trans heterozygous mutant c+/+d shows the d transformation. It is evident that the bithorax series shows a polarity in the sense that each locus tends to influence the other loci to the right, but not to the left. Operationally, what happens is that wild-type alleles to become expressed.

The bithorax series resembles the bacterial operon because of clustering of functionally related genes and because of the polarity effect. Moreover, the products of these genes may be rather directly responsible for the observed gross changes in morphology. Specifically, it has been postulated that the e^+ gene manufactures a substances S^e that suppresses the potential meso-like development of meta. The d^+ gene is thought to produce a difference substancy *sd* that suppresses the potential metalike development of *abd*. The a^+ gene would thus produce a substance *sa* that suppresses the potential meso-like development of meta. Common to these hypothesis is the assumption that wild-type alleles make a substance that the mutant alleles fail to make. If this were so, dosage effects might be demonstrable for wild-type genes ; but not for their mutant alleles. This can be tested by constructing fly genomes that contain one or more doses of the various alleles. Indeed, even several doses of a mutant allele do not yield a mutant phenotype as long as at least one wild-type allele is present in the genome.

The explanation offered is that some gene product is preventing, or suppressing, development. The d+ allele suppresses the development of an abdominal segment into a thoracic segment. The mutation d-d+ may be analogous to the mutations that transformed a multilegged ancestor into the six-legged insects of today. We are assuming here that abdominal segments, indeed, have the potential to form thoracic segments. On the other hand, the mutation a-a+ would have reduced the four-winged insect to a two-winged insect. In embryological terms, this transformation involves the reduction of the wing disc of the metathorax to a haltere disc. Early in evolution, the developmental potential of segment-forming primordia or imaginal discs must have been very great, but mutations have restricted the developmental capacities of these primordia.

One mechanism by which this could have been achieved is through mutations affecting the capability of some cells in a disc to divide. As we see from the fate map of the wing disc, blocking the proliferation of the posterior part would lead to a b-type animal. This hypothesis would be strongly supported if the presumed dormant cells could be stimulated to proliferate. A disc would then give rise to a structure that is does not normally form. We have already seen that such is not generally the case ; otherwise, fate maps could not have been established. But there are a few cases in which a disc has formed structures that it does not normally form. In fact, in one instance, a *Drosophila* disc has formed a structure that is not normally formed anywhere on the *Drosophila*. When a haltere disc is removed from a larva and injected into a host larva, it forms all the structures typical for a halters, plus a groups of adventitious bristles that are not normally found on any body segment of *Drosophila*. The only obvious disturbance of disc development in this case is the transplantation, which in many instances is known to induce some additional mitotic activity. It is therefore conceivable that a cluster of dormant cells was stimulated to proliferate and differentiate, thus giving rise to a new group of structures. This result supports the belief that at least a few cells of an imaginal disc do not differentiate during metamorphosis.

A quite different explanation of homoeosis and the formation of adventitious bristles assumes that the mutation or the surgical interference changes the fate of some cells of the disc. Changing the normal fate would mean acquiring a new kind of determination. This process has been termed *transdetermination*, and very good evidence has been obtained to show that it may indeed operate in the embryological experiments to be discussed now, and possibly also in the genetic experiments just described.

Transdetermination

Does a programmed cell located in one of the determined areas shown on fate maps of an imaginal disc retain its state of determination when provoked to divide more often than normally ? In other words, is determination maintained and transmitted through cell heredity : To answer this question, proliferation must be induced in a piece of imaginal disc for abnormally long periods of time ; then the cells must be induced to differentiate to assess the state of determination. Insects are ideal organisms for this kind of experiment. The hormonal milieu of larvae and adults permits imaginal discs to proliferate, but not to differentiate. Only in a pupal environment will cells undergo extensive differentiation. To stimulate proliferation, a disc is removed from its larval donor, cut in two and injected into a larva or adult, both of which support proliferation. Proliferation results in a steady increase in size of the implant. Several tissues have been maintained in such *in vivo* cultures for many years in a number of laboratories.

The developmental potentiality of the implants may be tested at any time by transplanting them into larvae. When the larvae metamorphose, the implants differentiate. In a typical experiment, one-half of a male genital disc was implanted into the abdomen of an adult fly. After 3 weeks the tissue was removed. It had been grown to several times the size of the original implant. This overgrown disc was cut into many fragments, and the fragments were forced through metamorphosis by implantation into larvae ready to pupate. The resulting differentiating structures were recovered from the abdominal cavity of the host flies and examined microscopically. Some of the fragments had developed chitinous structures, such as anal plates or claspers. Others had formed a sperm pump, and still others, paragonia. This behaviour is in accord with the fate maps.

The experimenter simply happened to isolate individual organ primoria from the tissue mass, and these developed according to their state of determination. This determination can be maintained during long periods of proliferation. In one experiment, and plate clones continued to give rise to only anal plates during more than 70 transfer generations, lasting several years. Occasionally, this strict cell heredity of determination is lost. An aliquot of anal plate primordium cells can suddenly give rise to a leg, or an antenna, or an eye, for example. Thus, in these cases genital disc tissue differentiates into structures that are not normally formed by the genital disc at all but by a disc of a different body segment. When this occurs, the genital disc has undergone transdetermination. The behaviour seems reminiscent of what we observed in homoeotic mutants. Clearly, cell lines derived from a single imaginal disc can be made to form almost any structure provided that they are kept in culture long enough.

Furthermore, there appears to be a definite order in which the various structures appear. The new program of development for a transdetermined structure can be maintained and propagated by cell heredity for many transfer generations. That is, tissue giving rise to legs, although formed from original anal plate material, will remember its new leg program indefinitely. Before we unreservedly accept these results we should rule out some obvious of jections to the experimental technique and to the interpretations. First, do the transdetermined cells originate in the implanted disc, or are they migrant cells from the host organism ? This question is easily answered by the use of genetic markers. Routinely, in these experiments the implant differs in genotype from the host. For example, ebony tissue may be cultured in yellow hosts. All host disc cells are, of course, yellow in genotype, and when differentiated, produce yellow chitinous products.

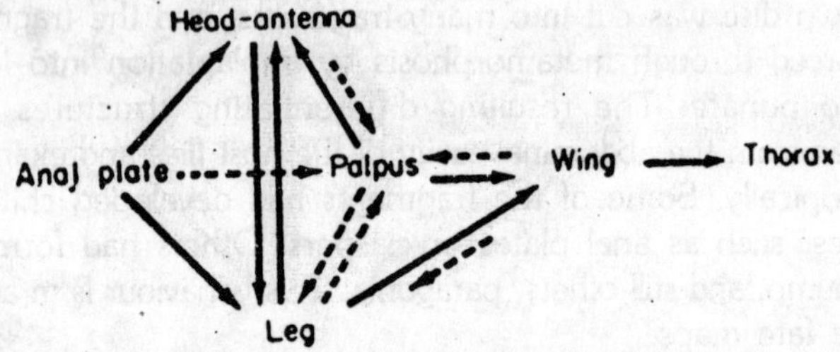

Fig. 10.7. Transdeterminations of imaginal discs.

All implanted disc cells on other hand are genetically ebony, and when differentiated, are black. Implants, whether transdetermined or not, all show the ebony colour. Second does the genital disc contain some leg cells that are revealed under the experimental conditions of forced proliferation? This possibility is more difficult to exclude. What is needed are clones derived from single cells previously proved to be determined as anal plate, for example. If such clones exhibited transdetermination, then the cryptic cell objection would be refuted. Although no one has succeeded in cloning single insect cells, a genetic trick has enabled investigators to examine a clone derived from a single cell.

The genetic trick takes advantage of rare instances of somatic crossing over, as described earlier in the discussion of cell lineage in eye development. Larvae heterozygous for yellow and singed were X-rayed. When somatic crossing-over occurs between the centromere and the marker genes, two cells differing in genotype result. One is now double homozygous for the two marker genes ; the other is double homozygous for their wild-type alleles. The double homozygous mutant cells are recognizable, after differentiation, by the colour and shape of bristles. Since somatic crossing-over is a rare event, it occurs only in randomly located cells dispersed throughout the tissue. Since this rare event is equivalent to a mutation, it is inherited by all progeny of the cell in which the event occurred. In a clone labeled this way, both normal and transdetermined structures were found. Therefore, transdetermination cannot be a consequence of selection of a cryptic cell carrying a different program, but must be due to a heritable change in program. Third, is somatic mutation responsible for transdetermination?

At first stage, mutation does seem a very plausible explanation, particularly in view of the bithorax series. A transdetermination from wing to haltere would be equivalent to the action of the b gene in the bithorax series. In fact, somatic mutation is difficult to exclude. The frequency of observed transdetermination is orders of magnitude higher than known mutation rates, but we should realize that the frequency of somatic mutation is proliferating disc cultures has not been determined. A more convincing argument against somatic mutation is that transdetermination occurs with a somewhat predictable frequency and in a predictable direction.

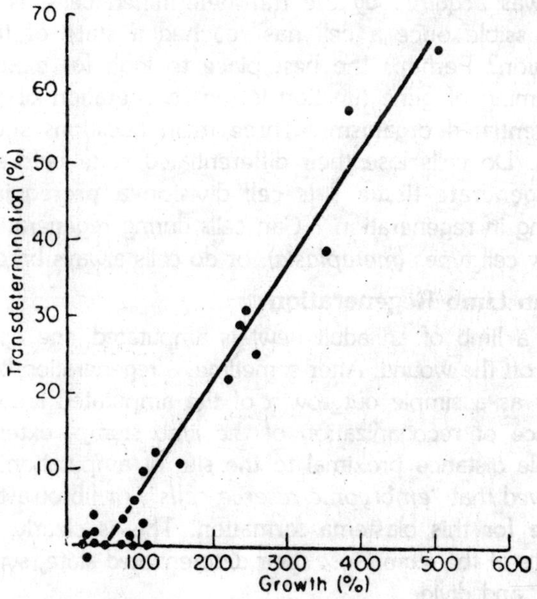

Fig. 10.8. The correlation of cell multiplication and transdetermination.

For example, a thorax forms more frequent from a wing disc than an eye forms from an anal plate disc. The mechanism of transdetermination is therefore not known. The only known prerequisite for transdetermination is cell proliferation. With so little known, hypotheses are not very restricted. Interpretation of the bithorax series postulated gene-controlled substances that prevented the development of wild-type discs in another direction. Perhaps similar substances become diluted during excessive proliferation, to depress hidden morphogenetic potentials. This raise a more general question of the stability of the differentiated state, which we shall examine in the following section on regeneration.

Regeneration and the Programming of Gene Function

The cellular differentiation occurring in normal development is brought about, in part at least, by sequential, selective, gene activation and repression. Whether this effect is achieved at the level of gene amplification, transcription, or translation is not important for the present discussion. What we want to discuss here is whether this sequence of events is irreversibly programmed for a given cell type. We do not think so. In the previous section, we described the loss of determination as a consequence of rapid cell proliferation in imaginal disc. Moreover, a new developmental program was acquired by the transdetermined cells. Is such a change possible once a cell has reached a state of terminal differentiation? Perhaps the best place to look for example of reprogramming of gene function is the regeneration of pairs of fully differentiated organisms. Three main questions should be considered. Do cells lose their differentiated state before giving rise to regenerate tissue ? Is cell division a prerequisite for participating in regeneration ? Can cells during regeneration give rise to new cell types (*metaplasia*), or do cells always breed true?

Amphibian Limb Regeneration

When a limb of an adult newt is amputated, the epidermis soon seals off the wound. After sometime, a regeneration blastema forms, not as a simple outgrowth of the amputated limb but as consequence of recoganization of the limb stump, extending a considerable distance proximal to the site of amputation. It was once believed that "*embryonic reserve cells*", or fibroblasts, were responsible for this blastema formation. This is clearly not so. Rather, cells of the stump lose their differentiated state, synthesize new DNA, and divide.

Electron microscope studies have shown that muscle cells of the stump gradually lose their myofibrils, their most characteristic differentiated feature. The use of tritiated thymidine and autoradiography shows that the nuclei of these same cells begin DNA synthesis. These cells, and not embryonic reserve cells, make up the bulk of the regeneration blastema. In larva newts after limb amputation, the cartilage cells proliferate, after losing the typical cartilaginous appearance. These cells then make up a substantial part of the blastema. Next, blastema cells differentiate into muscle cells, cartilage cells, and connective tissue cells of the regenerating limb. Epidermis, and also blood vessels and nerves, do not form from blastema cells but grow out from the intact tissues left behind in the stump. It is clear in this case that certain mature cells do lose their differentiated state before giving rise to the regenerate, and that they divide extensively. However, there is not direct evidence that metaplasia occurs. Each cell type in the regenerate may be derived only from antecedent cells of the same type.

Amphibian Lens Regeneration

When the lens of an adult salamander eye is removed, a transformation of cells of the dorsal iris margin is noticeable within a few days. Nuclei become rounded, and after about 5 days initiate DNA synthesis. A lumen forms between the two lamina of the iris, and the cells begin to lose their pigment. At this stage, about 10 days after lentectomy, DNA synthesis and cell proliferation are in full swing. The ventral cell layer of the iris becomes multilayered, and in about 3 weeks, forms a body of tissue within which lens fibers differentiate. Synthesis of DNA declines, and the alpha, beta, and gamma crystallins are synthesized. Thus, during lens regeneration, differentiated iris cells transform into lens cells. To accomplish this transformation, the iris cells must synthesize DNA and divide. Here, again regeneration is associated with a loss of the differentiated state and with proliferation. Moreover, in lens regeneration metaplasia seems to occur, for the normal lens is formed from epidermis, whereas the regenerated lens is formed from iris, which itself is of neural origin.

Mammalian Liver Regeneration

When a part of the liver of a mammal is removed surgically, rapid cell proliferation occurs in the remaining liver. These dividing cells make DNA, RNA, and also liver enzymes. Ultrastructurally, there is little evidence of dedifferentiation. However, there is excellent evidence to show that the regenerating cells pass through

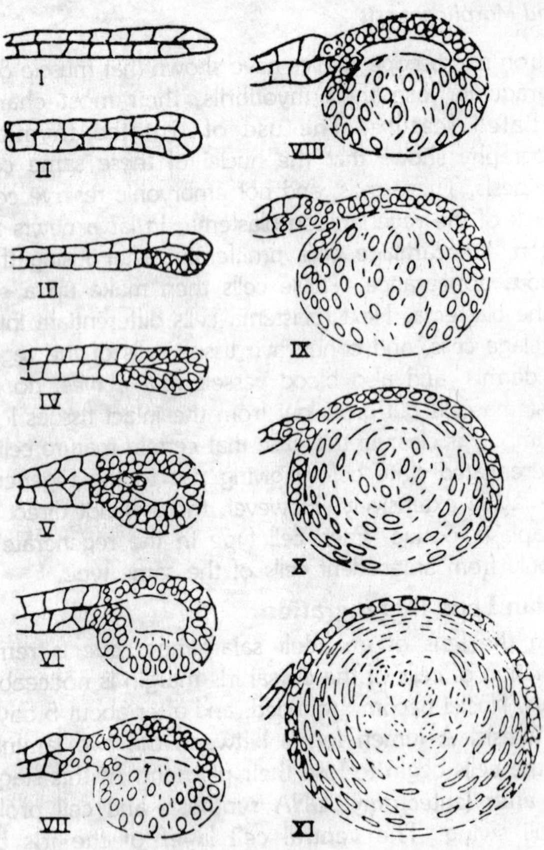

Fig. 10.9. Morphological stages of lens regeneration.

a new time sequence of gene function. The evidence for this is based upon nucleic acid hybridization. Labeled RNA precursor molecules were made available to regenerating livers, and at successive stage of regeneration, the newly formed RNA was analyzed by molecular hybridization. RNA from the same or from different stages of regeneration was used in competitive hybridization experiments. As the family of curves, the cells of the regenerate pass through a sequence of stages, each of which synthesizes different arrays of RNA; the final stage synthesizes RNA characteristic of the adult liver. Significantly, a similar sequence of RNA synthetic events occurs during fetal development of the liver; this was found when the RNA synthetic patterns were studied during normal liver ontogeny, again using the technique of nucleic acid hybridization. Clearly, regeneration of the liver proceeds through a programmed sequence of gene activities similar to that of normal liver ontogeny. Liver regeneration requires extensive proliferation but not much loss of differentiation. As might be expected, metaplasia does not occur. We might ask at this point whether cell proliferation is an absolute requirement for the cellular changes occurring during regeneration. Are there examples of differentiated cells losing their differentiated state and redifferentiating into new cells types without cell division?

The Lability of the Differentiated State in Hydra

Hydra is a simple coelenterate that presents a number of advantages for developmental studies. For example, it contains only about 10 different types of cells. In this organism, it is therefore easier to trace the fate of a particular cell during regeneration than in other organisms where hundreds of cells are involved in regeneration. Some cells of hydra have remarkable totipotency, notably the interstitial cells, which can give rise to virtually all cell types of the mature organism. Even fully differentiated cell types of hydra, when isolated, have the capacity to regenerate normal adults. The gastrodermis, as analyzed histologically is seen to consist of only two cell types ; digestive and gland cells. When these two cell types are cultured together in a series of appropriate culture media, they can form an entire hydra. Metaplasia clearly occurs in this case, and it is important to know whether the metaplasia required cell division.

It is impossible to follow the developmental path of each of these two cell types. The gland cells, as shown by electron microscope analysis, lose their endoplasmic reticulum and shed their lipid inclusions. These structures characterize the differentiated state of this cell type. After these structures are lost, they can no longer be distinguished from interstitial cells. This dedifferentiation occurs without cell division. Next, the cells divide and their progeny differentiate into typical cnidoblast cells and nematocysts, highly specialized cells quite different from the ancestral gland cells. When the digestive cells are isolated, the algal bodies within the cells break down. Next, muscle fibers begin to form and a mucous border characteristic of epidermal cells appears. These new cells now closely resemble normal epidermal cells. This transformation of digestive cells into epidermal cells can occur without cell division. Clearly metaplasia occurs in both cell types, but the metaplasia is not identical in both. The dedifferentiation of gland cells to form interstitial cells leads through cell division to the formation of very different cell types. In the transformation of the digestive cells into epidermal cells, on the other hand, no cell division is required. Rather, redifferentiation occurs directly. To our knowledge, this is the only example of direct metaplasia in the absence of cell division in a higher organism.

Lethal Cell Differentiation

One of the most remarkable types of cell differentiation leads to cell death—that is, the terminal developmental stage is lethal. Such cells are programmed to die and their death is a normal characteristic of development. Cell death occurs in many part of an organism, and in the embryo it is an essential feature of the normal morphogenesis of many organs. The union or detachment of parts of organs and the formation of a lumen in the center of an otherwise solid structure all require the death and disintegration of strategically placed cells. For example, cell death is essential to the separation of the fingers and toes from one another, to the separation of the lip from the gums, to the opening of the eyelids, and to the formation of the central canals of many ducts and organs. The failure of cell death to occur on schedule results in congenital abnormalities. One example, webbed toes, has already been described. Another is the occasional persistence of a small tail in human beings. In other locations, the survival of embryonic cells that are normally programmed to die may lead to the development of embryonic tumors. For example, the medulloblastomas derived from the rapidly growing external granular layer of the fetal cerebellum may represent surviving embryonic cells that should have died.

Other tumors of embryonic origin that may develop in a similar fashion are the sympathetic neuroblastomas, retinoblastomas, nephroblastomas, hepatoblastomas, and embryonic sarcomas. These tumors are composed of aberrant cells that originated either from transformed normal cells or from embryonic cells that failed to die on schedule. A conspicuous example of the role played by cell death in the normal development of organisms is the regression of larval organs during metamorphosis. The cells of the tail of the tunicate tadpole or the frog tadpole die and are resorbed as the larva transforms into an adult. Many other, less conspicuous, larval organs also regress during metamorphosis. Since all cells of the developing embryo have the same genetic makeup, the regression of certain cells and the persistence of others must flow from their state of differentiation on receiving an appropriate stimulus. A clear case involves the regression of cells in the Mullerian ducts of males during hormonal maturation.

Mammalian embryos are equipped with the rudiments of the sexual organs of both sexes ; one set is suppressed as a consequence of hormonal stimulation in accord with the genetic makeup of the individual. Likewise, during thyroxin-stimulated metamorphosis of frog larvae, some cells die and others proliferate. The response of the cell is thus highly specific to its state of differentiation. In the adult, too, cell differentiation can lead to cell death. The erythroblast of adult bone marrow passes through a sequence of developmental steps that result in the loss of its nucleus ; therefore, it soon dies. Likewise, the differentiation of the epidermis in vertebrates leads to a heavily keratinized cell that eventually dies and is sloughed off. Many secretory cells likewise die in the process of releasing their secretions. These are all normal types of cell differentiation that terminate in death. Similar sequences of differentiation can be induced experimentally.

Removal of embryonic limbs, for example, will bring about the death of many cells in the central nervous system that are normally responsible for innervating those limbs. Thus, these nerve cells require a highly specific functional relationship with the innervated cells to survive. Simple nutrition can scarcely be involved. The functional dependence can best be understood as that of maintaining the nerve cell in a stable equilibrium state of differentiation. Once this equilibrium is upset, the nerve cell begins to change and enters a lethal pathway of differentiation. Like all the steps in cellular differentiation, the time at which the differentiated step leading to cell death is taken occurs at a specific stage in the life of the cell. This has been clearly demonstrated in experiments on the wing bud of developing chick embryos.

The cells attaching the posterior margin of the wing bud to the body in the young chick embryo are normally destined to die as the wing bud separates from the body. These cells can be removed from the area in which they normally reside at an early stage of development and transplanted to the dorsal surface of the wing bud. In this new location, they do not die but become an integral part of the dorsal wing surface. If, however, these cells are allowed to remain in their original position beyond a critical developmental stage then when transplanted to a new site they nevertheless continue their differentiation to death. Thus, the initial steps of differentiation are not irreversible, but at some point along the pathway the lethal direction of differentiation becomes fixed, at least under all experimental conditions tested so far.

Like all other cell processes, the steps in cell differentiation leading to cell death are subject to gene control and can be changed by mutation. For example, *wedded* toes in human beings is due to a mutation that permits survival of the cells between the toes, even though genetically normal cells in the location would die. There are also many examples of mutations that induce cells in particular areas to die at specific stages in development ; cells with normal genotypes would have remained viable in those locations. For example, human beings with XO and XXY genotypes differentiate normal cells in most organs of the body, but viable gametes do not differentiate in the gonads. The developing germ cells die at an advanced stage of differentiation but other cells of the body develop essentially normally. In the XXY individuals, the germ cells die in early childhood or at an early stage in puberty.

The primordial germ cells with these abnormal karyotypes are incapable of completing their differentiation, although early steps in the pathway of gamete differentiation remain open to them. More severe *karyotypic* abnormalities—for example, trisomy for chromosome two—permits the embryonic cells to take part in the formation of an apparently normal placenta and certain fetal membranes, but these cells cannot form the body of an embryo. Single gene changes controlling cell death are also known. The tail less mutant of the mouse is a clear example. In such mice, a constriction appears in the proximal part of the tail at about 11 days of embryonic development, and a localized zone of cell death occurs.

The distal part of the tail then degenerates, and the mouse is born with only a short stump. *Rumplessness* is fowl embryos exhibits a similar form of degeneration. In these embryos, the presumptive tail tissue degenerates even before the onset of tail formation, and the newly hatched chick is completely without a tail. The gene for retinal degeneration in mice affects only certain specifically differentiated cells, the rod cells of the retina. These cells begin to differentiate, but before differentiation has been completed, the distal parts of the cells cease development and degenerative changes begin. It is important to note that these cells develop normally in the mouse throughout embryonic life and for the first 14 days after birth. No other type of cell appears to be affected by this gene.

INDEX

Cross-fertilization, 193

D

E

F

G

H